METHODS IN MOLECULAR BIOLOGY™

Series Editor
John M. Walker
School of Life Sciences
University of Hertfordshire
Hatfield, Hertfordshire, AL10 9AB, UK

For further volumes:
http://www.springer.com/series/7651

Computational Modeling of Signaling Networks

Edited by

Xuedong Liu, Ph.D.

Chemistry and Biochemistry, University of Colorado-Boulder, Boulder, CO, USA

Meredith D. Betterton, Ph.D.

Department of Physics, University of Colorado-Boulder, Boulder, CO, USA

Humana Press

Editors
Xuedong Liu, Ph.D.
Chemistry and Biochemistry
University of Colorado-Boulder
Boulder, CO, USA

Meredith D. Betterton, Ph.D.
Department of Physics
University of Colorado-Boulder
Boulder, CO, USA

ISSN 1064-3745 ISSN 1940-6029 (electronic)
ISBN 978-1-4939-5843-6 ISBN 978-1-61779-833-7 (eBook)
DOI 10.1007/978-1-61779-833-7
Springer New York Dordrecht Heidelberg London

Softcover re-print of the Hardcover 1st edition 2012

Printed on acid-free paper

Humana Press is part of Springer Science+Business Media (www.springer.com)

Preface

Cells sense their environment largely through the activities of signal transduction networks which regulate diverse aspects of cellular behavior. Signaling outputs are dynamic, extraordinarily complex, and yet highly specific. A comprehensive understanding of how signaling networks behave in space and time to generate specific biological responses is paramount to biology and medicine. Aberrations in signaling networks are associated with many human diseases such as cancer and diabetes. Better understanding of signaling mechanisms can potentially have a major impact on drug design and therapeutics.

Signaling networks are composed of numerous signaling pathways and each has its own intricate component parts. In the past three decades, the molecular biology approach has dominated the field of signal transduction research. Reductionism—with an emphasis on identifying and characterizing the components of each pathway—is the prevailing philosophy underlying much signaling research. The success of this approach has enabled biologists to enumerate a "parts list" for many signaling pathways. Robust molecular techniques and easy-to-standardize protocols researchers can follow have enabled the application of molec ular biology techniques to signaling. Automation of these techniques has led to sequencing of the entire genome of many organisms including humans. There has been an explosion of information gathered through genome wide technologies, such as measurements of gene expression through DNA microarray technology, assays of genome-wide protein–protein interaction maps, and proteomic measurements of posttranslational modifications. As the description of molecular components of signaling systems gets more detailed, it becomes more and more evident that the behaviors of signaling networks amount to more than the sum of their parts.

The ultimate goal of studying signaling networks is to predict cellular responses to external stimuli based on molecular mechanisms. Accurate predictions require quantitative understanding of the interactions of the signaling components. Mathematical and computational modeling techniques are emerging as powerful tools in cellular signaling research. There are several reasons behind the rapid rise in popularity of modeling in biological research. Firstly, the extraordinary complexity of signaling networks calls for the representation of intricate relationships in mathematical terms. Modeling helps to recognize patterns and recurring themes in signaling networks. Secondly, the modeling process requires critical evaluation of different types of biological data to get an integrated view of the system. Thus, modeling promotes critical thinking about the systems in question. Thirdly, once a model is built, it represents a sophisticated hypothesis that can be tested by running simulations and comparing the outputs to experimental data. The relative ease of running simulations can save time (relative to experiments). Finally, modeling can identify possible ways by which molecular mechanisms give rise to higher-order behavior and predict system behavior under conditions that have yet to be experimentally tested.

Despite these obvious advantages of modeling, significant challenges exist in order for experimentalists to adapt modeling into their daily research. Unlike molecular biology, where standard protocols for experimental design and manipulation are readily available from manuals like *Molecular Cloning* or *Current Protocols*, standard protocols that many experimentalists are accustomed to for mathematical and computational modeling are difficult to find. Perhaps this partly reflects the fact that modeling of signaling networks is still very much in its infancy. This book is written with the intention of helping the experimentalists to ease through this transition, while also enabling mathematical and physical scientists to learn how modeling is currently being applied to research in signaling. By no means will the current book be as comprehensive as manuals like *Molecular Cloning*, but we do hope that this book will serve as a catalyst for achieving this goal.

Chapters 1 and 2 by Iber, Fengos, and coauthors address two related topics in model formulation and analysis. Chapter 1 outlines the basic process that is the bread and butter of many computational systems biologists: it focuses on systems of ODEs and walks through the formulation and analysis of a model. Systems of ODEs are an appropriate model for situations where spatial variation is not important and the numbers of molecules is relatively large. A specific example of a simplified TGF-beta signaling model is used to illustrate the methods. This chapter introduces many useful concepts: in model formulation, they discuss the mathematical representation of some common biochemical reactions, including synthesis, degradation, shuttling, binding, and enzymatic activity (with or without inhibition). In the analysis section, Iber and Fengos discuss a number of useful tools in the analysis of dynamical systems, including nondimensionalization, phase-plane analysis, linear stability analysis, bifurcation diagrams, and sensitivity analysis. Familiarity with these tools of the trade is essential for anyone who wants to work (or understand the research) in this area. Chapter 2 by Geier et al. builds on Chap. 1 to discuss the parameterization of signaling models. This is an important issue that addresses questions ranging from the practical (how does one actually determine parameter values to use in day-to-day work in computational biology?) to the fundamental (how well can one constrain parameter values given limited data? How do uncertainties in the parameters affect conclusions one can draw from the model?).

Zi contributed the modeling tutorial in Chap. 3. This chapter is recommended for those new to the field who want an entry-level description of how to approach modeling of signaling pathways. Zi explains the differences between top-down approaches (using high-throughput data) and bottom-up approaches (using modeling). With a focus on ODE modeling of signaling pathways, the tutorial explains the commonly used types of kinetics (the law of mass action, the Hill equation, Michaelis–Menten kinetics) and explains how to go from a reaction to writing down an ODE. For the development of more complex models, Zi includes discussion of signal transduction steps including production/degradation of mRNA and protein, phosphorylation, and dephosphorylation, feedback, and signaling inputs. The chapter wraps up with a discussion of initial conditions and parameter estimation by least squares, and includes a workflow that goes through all the steps required in development and analysis of a model.

Chapter 4, Anderson, Liu and Ferrell discuss several important concepts critical for understanding and characterizing dynamical systems, including steady state, stability, reversibility, ultrasensitivity, bistability, bifurcations, and hysteresis. Using an example of a well-studied signaling pathway, they illustrate the use of purely graphical methods to determine the plausibility of two network behaviors, bistability and irreversibility.

In Chap. 5, Stites and Ravichandran explain the development and analysis of their model of Ras signaling using mass-action kinetics and ODEs. This chapter shows how the basic modeling techniques can be applied to address an important scientific question. The goal is to predict cellular levels of RasGTP, a key oncogenic protein. They use this signaling network as a prototype for a tutorial on the development, validation, and analysis of a mathematical model. The chapter gives a clear explanation of how the model was constructed, addressing the questions they wished to investigate, how they chose how much of the pathway to include, simplifications made in developing the model, and the parameter values used. They validated the model using predictions for experiments and sensitivity analysis. They conclude with new predictions for experiments and insight into possible mechanisms that affect RasGTP levels.

Chapter 6, by Lai, Wlkenhauer, and Vera, addresses the role of micro RNAs (miRNA) in regulating signaling. This work shows how experiments and modeling can be integrated in a single study, as well as a pedagogical explanation of different types of feedback in signaling networks. The chapter focuses in particular on the signaling module involving p53, Sirt1, and the micro RNA miR-34a, and investigates different silencing mechanisms of miR-34a on Sirt1. It is known that miRNAs can be involved in different types of feedback loop, including simple negative feedback, positive feedback involving an intermediate protein, and negative feedback involving an intermediate protein. Lai et al. develop an ODE model with rate equations based on power law terms, which also includes time delays; they discuss parameter estimation for their model. They explicitly consider 4 different models for repression of Sirt 1, including enhanced degradation of mRNA, reversible deactivation of mRNA, translation inhibition, and transcription inhibition. The model predictions are used to design an experiment that can distinguish the different proposed mechanisms; the comparison to experimental results suggests that translational repression seems the most likely mechanism.

In Chap. 7, authors Bandara and Meyer discuss a novel but increasingly important approach to modeling: how to use modeling to plan experiments. The chapter discusses three specific problems: first, parameter estimation—fitting a model to data to estimate the parameters of the model. Second, they discuss model discrimination—using models to design an experiment to determine which of two models is correct. Here the basic idea is to design an experiment that will make the difference between two model predictions very obvious—this can be done by maximizing a function that describes the difference between the two model predictions. The third problem is experimental design for model parameterization—the design of an experiment that will make the parameter estimation problem easier. Each of these problems is addressed in the context of least-squares fitting (minimizing the sum of squared errors between two variables). Optimization of an objective function based on the model predictions and data is the goal in all cases.

Chapter 8, by Maiwald, Eberhardt, and Blumberg, is a tutorial on using the software system PottersWheel. The focus of this software is ODE models, and PottersWheel is integrated with Matlab as a Matlab toolbox. The authors use the JAK/STAT pathway as a tutorial example and focus their discussion on model discrimination, parameter estimation, and experimental design. They walk through a series of steps in their tutorial, focusing on model creation, data import, fitting the model to the data, assessing goodness of fit, model refinement, analysis and model prediction, and design of new experiments. Some interesting features of PottersWheel may make it particularly attractive to those in the field. For example, models can be entered either with a GUI "model designer" or a text file. The software is designed to easily integrate experimental data; data from text or excel files can be loaded into the model.

In Chap. 9, Sekar and Faeder give a tutorial on rules-based modeling using the BioNetGen (BNG) language. The language is designed to handle combinatorial complexity in ODE-based signaling models, which occurs when the number of possible states and reactions becomes large. In this case, writing out every state and reaction by hand is tedious. Rules-based software packages based on BNG can start with a shorter list of reaction rules and generate the model states and equations automatically. Sekar and Fedar take a pedagogical approach, which will be useful for readers new to this type of modeling, and focus their discussion on how to deal with common challenges and potential pitfalls in constructing a rules-based model. The discussion of what a rules-based model is, how to represent the model in BNG syntax, and how to write model rules is useful for getting started with this approach and includes numerous examples. The tutorial on receptor–ligand interactions works through every step of the construction and analysis of an example model of the EGF receptor. More advanced topics are addressed in the sections on compartmental modeling and parameterization. Examples of rules written to describe common mechanisms in signaling pathways are included. The chapter concludes with some best practices for good modeling.

In Chap. 10, Prasad uses the example of the kinetics of thymocyte selection—the positive and negative selection of T cells based on the affinity of TCR–pMHC interaction—to illustrate the process of modeling, analysis of a model, and generating predictions for experiments. The author discusses basic reaction types, the steady state assumption and when to make it, basic reaction kinetics, and stochastic modeling. The example of thymocyte selection is a detailed, well-explained example. The chapter discusses how the model was formulated and the choice of simplifying assumptions. The discussion of equilibria of ODEs and bifurcation analysis is illustrated using the thymocyte kinetic model as an example. The discussion of the stoichiometric matrix, combinatorial complexity and rule-based simulation, and parameter sensitivity will be useful guides for others undertaking this analysis.

In Chap. 11, Song and You focus on modeling spatiotemporal dynamics using a synthetic bacterial ecosystem as an example. As the only chapter in this volume to include spatial variation in modeling, this chapter is a useful introduction to this large area. Song and You discuss model development for a PDE model, parameter estimation, and pattern formation. They discuss the use of finite element methods and comsol for numerically solving PDEs. The chapter works through the modeling process in detail using the example of a synthetic ecosystem, consisting of two engineered *E. coli* populations that act as predator and prey and communicate through quorum sensing. For low prey density, the predator cells die, while at high predator densities the preys die. The authors explain how they constructed their PDE model, with transport of a drift-plus-diffusion form with the drift determined by chemotaxis. They explain how to quantify biodiversity in their model and experimental system using the modified Simpson's biodiversity index.

Chapter 12, by Saadatpour and Albert, focuses on discrete (Boolean) modeling, giving a useful overview of this modeling approach with example models of the abscisic acid signal transduction in plants and the T-cell-survival signaling network in humans. The authors first address the advantages of discrete models, which are particularly valuable when qualitative results are acceptable and many model parameters are unknown. The authors walk through the steps in modeling, including (1) reconstructing the network from the literature; (2) identifying Boolean functions, which addresses how to turn the experimental information into logical rules; (3) implementing time, whether to use synchronous or asynchronous updates; (4) analyzing the dynamics of the system; (5) testing the validity and robustness of the dynamic model; and (6) analyzing the effect of node perturbations, by simulating experimental perturbations such as knockout or overexpression.

The Boolean approach to modeling is illustrated with two examples. The abscisic acid signal transduction network in plants is important in changing the size of holes in the outside of leaves, controlling how rapidly water is lost by transpiration. The other example is the T-LGL leukemia survival signaling network, which addresses a type of leukemia of cytotoxic T lymphocytes.

Chapter 13, by Walczak, Mugler, and Wiggins, is a useful complement to simulation-based approaches to modeling signaling networks: in this chapter, Walczak et al. review analytic approaches for stochastic models. Stochastic approaches are important when the numbers of molecules considered are relatively small, so that the inherent randomness of the dynamics is important. (By contrast, a deterministic model considers the average number of molecules only, and neglects fluctuations.) The methods discussed are a lightning tour of the three classic equations of stochastic modeling: the master equation, the Fokker-Planck equation, and the Langevin equation. The authors demonstrate a variety of methods for solving these equations, both analytic and numerical. Walczak et al. beautifully illustrate how understanding analytic approaches can inform numerical simulation, by allowing the construction of efficient simulation schemes. They use specific simple examples of models of production/degradation of a protein to illustrate general analytic methods. The authors start with a very simple model of production and degradation of a single protein, then add twists to describe self-activation or inhibition, transcriptional or translational bursting, and interaction of two genes. This chapter is a useful introduction to more mathematically sophisticated techniques in stochastic modeling and their application to signaling.

We thank the contributors of this book for their time and efforts devoted to share their state-of-the-art procedures and methods employed in their laboratories. We wish to express our deep appreciation for Dr. John Walker who inspired us to take on this project and supported us during its development.

Boulder, CO, USA

Xuedong Liu, Ph.D.
Meredith D. Betterton, Ph.D.

Contents

Contributors

RÉKA ALBERT • *Department of Physics, The Pennsylvania State University, University Park, PA, USA*
GRAHAM A. ANDERSON • *Chemical and Systems Biology, Stanford University Medical Center, Stanford, CA, USA*
SAMUEL BANDARA • *Chemical and Systems Biology, Stanford University Medical Center, Stanford, CA, USA*
MEREDITH D. BETTERTON • *Department of Physics, University of Colorado-Boulder, Boulder, CO, USA*
JULIE BLUMBERG • *Scientific Software, TIKANIS Solutions, Freiburg, Germany*
OLIVER EBERHARDT • *Scientific Software, TIKANIS Solutions, Freiburg, Germany*
JAMES R. FAEDER • *Department of Computational and Systems Biology, University of Pittsburgh School of Medicine, Pittsburgh, PA, USA*
FEDERICO FELIZZI • *Department of Biosystems, Science, and Engineering (D-BSSE) ETH Zurich, Basel, Switzerland*
GEORGIOS FENGOS • *Department of Biosystems, Science, and Engineering (D-BSSE) ETH Zurich, Basel, Switzerland*
JAMES E. FERRELL, JR • *Chemical and Systems Biology, Stanford University Medical Center, Stanford, CA, USA*
FLORIAN GEIER • *Department of Biosystems, Science, and Engineering (D-BSSE) ETH Zurich, Basel, Switzerland*
DAGMAR IBER • *Department of Biosystems, Science, and Engineering (D-BSSE) ETH Zurich, Basel, Switzerland*
XIN LAI • *Systems Biology and Bioinformatics Group, Department of Computer Science, University of Rostock, Rostock, Germany*
XUEDONG LIU • *Chemistry and Biochemistry, University of Colorado-Boulder, Boulder, CO, USA*
THOMAS MAIWALD • *Freiburg Center for Systems Biology, University of Freiburg, Freiburg, Germany; Scientific Software, TIKANIS Solutions, Freiburg, Germany*
TOBIAS MEYER • *Chemical and Systems Biology, Stanford University Medical Center, Stanford, CA, USA*
ANDREW MUGLER • *FOM Institute AMOLF, Amsterdam, The Netherlands*
ASHOK PRASAD • *Chemical and Biological Engineering, School of Biomedical Engineering, Colorado State University, Fort Collins, CO, USA*
KODI S. RAVICHANDRAN • *Department of Microbiology, Beirne B. Carter Center for Immunology Research, University of Virginia, Charlottesville, VA, USA*
ASSIEH SAADATPOUR • *Department of Mathematics, The Pennsylvania State University, University Park, PA, USA*
JOHN A.P. SEKAR • *Department of Computational and Systems Biology, University of Pittsburgh School of Medicine, Pittsburgh, PA, USA*
HAO SONG • *Division of Bioengineering, School of Chemical and Biomedical Engineering, Nanyang Technological University, Singapore, Singapore*

EDWARD C. STITES • *Medical Scientist Training Program, University of Virginia, Charlottesville, VA, USA*
JULIO VERA • *Systems Biology and Bioinformatics Group, Department of Computer Science, University of Rostock, Rostock, Germany*
ALEKSANDRA M. WALCZAK • *CNRS-Laboratoire de physique Theorique de l'Ecole Normale Superieure, Paris, France*
CHRIS H. WIGGINS • *Department of Applied Physics and Applied Mathematics, Center for Computational Biology and Bioinformatics, Columbia University, New York, NY, USA*
OLAF WOLKENHAUER • *Systems Biology and Bioinformatics Group, Department of Computer Science, University of Rostock, Rostock, Germany*
LINGCHONG YOU • *Department of Biomedical Engineering, Institute for Genome Sciences and Policy, Duke University, Durham, NC, USA*
ZHIKE ZI • *BIOSS Centre for Biological Signalling Studies, University of Freiburg, Freiburg, Germany*

Chapter 1

Predictive Models for Cellular Signaling Networks

Dagmar Iber and Georgios Fengos

Abstract

This chapter provides an introduction to the formulation and analysis of differential-equation-based models for biological regulatory networks. In the first part, we discuss basic reaction types and the use of mass action kinetics and of simplifying approximations in the development of models for biological signaling. In the second part we introduce phase plane and linear stability analysis to evaluate the time evolution and identify the long-term attractors of dynamic systems. We then discuss the use of bifurcation diagrams to evaluate the parameter dependency of qualitative network behaviors (i.e., the emergence of oscillations or switches), and we give measures for the sensitivity and robustness of the signaling output.

Key words: Mathematical modeling, ODE models, Signaling models

1. Purpose of Modeling

Biological functionality arises from the complex interactions of simple components. Decades of intense research in molecular and cellular biology have led to detailed wiring diagrams for the regulatory interactions in many important signaling pathways. Detailed biochemical and structural studies provide us with further information about the mode of interaction as well as about the kinetics of catalyzed reactions. Biologists typically describe the complex interactions in signaling networks with cartoons (Fig. 1). The molecular species (genes, proteins, lipids, metabolites, etc.) and their complexes are depicted by icons; arrows indicate transformations, i.e., production, degradation, binding, dissociation, and chemical modifications (phosphorylation, etc.). For simple regulatory networks these cartoons are, in general, sufficient to grasp the regulatory logic of the circuit.

With the recent rise of high-throughput technologies and the increased sophistication of computational tools an increasing regulatory complexity is being revealed with many more cross-interactions

Xuedong Liu and Meredith D. Betterton (eds.), *Computational Modeling of Signaling Networks*,
Methods in Molecular Biology, vol. 880, DOI 10.1007/978-1-61779-833-7_1,

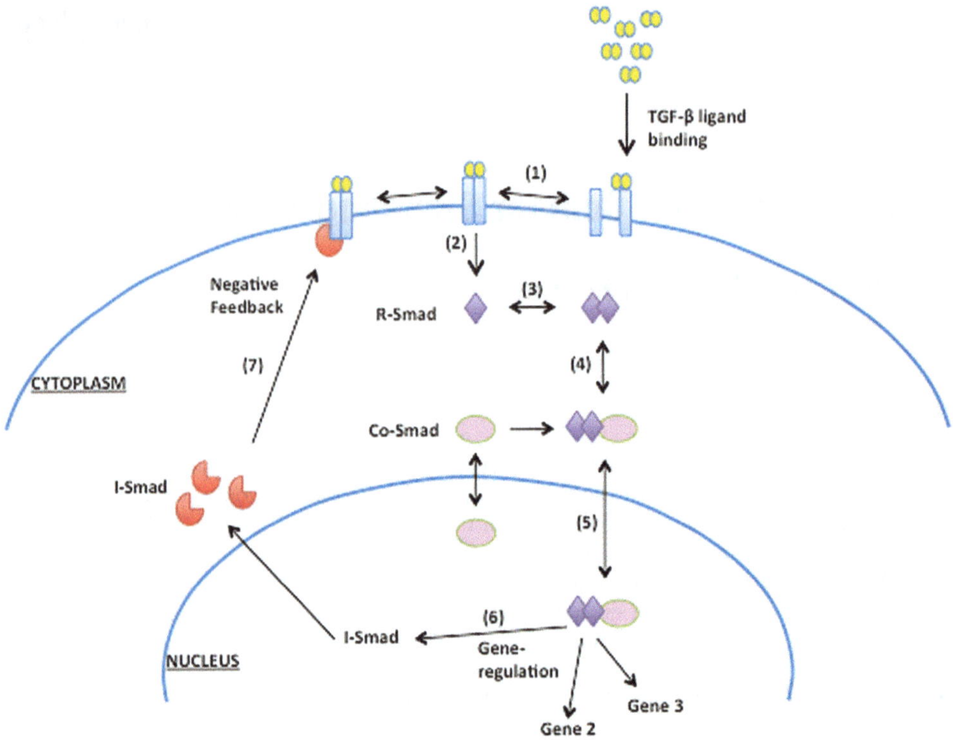

Fig. 1. A cartoon description of the TGF-beta signaling network. TGF-beta is a soluble secreted protein, that signals by binding to the TGF-beta receptor (1). The ligand-bound receptor phosphoryates the regulatory Smad (R-Smad) (2). After dimerization (3), phosphorylated R-Smads bind a Co-Smad (4) and enter the nucleus (5) where they regulate a wide range of genes. One of the genes that is upregulated encodes an inhibitory Smad (I-Smad) (6) that downregulates TGF-beta signaling by interfering with the receptor-dependent phosphorylation of the R-Smads (7).

and interdependencies. As the wiring is getting more complex, verbal reasoning quickly reaches its limitations in delineating the regulatory logic. Mathematical models are the method of choice to enable such an analysis.

The type of mathematical model depends on the particular setting. This chapter will focus on ordinary differential equation (ODE) models. These can be used when there are no spatial inhomogeneities, i.e., the components are "well-stirred," and the concentration of all components is sufficiently high such that a continuous description is applicable. If spatial inhomogeneities are important then partial differential equations (PDEs) need to be used. If particle numbers are low, stochastic effects are playing a role and the system has to be modeled with stochastic differential equations or with Monte Carlo methods.

Models cannot but approximate reality. The level of detail of a model is dictated by the question under investigation and by the type of data that is available to test the model with. Biological questions may concern qualitative properties of the regulatory system, i.e., its ability to show transient, sustained, or oscillatory responses, its sensitivity to a given input, or its robustness to noise. A quantitative comparison of model predictions and experiments may be necessary to distinguish between competing hypotheses and to judge whether a certain regulatory behavior is at all possible and plausible given the observed type and strengths of interactions and the speed of reactions.

2. Part I: Development of Mathematical Models

To translate the cartoons into a set of ODEs we need to assign a single state variable $x_i(t)$ to each icon. Figure 1 shows a typical depiction of the TGF-beta network. One state variable would be the unbound ligand, $x_1(t)$. Another state variable would be the ligand–receptor complex, $x_2(t)$, and so on. The set of values of all state variables $\{x_1(t), x_2(t), \ldots\}$ at a given time point t constitutes the state of the system at time t. We are interested how the system evolves over time and what equilibrium it assumes. If mass is conserved in the system then the time evolution of $x_i(t)$ can be formulated as a differential equation of the form

$$\frac{dx_i}{dt} = \textit{synthesis} - \textit{degradation} \pm \textit{shuttling} \pm \textit{complex formation} \pm \textit{chemical modification}. \quad (1)$$

The positive kinetic terms can be combined in a "gain" rate, the negative kinetic terms can be combined in a "loss" rate. When gain and loss rates balance the variable no longer changes with time, i.e., $\frac{dx_i}{dt} = 0$. When the gain and loss rates of all variables balance then the system reaches an equilibrium point, also referred to as steady state or fixed point. In general, the gain and loss rates change as the values of the state variables change. This is the basis of all feedback regulation. Mathematically, we say that the system of ODEs is coupled, i.e., the differential equations for the different variables depend on each other

$$\frac{dx_i}{dt} = f(x_1, x_2, \ldots, x_n). \quad (2)$$

This means that we need to consider the entire set of equations simultaneously and cannot solve the different equations separately. Software packages (such as Matlab or Mathematica) are available that provide algorithms to solve these sets of equations numerically. In the following, we will discuss how the rates of some typical biological reaction types are affected by changes in the values (i.e., concentrations) of state variables.

2.1. Basic Reaction Types

The most accurate model can be obtained when the law of mass action is used to formulate kinetic laws for all elementary reactions in Eq. 1. The *law of mass action* states that the rate of a reaction is proportional to the concentrations of the participating molecules.

Based on this general approach, we can distinguish the following basic reaction types that are frequently found in biological settings.

2.1.1. 0th Order Reactions: Constant Reaction Rates

0th order reactions are the simplest of all reactions because the rate of the reaction is constant and does not depend on any time-varying species. This kinetic law is used frequently to describe the synthesis of a molecular component.

Constant Synthesis

Assuming that the species X is produced at a constant rate k_{prod} we write for the concentration of X, $[X]$,

$$\frac{d[X]}{dt} = k_{prod}. \tag{3}$$

This equation can be solved as $[X(t)] = [X(t_0)] + k_{prod}(t - t_0)$ and we note that the concentration of X at time t depends only on the initial value of X at time t_0 and on the time interval $t - t_0$ that has passed. Accordingly the rate at which X is produced does not change when the concentration of X is changed (Fig. 2A, a).

2.1.2. 1st Order Reactions: Monomolecular Reactions

Most biological reactions are catalyzed or affected by components whose concentrations vary with time. Reactions that only depend on one such component are referred to as monomolecular reactions. Important examples include the decay of a molecular species or its transport between compartments, i.e., cytoplasm and nucleus.

Linear Degradation

The rate at which a protein, mRNA, or similar is removed or inactivated is often proportional to its own abundance, i.e., it changes linearly with its own concentration (Fig. 2A, b). We write for the concentration of such a component X

$$\frac{d[X]}{dt} = -k_{deg}[X]. \tag{4}$$

This equation can be solved as $[X(t)] = [X(t_0)]\exp(-k_{deg}(t - t_0))$ and we note that the concentration of X decays exponentially over time. An important measure is the characteristic time $t_{1/2} = \frac{\ln(2)}{k_{deg}}$ by which the initial concentration $[X(t_0)]$ has decreased by half.

Shuttling Between Compartments

Similarly the shuttling between two compartments (i.e., nucleus and cytoplasm) can be described by two coupled differential equations for the concentrations of X in the nucleus, $[X_{\text{n}}]$, and in the cytoplasm $[X_{\text{c}}]$. If X is exported from the nucleus at rate k_{out} and imported from the cytoplasm at rate k_{in} then we have

$$\begin{aligned} \frac{d[X_n]}{dt} &= k_{in}[X_c] - k_{out}[X_n], \\ \frac{d[X_c]}{dt} &= -\frac{d[X_n]}{dt} = -k_{in}[X_c] + k_{out}[X_n]. \end{aligned} \tag{5}$$

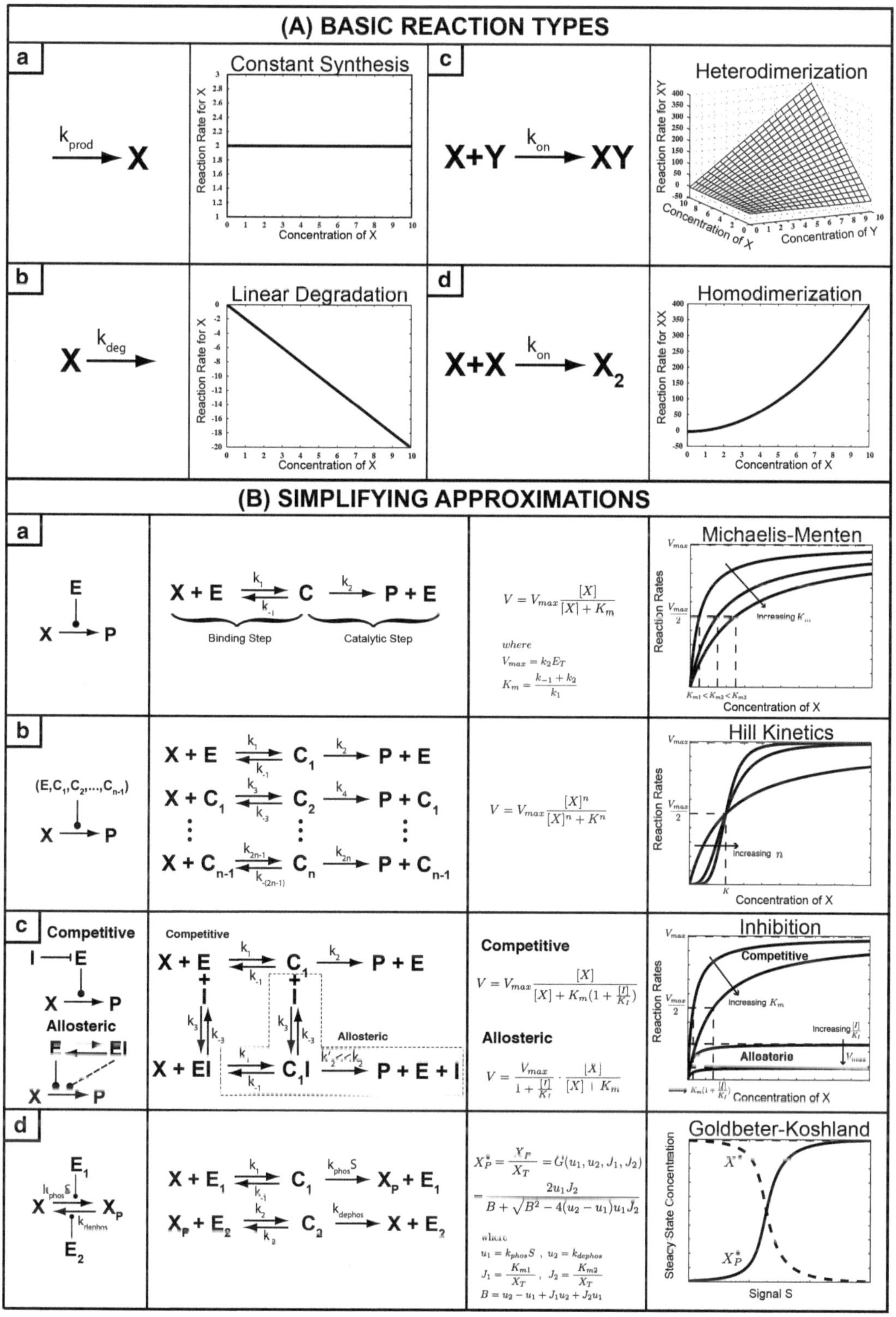

Fig. 2. Basic reaction types (A): (**a**) Constant synthesis. (**b**) Monomolecular reactions: Linear degradation. (**c**) Bimolecular reactions: Heterodimer formation. (**d**) Bimolecular reactions: Homodimer formation. Simplifying approximations (B): (**a**) Michaelis–Menten kinetics. (**b**) Hill kinetics. (**c**) Hill kinetics with allosteric or competitive inhibition. (**d**) Goldbeter–Koshland kinetics. The reaction scheme and a plot of the representative reaction rate versus the concentration of the reactant *X* (A) or the enzyme *E* (B) are depicted. In (B, d) the steady state concentrations of X_p and *X* are plotted versus the signal strength *S*.

We can reformulate this set of equations by observing that the total amount of X is conserved, i.e., $[X_c]V_c + [X_n]V_n = T = const.$ V_c and V_n refer to the different volumes of cytoplasm and nucleus. These volume factors are important since only the number of proteins is conserved (according to conservation of mass). Given the difference in volume the concentrations are, however, not conserved. We can then write $[X_c] = \frac{T-[X_n]V_n}{V_c}$ and obtain a differential equation that is similar to Eq. 1.4 except for an additional constant term, $k_{in}\frac{T}{V_c}$,

$$\begin{aligned} \frac{d[X_n]}{dt} &= k_{in}\frac{T-[X_n]V_n}{V_c} - k_{out}[X_n] \\ &= k_{in}\frac{T}{V_c} - \left(k_{in}\frac{V_n}{V_c} + k_{out}\right)[X_n]. \end{aligned} \tag{6}$$

2.1.3. 2nd Order Reactions: Bimolecular Reactions

Most reactions in biology involve some form of complex formation and therefore depend on the interaction of more than one time-varying component. Here it is important to distinguish between homo- and heterodimerization.

Complex Formation: Heterodimers

The formation of heterodimers, XY, is the result of the interaction of two components X and Y (Fig. 2A, c). The reaction rate depends linearly on both the concentrations of X and of Y. Assuming that the reaction proceeds at rate k_{on} and that the total concentrations of both components is constant we have

$$\frac{d[XY]}{dt} = k_{on}[X][Y] = k_{on}(X_T - [XY])(Y_T - [XY]), \tag{7}$$

where $X_T = [X] + [XY]$ and $Y_T = [Y] + [XY]$ are the total concentrations of X and Y, respectively. This equation can be solved to give $[X](t) = X_T - [XY(t)] = \frac{Y_T - X_T}{\frac{Y_T}{X_T}\exp((Y_T - X_T)kt) - 1}$.

Complex Formation: Homodimers

Similarly, the kinetics of homodimer formation between two X components can be described by the following quadratic rate law:

$$\frac{d[X_2]}{dt} = k_{on}[X]^2 = k_{on}(X_T - 2[X_2])^2, \tag{8}$$

where X_T is the total amount of X which we again assume to be constant. Here the rate of homodimer X_2 formation depends non-linearly on the concentration of the monomers X (Fig. 2A, d). The equation can be solved to give $[X](t) = \frac{X_T}{2X_T kt + 1}$.

We note that there are many cases in which the total concentrations are not constant. The above simplification would then not apply and a set of coupled ODEs for the monomers and the dimers would then need to be solved. In case of higher order complexes the formation can, in general, be modeled as a sequential step of bimolecular reactions.

2.2. Simplifying Approximations

If we formulate the kinetics of large networks based on first principles, then the description becomes very complex and will be accurate only if we are able to determine a large number of parameters with high accuracy. In particular, in case of cooperative enzymes, it can be very difficult to obtain accurate data on the reaction rates of all intermediate complexes. Most of the times we do not know all elementary/molecular interactions that regulate a particular reaction. Therefore there are many situations in which simplifications are sufficient and in fact preferable. Even from a computational point of view, it can make calculations more efficient.

2.2.1. Michaelis–Menten Kinetics

One frequently used approximation is quasi-stationarity of a reaction. This is used in the derivation of Michaelis–Menten kinetics for the enzymatic turnover of a substrate (Fig. 2B (a) LHS). Here the different timescales are exploited on which reactions proceed. If some reactions proceed much faster than others then certain concentrations are constant at the beginning while other concentrations barely change at a later time. To derive the Michaelis–Menten kinetics for the enzymatic reaction in the second column in Fig. 2B (a), it is assumed that the substrate concentration, $[X]$, is much larger than the total amount of the enzyme, $E_T = [E] + [C]$, and that the formation of the complex, C, is fast in comparison to the formation of product P. The concentration of the complex then quickly reaches a quasi-stationary concentration $[C] = \frac{k_1[X]}{k_1[X]+k_{-1}+k_2} E_T$. The rate at which the product P is formed is then given by

$$\frac{d[P]}{dt} = k_2 C = k_2 E_T \frac{k_1[X]}{k_1[X] + k_{-1} + k_2} = v_{max} \frac{[X]}{[X] + K_m}. \quad (9)$$

$v_{\mathrm{max}} = k_2 E_{\mathrm{T}}$ is the maximal rate at which this reaction can proceed when the substrate concentration is large ($[X] \gg K_{\mathrm{m}}$). $K_m = \dfrac{k_{-1} + k_2}{k_1}$ is the Michaelis–Menten constant and specifies the substrate concentration at which the reaction proceeds at half-maximal rate. For a more detailed discussion of the derivation of Michaelis–Menten kinetics, see standard text books in Mathematical Biology (1; 2). Importantly, the rate at which product is formed versus the substrate concentration yields a hyperbolic graph (Fig. 2B (a) RHS). While the conditions for Michaelis–Menten kinetics do not always strictly apply, such dependency of the reaction rate on the substrate concentration is observed more generally. In such cases, the reaction rate ν can be approximated by $v = v_{max} \frac{[X]}{[X]+K_m}$.

2.2.2. Hill Kinetics: Cooperativity

Many proteins have more than one binding site for their interaction partners (Fig. 2B (b) LHS). Binding of the first ligand can trigger a conformational change that alters the binding characteristics at all binding sites (Fig. 2B (b) 2nd column). The detailed modeling of all interactions and transitions is tedious. It can be shown (2) that if the first ligand binds with very low affinity (i.e., large $K_1 = \frac{k_{-1}+k_2}{k_1}$), and all subsequent ligands $i = 2 \ldots n$ bind with an increasing affinity (i.e. smaller K_i), then

$$\frac{d[P]}{dt} = v_{max} \frac{[X]^n}{K^n + [X]^n}. \tag{10}$$

Strictly speaking this formula is obtained in the limit $K_1 \to \infty$ and $K_n \to 0$ while keeping $K_1 K_n$ finite. $\frac{[X]^n}{K^n+[X]^n}$ is referred to as Hill function with Hill constant $K = (\prod_{i=1}^{n} K_i)^{\frac{1}{n}}$ and Hill coefficient n. If we plot the rate at which product is formed versus the substrate concentration, we obtain a sigmoid graph (Fig. 2B (b) RHS). The Hill constant K corresponds to the concentration at which the reaction proceeds at half-maximal speed. The Hill factor n determines the steepness of the response. Typically n is smaller than the total number of binding sites because the idealized limits from above do not apply.

2.2.3. Inhibitory Interactions

Inhibitors (I) of a chemical reaction either fully prevent a reaction or reduce the reaction rate. When the effect of an inhibitor is reversible, the steady state of the inhibited species is reduced, whereas in the case of irreversible inhibition the steady state is zero. Here we will only focus on reversible inhibitions. An important regulatory paradigm is the use of inhibitors and activators to modulate the speed of reactions. Inhibitors can either compete with the substrate for the catalytic cleft (competitive inhibition) or alternatively inhibitors can induce a conformational change that alters the activity of the enzyme (allosteric inhibition).

Competitive Inhibition

Inhibitors that bind to the active site of an enzyme and compete with substrate for access are termed competitive inhibitors (Fig. 2B (c), excluding dotted part). As a result a higher amount of substrate is required to achieve a particular reaction rate. It can be shown (2) that this competition is reflected in an increase in the Michaelis–Menten constant from K_m to $K_m(1 + \frac{I}{K_I})$, where K_I is the dissociation constant for the enzyme–inhibitor interaction. The velocity of the reaction is then

$$\frac{d[P]}{dt} = k_2[C_1] = v_{max} \frac{[S]}{[S] + K_m(1 + \frac{[I]}{K_I})}. \tag{11}$$

Similarly, in case of Hill kinetics competitive inhibition is modeled by an increase in the Hill constant K by a factor of $1 + \frac{[I]}{K_i}$.

Allosteric Inhibition

Allosteric (or noncompetitive) inhibitors do not bind to the substrate binding site but affect the reaction rate by binding to a different site where they may induce a conformational change (Fig. 2B (c), including dotted part). While this conformational change can, in principle, also affect the binding affinities in the active site, allosteric inhibitors, in general, reduce the maximal velocity of the reaction v_{max} (i.e., $k_2' \ll k_2$ in Fig. 2B, c), and we have

$$v = \frac{v_{max}}{1 + \frac{[I]}{K_I}} \frac{[X]}{K_m + [X]}. \tag{12}$$

The above approximations can be combined in more complex reaction schemes. Thus if multiple inhibitors bind to the enzyme in a cooperative way, i.e., the first inhibitor binds weakly (with large dissociation constant $K_{I1} \rightarrow \infty$) and the subsequent inhibitors $i = 2 \ldots p$ with higher affinity (decreasing dissociation constant $K_{Ii} \rightarrow 0$) while $K_{I1}K_{Ip}$ remains finite, then in line with Subheading 2.2.2

$$v = \frac{v_{max}}{1 + \left(\frac{[I]}{K_I}\right)^p} \frac{[X]}{K_m + [X]} \tag{13}$$

where $K_I^p = \prod_{i=1}^{p} K_{Ii}$ and p is analogous to the Hill constant.

2.2.4. Goldberger-Koshland Kinetics

The biological activity of signaling proteins is often controlled by a reversible chemical transformation, e.g., phosphorylation, methylation etc. If we were to model all steps explicitly the models would again be complex (Fig. 2B, d), and experimental data may be missing to parameterize the model. These enzymatic reactions are therefore often approximated with Michaelis–Menten reactions, and we have for the kinetics of the phosphorylated and unphosphorylated forms, X_p and X, respectively,

$$\frac{d[X_p]}{dt} = -\frac{d[X]}{dt} = k_{phos} S \frac{X_T - [X_p]}{K_{M1} + X_T - [X_p]} - k_{dephos} \frac{[X_p]}{K_{M2} + [X_p]}, \tag{14}$$

where k_{phos} and k_{dephos} are the rates of phosphorylation and dephosphorylation, respectively. The concentrations of the enzymes E_1 and E_2 do not explicitly appear here as they are considered to be approximately constant and therefore are lumped into the rate constants. K_{M1} and K_{M2} are the Michaelis–Menten constants of the respective enzymatic reactions. Here S refers to an external signal that is assumed to only affect the kinase and thus the phosphorylation reaction. In equilibrium $\frac{d[X_p]}{dt} = \frac{d[X]}{dt} = 0$ and we obtain the Goldberger-Koshland formula

$$\begin{aligned} X^* &= \frac{[X_p]}{[X_T]} = G(u_1, u_2, J_1, J_2) \\ &= \frac{2u_1 J_2}{B + \sqrt{B^2 - 4(u_2 - u_1)u_1 J_2}}, \end{aligned} \tag{15}$$

where $u_1 = k_{phos}$ S, $u_2 = k_{dephos}$, $J_1 = \frac{K_{M1}}{X_T}$, $J_2 = \frac{K_{M2}}{X_T}$, and $B = u_2 - u_1 + J_1 u_2 + J_2 u_1$. X_T refers to the total concentration of the signal protein X, i.e., $X_T = [X] + [X_p]$. In the context of larger regulatory networks with such regulatory motif (Fig. 2d), the Goldberger-Koshland formula can be used to approximate the

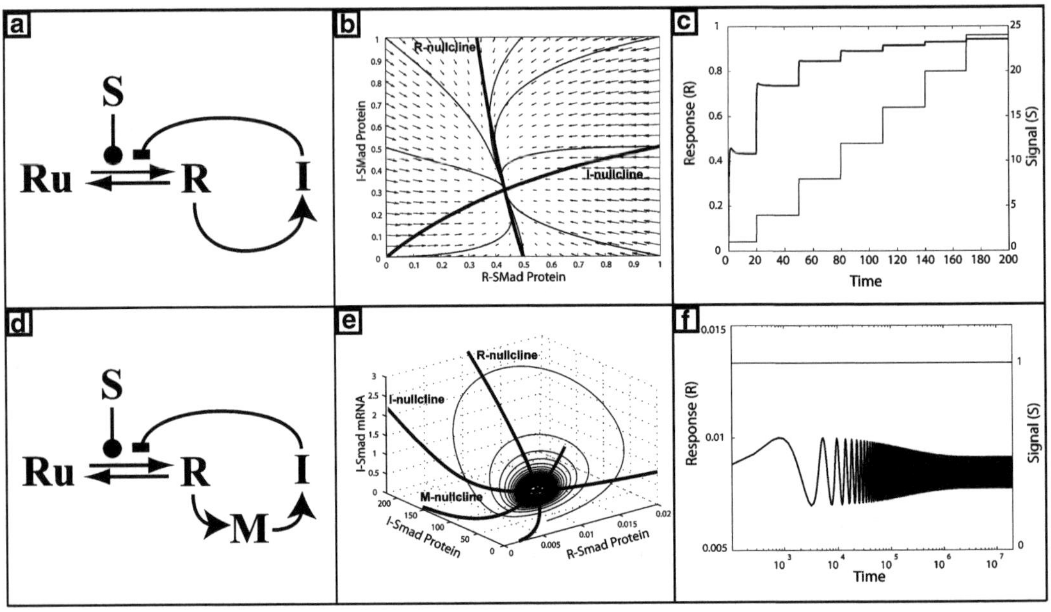

Fig. 3. Modeling examples The wiring diagram (**a, d**), the phase plane (**b, e**), and the time-dependent evolution of activated R-Smad levels (R) in response to signal pulses (**c, f**) for a 2-component (**a–c**) or a 3-component (**d–f**) model for TGF-beta signaling.

fraction of active enzyme dependent on the input signal S as long as quasi-stationarity for the reaction that regulates the enzyme relative to the rest of the network is a reasonable assumption.

2.3. Example: TGF-Beta Signaling

As an example we develop a model for TGF-beta signaling. In order to illustrate the use and limitations of models and to explain how models must and can be extended to analyze biological questions, we will start with one of the simplest possible models. The considered components and interactions are graphically summarized in Fig. 3a. The model considers only two time-dependent variables that describe the dynamics of the concentrations of phosphorylated R-Smad, $[R]$, and for the concentration of the inhibitory I-Smad, $[I]$. These two components are part of five reactions: (1) the signal-dependent activation of the R-Smad, (2) the subsequent induction of I-Smad production, and (3) the negative feedback of the I-Smad back on the R-Smad. Both proteins are turned over in reactions (4) and (5). After having decided the topology and the reactions of the network we need to define the kinetic laws for the gain and the loss rates. We assume that the rate at which phosphorylated R-Smad is formed depends linearly on the concentration of unphosphorylated R-Smad, $[R_u]$. If we assume that the total concentration of R-Smad, R_T, does not change on the considered timescale, i.e., there is no expression of degradation of R-Smads or the two processes perfectly balance, then we can write $[R] + [R_u] = R_T$.

The gain rate could then be formulated as $v_+ = k_1(R_T - [R])$. If we further assume that the maximal rate of R-Smad phosphorylation depends linearly on the signal strength, S, then we need to extend our formulation of the gain rate and write $v_+ = k_1 S(R_T - [R])$. The I-Smad inhibits phosphorylation of the R-Smad by binding to the free receptor as well as to the receptor–ligand complex. Since the I-Smad and the ligand bind the receptor at different sites we assume allosteric cooperative inhibition of the ligand-induced activation. In analogy to Subheading 2.2.3, we then have for the rate of R-Smad phosphorylation, $v_+ = \frac{k_1 S}{1+\left(\frac{[I]}{K_I}\right)^p}(R_T - [R])$. As regards the loss rate, we assume that the rate of R-Smad dephosphorylation depends linearly on the concentration of phosphorylated R-Smad ($v_- = k_2[R]$). This implies that it is the availability of the substrate $[R]$ rather than the availability of the phosphatase that is limiting. The concentration of the phosphatase can therefore be considered to be constant and can be lumped into the reaction constant. If the phosphatase was limiting we would need to use a Michaelis–Menten kinetics or Hill kinetics for the reaction as discussed in connection with the Goldbeter–Koshland kinetics. If both concentrations were not limiting then the reaction would proceed at a constant rate over time. We thus have for the kinetics of the active R-Smad, R,

$$\frac{d[R]}{dt} = v_+ - v_- = \frac{k_1 S}{1+\left(\frac{[I]}{K_I}\right)^p}(R_T - [R]) - k_2[R]. \quad (16)$$

The production of the I-Smad depends on the concentration of the active R-Smad, a transcription factor. R-Smad most likely binds to DNA in a cooperative fashion. Accordingly, we model the rate of I-Smad production (gain rate) by a Hill function, $\frac{[R]^q}{[R]^q + K_R^q}$, with Hill constant K_R and Hill factor q. k_3 is the maximal rate at which the I-Smad can be produced when R-Smad is abundant ($[R] \gg K_R$). We assume linear decay of the I-Smad at rate k_4, i.e., we assume that the concentration of the protease that degrades the I-Smad is not limiting. We then have

$$\frac{d[I]}{dt} = k_3 \frac{[R]^q}{[R]^q + K_R^q} - k_4[I]. \quad (17)$$

3. Part II: Model Analysis

3.1. Nondimensionalization

Before we continue with the analysis of this model it is sensible to simplify the mathematical formulation. We will rescale all variables and parameters in a way that the new variables and parameters have no physical dimensions and can therefore more easily be compared. Moreover, we will reduce the total number of parameters in the model by combining parameter values. There is no standard method to nondimensionalize a model, and some consider the procedure an art because a clever nondimensionalisation can sometimes greatly facilitate the subsequent analysis of the problem. There are some guidelines, however: (1) If there is a maximal value that a variable can attain it is sensible to normalize with respect to this maximal value. (2) If a variable is linked to a certain parameter inclusion of this parameter in the normalization can reduce the total number of parameters in the model. (3) If possible, parameters should be combined to obtain small and large parameters as this enables the use of perturbation methods. Keeping all this in mind, we rewrite the model in dimensionless form by making the following substitutions:

$$R = \frac{[R]}{[R_T]} \Leftrightarrow [R] = R[R_T], \tag{18}$$

$$I = \left(\frac{[I]}{K_I}\right)^p \Leftrightarrow [I] = IK_I, \tag{19}$$

$$\tau = k_2 t \Leftrightarrow t = \frac{\tau}{k_2}. \tag{20}$$

The nondimensionalized model then reads

$$\frac{dR}{d\tau} = \sigma \frac{1 - R}{1 + I^p} - R = f(R, I), \tag{21}$$

$$\frac{dI}{d\tau} = \kappa_1 \frac{R^q}{R^q + K^q} - \kappa_2 I = g(R, I) \tag{22}$$

with $\sigma = \frac{k_1}{k_2} S$, $\kappa_1 = \frac{k_3}{k_2 K_I}$, $\kappa_2 = \frac{k_4}{k_2}$, and $K = \frac{K_R}{R_T}$.

3.2. Equilibrium Concentrations

The concentrations of the R-Smad and the I-Smad in the limit of long times, i.e., when the system has attained its equilibrium, are obtained by setting the time derivatives to zero, i.e.,

$$\frac{dR}{dt} = 0 \Rightarrow R(I) = \frac{\sigma}{\sigma + 1 + I^p} \qquad R - nullcline, \tag{23}$$

$$\frac{dI}{dt} = 0 \Rightarrow I(R) = \frac{\kappa_1}{\kappa_2} \frac{[R]^q}{[R]^q + K^q} \qquad I - nullcline. \tag{24}$$

The two functions $R(I)$ and $I(R)$ are referred to as *nullclines*. Since the nullclines are those curves for which the respective time derivatives are zero, on the R-nullcline R does not change with time ($\frac{dR}{dt} = 0$), while on the I-nullcline I does not change with time ($\frac{dI}{dt} = 0$). An algebraic calculation of the steady-state values of R and I is tedious. It is easier to determine the steady state graphically as the intersection of the two nullclines, i.e. the point where all the temporal derivatives are zero (Fig. 3b).

3.3. Phase Plane Analysis

Figure 3b is referred to as *phase plane*. We can use the phase plane to understand how the dynamic system evolves in time. For this, we plot the trajectories

$$\vec{x}(t) = \begin{pmatrix} R(t) \\ I(t) \end{pmatrix}, \tag{25}$$

a parametric representation of R and I with parameter t, in the I-R plane over time t (Fig. 3b, gray lines). As we can see all trajectories meet in a common point, the intersection point of the two nullclines. To see how the system develops from any point (R^*, I^*) in the phase plane we plot small arrows that represent the vector field of tangents

$$\vec{t} = \begin{pmatrix} dR/dt \\ dI/dt \end{pmatrix} \tag{26}$$

to the trajectory $\vec{x}(t)$. These tangent vectors $\vec{t}$ point in the direction in which the system develops from a point (R^*, I^*). The length of the tangent vector indicates the speed with which the system will change. Accordingly the arrows are of zero length in the steady state. As the time derivative along the nullclines is zero, the system is unable to evolve in their direction. Therefore the trajectories can only cross the nullclines perpendicularly. If all vectors point to the steady state then the steady state is globally stable, i.e., the system returns to the steady state after each perturbation. If only the arrows in the vicinity of the steady state all point to the steady state then the steady state is said to be locally stable. The phase plane with the trajectories and phase vector field is called a *phase portrait*.

In the phase portrait for the 2-dimensional TGF-beta model, all trajectories converge in the steady state (Fig. 3b). The steady state is therefore stable. It should be noted that despite this stability, an increase in the signaling strength still leads to an increase in the steady-state concentration of the active R-Smad. The stable steady state moves in the phase plane (the R-nullcline is being shifted to the right), but it remains stable and the system does not overshoot as the steady state moves (Fig. 3c). There is therefore neither a transient response nor perfect adaptation in response to a change in the signaling strength. For a more detailed discussion for conditions of adaptation, see (3, 4).

3.3.1. Oscillation

We can introduce sustained oscillations by including a further intermediate step, the formation of I-Smad mRNA (M) in the model for TGF-beta signaling (Fig. 3d). By introducing an intermediate step we introduce a time delay. This *time delay* together with the *nonlinearity* as introduced by the Hill factors and the *negative feedback* of I on R are necessary to obtain oscillations. For a more detailed discussion of oscillators, see reviews by Novak and Tyson (5; 6). If it is assumed that the translation of mRNA into protein as well as all decay processes are linear processes, the model reads in its nondimensional form

$$\begin{aligned}\frac{dR}{d\tau} &= \sigma(1-R)\frac{1}{1+I^p} - \delta_1 R = F(R,M,I),\\ \frac{dM}{d\tau} &= \mu\frac{R^q}{R^q+K^q} - \delta_2 M = G(R,M,I),\\ \frac{dI}{d\tau} &= M - \delta_3 I = H(R,M,I).\end{aligned} \tag{27}$$

For appropriate *parameter values*, we now obtain a limit cycle around the steady state (Fig. 3e) and sustained oscillations emerge (Fig. 3f).

3.4. Linear Stability Analysis

For sufficiently simple systems, a graphical analysis can be employed to judge the stability of a steady state and to reveal oscillations, adaptation, or switches. In the following, we discuss a more generally applicable method, the linear stability analysis, to evaluate the (local) stability of steady states. The idea behind a linear stability analysis is to introduce a small perturbation at the steady state and to study whether this perturbation grows or decays over time. In the first case, the steady state is said to be unstable while in the latter case the steady state is stable. Linearization of the system of differential equations at the steady state greatly facilitates the analysis but means that our results apply only locally, i.e., in the vicinity of the studied steady state. Since we will deal with linear systems we start by combining our set of state variables in a vector, i.e.,

$$\vec{x} = \begin{pmatrix} R(t) \\ I(t) \end{pmatrix} \quad \textit{and} \quad \frac{d\vec{x}}{dt} = \begin{pmatrix} \dot{R}(t) \\ \dot{I}(t) \end{pmatrix} = \begin{pmatrix} f(R,I) \\ g(R,I) \end{pmatrix}, \tag{28}$$

where the dots denote time derivatives The functions f and g are in general nonlinear functions. Let $\vec{w} = \vec{x} - \vec{x}_s = (R - R_s, I - I_s)^T$. We can approximate the values of $f(R, I)$ and $g(R, I)$ close to this steady state (R_s, I_s) using Taylor series expansion, i.e.,:

$$\begin{aligned} f(R,I) = f(R_s,I_s) &+ \frac{1}{1!}\frac{\partial f}{\partial R}|_s(R-R_s) + \frac{1}{2!}\frac{\partial^2 f}{\partial R^2}|_s(R-R_s)^2 \\ &+ \ldots + \frac{1}{1!}\frac{\partial f}{\partial I}|_s(I-I_s) + \frac{1}{2!}\frac{\partial^2 f}{\partial I^2}|_s(I-I_s)^2 + \ldots \end{aligned} \tag{29}$$

and likewise for $g(R, I)$. We next linearize the system by ignoring all terms that are of order two and higher, i.e.,

$$\begin{aligned} f(R,I) &= f(R_s,I_s) + \frac{\partial f}{\partial R}|_s(R-R_s) + \frac{\partial f}{\partial I}|_s(I-I_s),\\ g(R,I) &= g(R_s,I_s) + \frac{\partial g}{\partial R}|_s(R-R_s) + \frac{\partial g}{\partial I}|_s(I-I_s). \end{aligned} \tag{30}$$

The steady state is stable if the perturbation decays to zero for long times t, i.e., $\vec{w} \to 0$ as $t \to \infty$. To express the differential equation in terms of $\vec{w}$ we use the linearized system of equations, and we write

$$\frac{d\vec{x}}{dt} = \frac{d\vec{x}_s}{dt} + J(\vec{x} - \vec{x}_s) \Rightarrow \frac{d\vec{w}}{dt} = J\vec{w}. \quad (31)$$

J is referred to as Jacobian, and represents the first-order expansion of the differential equations at the steady state (R_s, I_s), i.e.,

$$J = \begin{pmatrix} \frac{df}{dR}|_{Rs,Is} & \frac{df}{dI}|_{Rs,Is} \\ \frac{dg}{dR}|_{Rs,Is} & \frac{dg}{dI}|_{Rs,Is} \end{pmatrix} = \begin{pmatrix} f_R\, f_I \\ g_R\, g_I \end{pmatrix}. \quad (32)$$

For the simple TGF-beta model, we obtain from Eqs. 21 and 22

$$J = \begin{pmatrix} -\frac{\sigma}{1+I_s^p} - 1 & -\sigma\frac{pI_s^{p-1}}{(1+I_s^p)^2} \\ \kappa_1\frac{qR_s^{q-1}K^q}{(R_s^q+K^q)^2} & -\kappa_2 \end{pmatrix} = \begin{pmatrix} -- \\ +- \end{pmatrix}. \quad (33)$$

The sign of the entries in the Jacobian reflect, the type of interactions. Since both R and I affect their own concentration negatively, $\frac{df}{dR}|_{Rs,Is}<0$, $\frac{dg}{dI}|_{Rs,Is}<0$. Furthermore R has a positive impact on I, and accordingly the entry J_{21} is positive, while I has a negative impact of R, and accordingly $J_{12} < 0$.

The differential equation for $\vec{w}(t)$, $\frac{d\vec{w}}{dt} = J\vec{w}$, can be solved as $\vec{w}(t) = \sum_i \beta_i \vec{v}_i \exp(\lambda_i t)$. The β_i are constants and are determined by the initial conditions. v_i and λ_i are the eigenvectors and eigenvalues of J, and follow from the relation $J\vec{v} = \lambda\vec{v}$ which can be rewritten as $(J - \lambda I)\vec{v} = 0$ (where I is the identity matrix). Unless $\vec{v} = 0$ we must have

$$\det(J - \lambda I) = 0. \quad (34)$$

For a system with n coupled ODEs Eq. 34 leads to the *characteristic polynomial* of degree n,

$$P(\lambda) = \sum_{i=0}^{i=n} \alpha_i \lambda^i = 0. \quad (35)$$

For a system with two coupled ODEs we have $P(\lambda) = \lambda^2 - tr(J) + \det(J) = 0$, where $tr(J)$ refers to the trace of J and $\det(J)$ to the determinant of J; accordingly

$$\lambda_{1,2} = \frac{tr(J) \pm \sqrt{tr(J)^2 - 4\det(J)}}{2} \quad (36)$$

Even though $\vec{w}(t)$ is a real-valued vector we notice from Eq. 36 that the eigenvalues (as well as the eigenvectors and the β_i) can be complex numbers, i.e. $\lambda_i = \Re(\lambda_i) + i\Im(\lambda_i)$ and thus

$$\vec{w}(t) = \sum_i \beta_i \vec{v}_i \exp(\lambda_i t) = \sum_i \beta_i \vec{v}_i \exp(\Re(\lambda_i)t)\exp(i\Im(\lambda_i)t). \tag{37}$$

The stability of the steady-state depends on whether or not the initial perturbation $\vec{w}(0)$ decays to zero as $t \to \infty$. The absolute value of $\exp(i\Im(\lambda_i)t) = \cos(\Im(\lambda_i)t) + i\sin(\Im(\lambda_i)t$ is one for all times t. The long-term behaviour of $\vec{w}(t)$ therefore depends on $\exp(\Re(\lambda_i)t)$. If the real parts of all eigenvalues are negative ($\Re(\lambda_i) < 0 \ \forall i$) then the exponential functions $\exp(\Re(\lambda_i)t)$ decay to zero for large times and so does the perturbation $\vec{w}$. We thus require that the real parts of all eigenvalues are negative for the steady state to be linearly (locally) stable. Contrary if all real parts of the eigenvalues are positive ($\Re(\lambda_i) > 0 \ \forall i$) then the steady state is (locally) unstable. If some of the real parts are positive and others are negative then we speak of a (locally) half-stable steady state.

In case of a 2-component model we require $tr(J) < 0$ and $\det(J) > 0$ for the real parts of all eigenvalues to be negative (Eq. 36). By inserting the entries from the Jacobian in Eq. 38 we obtain for our system of interest $tr(J) = f_R + g_I = -\frac{\sigma}{1+I_s^p} - 1 - \kappa_2$ and $\det(J) = f_R g_I - f_I g_R = (-\frac{\sigma}{1+I_s^p} - 1)(-\kappa_2) - (-\sigma\frac{pI_s^{p-1}}{(1+I_s^p)^2})(\kappa_1 \frac{qR_s^{q-1}K^q}{(R_s^q+K^q)^2})$. We thus have $tr(J) < 0$ and $\det(J) > 0$ for all σ and the steady state is therefore stable for all signal strengths $\sigma \geq 0$.

If (some of) the eigenvalues have non-zero imaginary part ($\Im(\lambda_j) \neq 0$) then $\vec{w}(t)$ oscillates in time. If the eigenvalues (λ_j) with non-zero imaginary part have positive real part ($\Re(\lambda_j) > 0$) then the amplitude of the oscillations grows with time and the trajectory is said to spiral out from the steady state (unstable steady state). If $\Re(\lambda_i) < 0$ then the oscillations dampen with time and the trajectories spiral into the steady state (stable steady state). If $\Re(\lambda_i) = 0$ then we have a center solution that gives rise to sustained oscillations, i.e. the amplitude of the oscillations does not change with time. The amplitude depends on the initial conditions and can thus change in response to perturbations.

In case of the 3-component model (Eq. 27) we have

$$J = \begin{pmatrix} -\frac{\sigma}{1+I_s^p} - \delta_1 & 0 & -\sigma\frac{pI_s^{p-1}}{(1+I_s^p)^2} \\ \mu\frac{qR_s^{q-1}K^q}{(R_s^q+K^q)^2} & -\delta_2 & 0 \\ 0 & 1 & -\delta_3 \end{pmatrix} = \begin{pmatrix} - & 0 & - \\ + & - & 0 \\ 0 & + & - \end{pmatrix} \tag{38}$$

and the coefficients of the characteristic polynomial (Eq. 35) are thus given as $\alpha_3 = 1 > 0$, $\alpha_2 = -tr(J) > 0$, $\alpha_1 = (\delta_2 + \delta_3)(\frac{\sigma}{1+I_s^p} + \delta_1) + \delta_2\delta_3 > 0$, and $\alpha_0 = -\det(J) > 0$ for all signal strength $\sigma \geq 0$. We can determine the stability of the steady states without solving this 3rd order polynomial by using the *Ruth-Hurwitz criterion* and *Descartes' Rule of Signs*. According to the Ruth-Hurwitz criterion all roots of this 3rd order characteristic polynomial have negative real parts if all coefficients are positive and $\alpha_2\alpha_1 > \alpha_3\alpha_0$. While the first condition is met for all physiological parameter sets, the second condition is met only for some of the possible parameter sets. We can further consider Descartes' Rule of Signs.

3.4.1. Descartes' Rule of Signs

Consider the polynomial $P(\lambda) = \sum_{i=0}^{n} \alpha_i \lambda^i = 0$. Let N be the number of sign changes in the sequence of coefficients $\{\alpha_i, \alpha_{i-1}, \ldots, \alpha_0\}$, ignoring any that are zero. Then there are at most N roots of $P(\lambda)$ which are real and positive, and further, there are N, $N-2$, or $N-4$, real positive roots. By setting $\omega = -\lambda$ and again applying this rule, information is obtained about possible negative real roots.

We notice that there are no sign changes in the coefficients for λ and three sign changes for $\omega = -\lambda$. There are thus no positive, real roots, and either three or one negative, real root(s), and accordingly either three negative real eigenvalues, or one negative real eigenvalue and two conjugate complex eigenvalues. Depending on the choice of parameters (in particular, the signal strength σ), we thus expect either a stable steady state, a center solution (sustained oscillations), or a spiral. Figure 3e shows the phase plane for a particular choice of parameters and Fig. 3f records the values of R over time. The steady state is an unstable spiral but the oscillations appear to assume constant amplitudes after some time and the trajectories all evolve toward a single attractor. This attractor is referred to as a *limit cycle*.

3.4.2. Limit Cycles

Limit Cycles are isolated closed trajectories that can act as attractors to neighboring trajectories. Trajectories that emerge from a locally unstable steady state can thus be attracted to a limit cycle and give rise to sustained limit cycle oscillations. Limit cycles are inherently nonlinear phenomena that cannot be analyzed or predicted with linear methods. Linear methods only predict an unstable steady state but do not reveal the limit cycle attractor. In two dimensions the *Poincaré–Bendixson Theorem* can be used to identify a region in the phase plane that contains a limit cycle. Typically sustained oscillations in biology do not correspond to center solutions but to limit cycles.

3.5. Bifurcation Diagrams

Often the stability of steady states changes as parameter values are altered. Bifurcation diagrams are employed to analyze how the values and the stability of equilibrium points depend on a regulatory control parameter, the bifurcation parameter. Those points at which the stability of an equilibrium point changes or new steady-state solutions appear or disappear are called bifurcation points. Particularly important bifurcation behaviors in biological models include *transcritical bifurcations*, *saddle-node bifurcations*, and *Hopf bifurcations*. The first two bifurcations generate so-called switches where the system switches from one stable steady-state solution (which might correspond to inactive) to another stable steady-state solution (which might correspond to active), while the Hopf bifurcation gives rise to oscillatory solutions. Several software packages are available to draw bifurcation diagrams, including matcont (http://www.matcont.ugent.be) and auto (http://cmvl.cs.concordia.ca/auto).

3.5.1. Oscillations

Network topologies that permit oscillations have been reviewed by Novak and Tyson (5; 6). One such possible architecture, the negative feedback oscillator, is depicted in Fig. 4a. The middle column shows the bifurcation diagram with the signal strength S as bifurcation parameter. Stable equilibria are depicted as solid lines while unstable equilibria are depicted as dotted line. There are two Hopf bifurcation points. At Hopf bifurcations oscillatory solutions emerge (region (1)). In terms of the linear stability analysis a pair of complex conjugate eigenvalues of the linearized system crosses

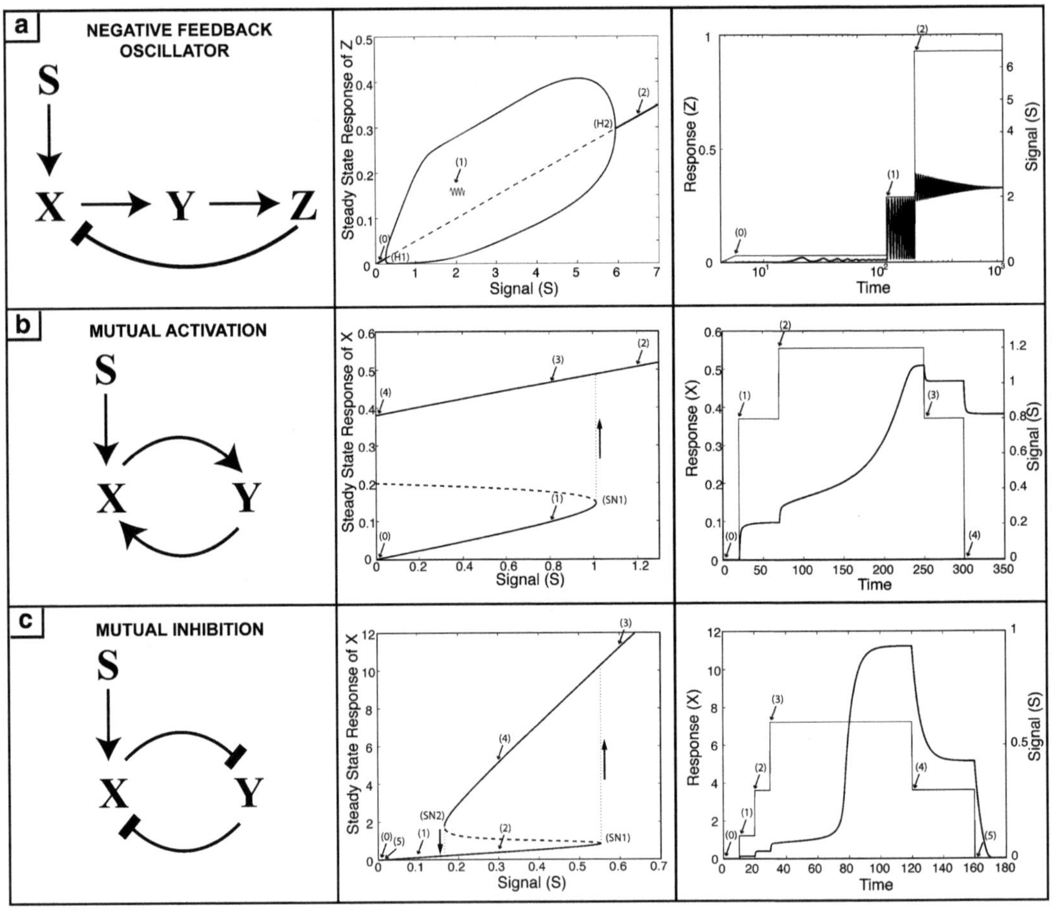

Fig. 4. Bifurcation diagrams. (**a**) Negative feedback oscillator, (**b**) Toggle switch (**c**) Buzzer. (*left*) Network motifs. (*center*) Bifurcation diagrams with bifurcation parameter *S*. Stable steady states are denoted by *solid lines*, unstable steady states by *dotted lines*. Hopf bifurcations are denoted by H, saddle-node bifurcations by SN. (*right*) Time-dependent responses to changes in signal strength *S*.

the imaginary axis of the complex plane. The equilibrium point thus becomes unstable (dotted lines) and the limit cycle becomes the new attractor. The solid lines in region (1) depict the extreme values that the limit cycle attains for component Z. The rightmost column shows the behavior of the system over time. Starting with a signal strength in region (0), we notice that oscillations decay away quickly to the stable steady state. As the signal strength is increased to fall into region (1) we obtain sustained oscillations; the amplitude depends on the signal strength. As the signal strength is increased further to fall in region (2), the oscillations again dampen away to the stable steady state.

3.5.2. Switches

Switches appear when an equilibrium point becomes unstable and the system jumps to another equilibrium as the bifurcation parameter passes the bifurcation point. When two equilibrium points collide and annihilate each other we speak of a saddle-node bifurcation (Fig. 4b, c). Saddle-node bifurcations generate switches that are robust to fluctuations, because they enable systems which have some kind of memory effect. In biological settings, we distinguish one-way switches (Fig. 4b) and toggle switches (Fig. 4c) which are created by mutual activation and mutual inhibition, respectively. Both bifurcation diagrams contain a region with three steady states (2 stable and 1 unstable). If the system is started on the lower branch at low signaling strength (point 0), it will follow this branch as the signal strength S is increased (point 1) until the system reaches the saddle-node bifurcation point (SN1) where the stable equilibrium branch collides with the unstable equilibrium branch. As the two steady states are annihilated at the bifurcation point, a further increase in the signal strength S results in a jump to the remaining equilibrium (point 2). If the signal strength is reduced again, the system continues to follow the new equilibrium branch (point 4) and does not switch back at the previous bifurcation point. The equilibrium that the system attains at a given signal strength S therefore depends on the history of the system, a phenomenon referred to as hysteresis. The new stable equilibrium branch meets the unstable branch at yet another saddle-node bifurcation point (SN2) where both equilibria are annihilated and the system jumps back to the initial stable equilibrium branch. If this second bifurcation point lies within the physiological range of the bifurcation parameter (here the signaling strength S), then the system can return to the previous steady state (point 5) and we speak of a toggle switch (Fig. 4c). If the second bifurcation point lies outside the range (i.e., on the negative axis) then the system cannot return, the switch is stable, and we speak of a one-way switch (Fig. 4b).

3.5.3. Ultrasensitivity

Switch-like changes in the equilibrium concentrations of a system can also be observed without a bifurcation, as long as the equilibrium point is very sensitive to a control parameter. Reactions that respond with greater sensitivity to a signal X then expected from a Michaelis–Menten kinetics are termed ultrasensitive. Ultrasensitivity is achieved in the Goldbeter–Koshland kinetics if the Michaelis–Menten constants are small and the enzymes are saturated. In the limit, we have zeroth-order kinetics and thus zeroth-order ultrasensitivity. A small change in $v_{\max}$ due to stimulation then leads to an immediate increase in the signaling component.

3.6. Sensitivity and Robustness

The sensitivity of dynamic systems is key to its regulatory control. It has been argued that life is positioned at the edge of chaos; cellular control needs to be stable enough to function but unstable enough to respond and evolve (7). It is therefore important to investigate the sensitivity and robustness of signaling models. A bifurcation analysis can be used to investigate how the stability and position of a steady states depends on a particular parameter value. In general, however, it is computationally infeasible to carry out a complete bifurcation analysis with respect to all parameter combinations and for all steady states. Instead, the response of the steady states to perturbations is evaluated to predict the qualitative behavior of the model. For large models, it is unfeasible to determine the sensitivity of all steady states with respect to all possible combinations of perturbations in parameter values. The analysis is therefore, in general, restricted to single parameter perturbations, and often only local parameter perturbations are considered.

3.6.1. Steady-State Sensitivity

In steady state, we have $\frac{dx}{dt} = f(\vec{x}(p), p, t) = 0$. The sensitivity of the steady-state value to changes in the parameter values, i.e., $\frac{dx}{dp}$ can be calculated from

$$\frac{d}{dp}\frac{dx}{dt} = \left(\frac{\partial f}{\partial x}\right)\frac{dx}{dp} + \left(\frac{\partial f}{\partial p}\right) = 0 \quad \Rightarrow \quad \left.\frac{dx}{dp}\right|_s = -\left.\left(\frac{\partial f}{\partial x}\right)\right|_s^{-1}\left.\left(\frac{\partial f}{\partial p}\right)\right|_s. \tag{39}$$

In general, x refers to a vector of the state variables, and p is a vector with the parameter values. In that case, $\left.\left(\frac{\partial f}{\partial x}\right)\right|_s$ refers to the Jacobian of the dynamical system, evaluated at the steady state. The inverse of the Jacobian $\left(\frac{\partial f}{\partial x}\right)^{-1}$ exists for most of the cases typically studied in biology. While this method is most precise, the sensitivity, $S_i = \frac{dx_s}{dp_i}$, of the steady state x_s of the dynamical system to perturbations in parameter p_i can also be assessed by numerically determining the finite differences $S_i = \frac{\Delta x_s}{\Delta p_i}$. Typically, however, normalized sensitivities, so-called control coefficients, are analyzed

$$C_i = \frac{\Delta x \setminus x_s}{\Delta p_i \setminus p_i}. \tag{40}$$

Typically parameters are perturbed by 1% ($\Delta pi = 0.01 p_i$), but other perturbations can be used.

3.6.2. Sensitivity of Trajectories

To explore the sensitivity of the trajectories to changes in the parameter values, we can calculate similar finite differences for the trajectories. A better approach is to calculate the sensitivity, $\frac{dx}{dp}(t)$, as we solve the dynamical system by simultaneously integrating

$$\frac{d}{dt}\frac{dx}{dp} = \frac{\partial f}{\partial x}\frac{dx}{dp} + \frac{\partial f}{\partial p}. \tag{41}$$

with initial condition $\frac{dx}{dp}(0) = 0$. This can be used to assign confidence intervals to the predicted trajectories based on confidence intervals on the parameter values as discussed in the next chapter.

4. Summary

Simple models of biological signaling networks provide powerful tools to understand and detect qualitative dynamical behaviors such as oscillations, adaptation, and switches. While such models are invaluable the level of simplification needs to be chosen carefully to not lead to wrong conclusions. By example of the model for TGF-beta signaling, we show how the level of detail affects the predicted qualitative behavior (stable steady state versus sustained oscillations). Conditions for which oscillations, adaptation, or switches can be observed in biological signaling networks are beyond the scope of this chapter, but have been reviewed and discussed extensively (3; 4; 5; 6). Equally, it is important to consider the sensitivity and robustness of the model, both with regard to parameter values and concentrations. A high sensitivity to parameter values implies that a mechanism may not exist as such in biology or that the parameters may be tuned and controlled by further sophisticated regulatory interactions. Great sensitivity to fluctuations in the components' concentrations can give rise to a large heterogeneity in a cellular population.

The impact of parameter values on the dynamics of the system can be predicted by analyzing the attractors of the dynamical systems: Phase plane analysis and a linear stability analysis can be employed to evaluate the local stability of equilibrium points and to predict the dynamic behavior. How the stability and position of a steady state depends on control parameter, so-called bifurcation parameters, can be visualized in bifurcation diagrams. There are three particularly important types of bifurcations in biological signaling models: transcritical, saddle-node, and Hopf bifurcations. The first two bifurcations generate a switch while the Hopf bifurcation leads to oscillations.

To integrate available experimental data in the model and to predict which behavior the system actually assumes under physiological conditions we, in general, require more detailed descriptions. As larger models are considered it becomes important to have efficient methods to encode these models. There are several software packages to assist. These include CellDesigner (www.celldesigner.org), Systems Biology Toolbox 2 (www.sbtoolbox2.org), and PottersWheel (www.potterswheel.de). CellDesginer allows models to be formulated graphically (see Fig. 2.1 in the next chapter) and to have the chart to be translated into a set of equations. While the equations can be directly simulated in Cell Designer, the Systems Biology Toolbox and PottersWheel offer much more powerful methods for analysis. It is possible to exchange the models between these different softwares using SBML (Systems Biology Markup Language, www.sbml.org) and it is advisable to analyze models with the help of software packages such as SBT, PottersWheel, or similar. Larger models comprise many more parameters and it is important to constrain these based on experimental information in order to arrive at a predictive power of the model. All the packages also offer methods to estimate parameter values based on experimental data. Details on parameter estimation will be provided in the next chapter.

Acknowledgments

We thank members of the Iber group for the critical reading of the manuscript. This work was financially supported by SystemsX, the Swiss Initiative for Systems Biology, with an iPhD grant and an RTD grant through the InfectX project.

References

1. Murray JD (2003) Mathematical biology. Springer, New York
2. Keener J, Sneyd J (2001) Math Physiol
3. Behar M, Hao N, Dohlman HG, Elston TC (2007) Mathematical and computational analysis of adaptation via feedback inhibition in signal transduction pathways. Biophys J 93(3): 806–821
4. Ma W, Trusina A, El-Samad H, Lim WA, Tang C (2009) Defining network topologies that can achieve biochemical adaptation. Cell 138(4):760–773
5. Novák B, Tyson JJ (2008) Design principles of biochemical oscillators. Nat Rev Mol Cell Biol 9(12):981–991
6. Tyson JJ, Chen KC, Novak B (2003) Sniffers, buzzers, toggles and blinkers: dynamics of regulatory and signaling pathways in the cell. Curr Opin Cell Biol 15(2):221–231
7. Kauffman SA (1993) The origins of order: self organization and selection in evolution. Oxford University Press, Oxford

Chapter 2

Analyzing and Constraining Signaling Networks: Parameter Estimation for the User

Florian Geier, Georgios Fengos, Federico Felizzi, and Dagmar Iber

Abstract

The behavior of most dynamical models not only depends on the wiring but also on the kind and strength of interactions which are reflected in the parameter values of the model. The predictive value of mathematical models therefore critically hinges on the quality of the parameter estimates. Constraining a dynamical model by an appropriate parameterization follows a 3-step process. In an initial step, it is important to evaluate the sensitivity of the parameters of the model with respect to the model output of interest. This analysis points at the identifiability of model parameters and can guide the design of experiments. In the second step, the actual fitting needs to be carried out. This step requires special care as, on the one hand, noisy as well as partial observations can corrupt the identification of system parameters. On the other hand, the solution of the dynamical system usually depends in a highly nonlinear fashion on its parameters and, as a consequence, parameter estimation procedures get easily trapped in local optima. Therefore any useful parameter estimation procedure has to be robust and efficient with respect to both challenges. In the final step, it is important to access the validity of the optimized model. A number of reviews have been published on the subject. A good, nontechnical overview is provided by Jaqaman and Danuser (Nat Rev Mol Cell Biol 7(11):813–819, 2006) and a classical introduction, focussing on the algorithmic side, is given in Press (Numerical recipes: The art of scientific computing, Cambridge University Press, 3rd edn., 2007, Chapters 10 and 15). We will focus on the practical issues related to parameter estimation and use a model of the TGFβ-signaling pathway as an educative example. Corresponding parameter estimation software and models based on MATLAB code can be downloaded from the authors's web page (http://www.bsse.ethz.ch/cobi).

Key words: Parameter estimation, Dynamics models, User guide

1. Pre-regression Diagnostics

Typically, modeling efforts are started after some experimental data has already been obtained, and based on these data a model is developed and parameterized. Initially, some (or most) parameter values will be unknown or can only be constrained to a biologically relevant range. Due to this incomplete information, a pre-regression analysis of the model is important to evaluate which and how

Xuedong Liu and Meredith D. Betterton (eds.), *Computational Modeling of Signaling Networks*, Methods in Molecular Biology, vol. 880, DOI 10.1007/978-1-61779-833-7_2,

model parameters can be estimated from future data. To estimate a parameter, changes in this parameter value need to affect the prediction of the model with regard to a state value for which there is data. In other words: the measured output must be sensitive to each parameter that we seek to estimate. Therefore, pre-regression analysis of the model is centered around sensitivity analysis, answering the question: "How does a system output depend of a certain parameter value?".

1.1. Sensitivity Analysis

A sensitivity analysis evaluates the dependence of a system output on a certain set of model parameters. Let $m = \{ m_1, m_2, \ldots, m_M\}$ be a set of M measurable output and $p = \{ p_1, p_2, \ldots, p_P\}$ the set of P unknown model parameters. The matrix of sensitivity coefficients $S_p{}^m$ (see Chapter 1) of the output with respect to the parameters is defined for each entry as

$$S_{p_j}^{m_i} = \frac{\partial m_i}{\partial p_j}, \quad i = 1, \ldots, M \quad j = 1, \ldots, P. \tag{1}$$

To optimally identify a model parameter, its corresponding sensitivities should be large and distinct from the sensitivities of all the other parameters. In other words, each column of the sensitivity matrix as defined by Eq. 1 should have at least one large entry and all columns must be linearly independent. If the latter does not hold, certain parameter changes can compensate for each other leaving the model output unchanged. As a result, parameter estimates will be correlated when inferred from the measurements. Since the sensitivities are state and time dependent, the way that the system is measured has a strong impact on the calculated sensitivities. Accordingly, sensitivity analysis can be used to plan new informative experiments with the aim to maximize the identifiability of the model parameters (2). Note that this type of analysis is local, i.e., applies only for a specific set of (a priori defined) parameter values. For a generally valid conclusion, the analysis may have to be carried out over a large set of biologically plausible parameter values.

2. Parameter Estimation

We now turn to the problem of parameter estimation. We highlight the major steps in the optimization of parameters with a special emphasis on gradient-based optimization methods.

2.1. The Model

Our dynamical system with N state variables can be described by a set of ordinary differential equations. If we write $x(t)$ for a vector with all state variables, k for a vector with all parameters, and x_0 for the vector for all initial expressions, this set of differential equations can be expressed as

$$\frac{dx(t)}{dt} = f(x(t), t, k), \qquad x(t_0) = x_0. \tag{2}$$

Often, the state variables cannot be directly observed, and there are combinations of state variables, or relative quantities (or possibly even more complicated functions of the state variables) that are measured in experiments. We therefore specify an observation function $g : \mathbb{R}^N \rightarrow \mathbb{R}^M$ which maps the state variables x to a set of M observables,

$$y(t) = g(x(t), s) \tag{3}$$

We require both $f(\cdot)$ and $g(\cdot)$ to be continuously differentiable functions with respect to their parameters. The vector s comprises the parameters of the observation function. Note that our formulation includes measurement settings where we can only partially observe the system such that $M < N$. The set of problem-specific parameters p includes the initial conditions, the model parameters, and the parameters that are specific to the measurements, i.e., $p = \{x_0, k, s\}$. The initial values of the dynamical system are also parameters as they are usually unknown.

2.2. Measurement Data

We denote the measured data by y_{ij}. Generally, these measurements y_{ij} are subject to error, i.e., they are the sum of the observables $y_j(t_i)$ and a measurement error, ε_{ij}.

$$y_{ij} = y_j(t_i) + \epsilon_{ij}. \tag{4}$$

In the following, measurement errors are assumed to be independent across all observations and all time points and follow independent Gaussian distributions with zero mean and state and time-dependent variance σ_{ij}^2. Due to the law of large numbers these assumptions apply in many practical settings. However, other distributions, in particular log-normal distributions, are also encountered, for instance when protein concentrations are low. Therefore, independence and normality should be checked in the course of data pre-processing. Data should be transformed appropriately in case of deviations from normality as discussed in (3). The observation function $g(\cdot)$ can take account for this transformation.

2.3. Nonlinear Regression

Intuitively, an optimal model should minimize the deviation between model prediction and data and thus make the measurements most likely given the model. In other words, an optimal parameter set is obtained by maximizing the likelihood L of the data y with respect to the parameter set p. The likelihood L takes the following form given our assumptions above:

$$L(y|p) = \prod_{i=1}^{T}\prod_{j=1}^{M}\frac{1}{\sigma_{ij}\sqrt{2\pi}}\exp\left(-\frac{1}{2}\frac{(y_{ij}-g_j(x(t_i,p),p))^2}{\sigma_{ij}^2}\right). \quad (5)$$

Due to the asymptotic properties of the maximum likelihood principle, it occupies a central position in estimation theory. In the limit of infinitely many data, it yields an unbiased, normally distributed parameter estimate with a minimal variance (4). However, in many practical settings, these properties are not matched due to limited amounts of data. Still, maximum likelihood is the most commonly used estimation principle because it is rather easy to implement. In practical terms, to find the maximum of the likelihood function the negative log likelihood is minimized.

$$-\log[L(y|p)] = \sum_{i=1}^{T}\sum_{j=1}^{M}\frac{1}{2}R_{ij}(p)^2 + c_{ij},$$

$$R_{ij}(p) = \frac{y_{ij}-g_j(x(t_i,p),p))}{\sigma_{ij}}, \qquad c_{ij} = \log\left[\sigma_{ij}\sqrt{2\pi}\right]. \quad (6)$$

The term c_{ij} in Eq. 6 is independent of p, and can be left out of the minimization. The maximum likelihood estimator for the model parameters is thus given by

$$p^* = \arg\min_{p} \sum_{i=1}^{T}\sum_{j=1}^{M}\frac{1}{2}R_{ij}(p)^2. \quad (7)$$

Note that Eq. 7 is essentially a least squares minimization problem (5).

2.4. Gradient Calculation

Common methods to minimize Eq. 7 are gradient based such as the classical Gauss–Newton or Levenberg–Marquardt methods (5). Gradient-based methods follow an iterative procedure in order to minimize Eq. 7 where in each step the gradient of the residuals $R_{ij}(p)$ is used to calculate a parameter update.

$$\begin{aligned}\frac{\partial}{\partial p_l}R_{ij}(p) &= -\frac{1}{\sigma_{ij}}\frac{\partial}{\partial p_l}g_j(x(t_i,p_l),p_l))\\ &= -\frac{1}{\sigma_{ij}}\left(\sum_{n=1}^{N}\left.\frac{\partial g_j}{\partial x_n}\right|_{t_i}\left.\frac{dx_n}{dp_l}\right|_{t_i}+\left.\frac{\partial g_j}{\partial p_l}\right|_{t_i}\right)\end{aligned} \quad (8)$$

$\frac{\partial g_j}{\partial x_n}$ and $\frac{\partial g_j}{\partial p_l}$ in Eq. 8 are the Jacobians of the differential equation system with respect to the state variables and with respect to the parameters. $S_{p_l}^n = (dx_n)/(dp_l)$ in Eq. 8 are the so-called sensitivities of the state variables to changes in the parameter values that we discussed in the previous chapter. In general there is no analytic solution to the trajectories $x(t, p)$ and therefore the sensitivities

$S^n_{p_l} = (dx_n)/(dp_l)$ in Eq. 8 have to be calculated numerically. Naively, one may approximate them by finite differences, which is also the default in many optimization software packages. However, this approximation is numerically unstable and becomes computationally very expensive in case of a high-dimensional parameter space as it demands many integrations of the differential equations (Eq. 2) (6). Alternatively, the sensitivities can be computed by an integration of the sensitivity equations (as discussed in the previous chapter) in parallel with Eq. 2.

$$\frac{dS^n_{p_l}}{dt} = \frac{d}{dt}\frac{dx_n}{dp_l} = \frac{d}{dp_l}\frac{dx_n}{dt} = \frac{df(t, x(t), k)}{dp_l} = \sum_{q=1}^{N} \frac{\partial f_n}{\partial x_q}\frac{dx_q}{dp_l} + \frac{\partial f_n}{\partial p_l}$$
$$S^n_{p_l}(0) = \left(\frac{dx_q}{dp_l}\right)(0) = \begin{cases} 1: & p_l \in \{x_0\} \\ 0: & p_l \in \{s, k\} \end{cases} \tag{9}$$

Note that $\partial f/\partial x_0 = 0$ and $\partial g/\partial x_0 = 0$. Since $f(\cdot)$ and $g(\cdot)$ are specified beforehand, their derivatives with respect to parameters and state variables can also be computed beforehand, either manually or for more complex systems by applying symbolic computations. In case of using implicit ODE solvers, the Jacobian $\partial f/\partial x$ should also be provided to the solver as it speeds up the calculations considerably. In our experience, an efficient and reliable computation of the sensitivities and gradients of the residuals is absolutely crucial for the success of a gradient-based minimization in high-dimensional parameter space. The parallel solution of the sensitivity equation has also the advantage of getting the sensitivity of certain system properties, defined by an appropriate observation function, in parallel with the solution of Eq. 2.

2.5. Minimization Process

A brief workflow of the gradient-based optimization procedure is given in Table 1. The procedure starts with an initial set of parameter values. In each cycle of the iteration, the ODE systems (Eqs. 2 and 9) are solved and the residuals and gradient of the residuals (Eq. 8) are calculated. Based on the current parameter values and the gradient, new parameter values that minimize the sum of squared residuals (Eq. 7) are minimized. It is noteworthy that the minimization step itself is only a minor contributor to the total computation time. The main computational burden is created by the need to solve the system of ODEs for the state variables and for the sensitivities many times. There are three commonly used procedures to update the parameters. We will only briefly mention them and refers the more interested reader to the discussions and algorithmic implementations presented in (5). In the Gauss–Newton procedure, the update step is calculated by solving by a linear regression in the unknown parameter increments. The Levenberg–Marquardt procedure adds considerable robustness to

Table 1
General workflow of gradient-based minimization procedures. In each parameter update step, the system of ODEs and the sensitivity equations are integrated

```
Initialize model system and parameters
LOOP
    Integrate ODE (Eq. 2.2) and sensitivity equations (Eq. 2.9) based on
    current parameter vector
    Calculate residuals defined in Eq. 2.7 and Jacobian of residuals based
    on sensitivities and Eq. 2.8
    Use gradient-based technique such as Trust-Region to calculate
    parameter increment utilizing
        Jacobian of residuals
    IF convergence criteria fulfilled
        BREAK
    ELSE
        Update parameter vector
    ENDIF
ENDLOOP
Calculate fit statistics, parameter variances and confidence limits
```

Gauss–Newton by an adaptive regularization of the linear regression problem to catch ill-conditioned cases, e.g., if some parameters are nonidentifiable. The third commonly used update scheme approximates the optimized function (Eq. 7) in a local region by a simpler, possibly lower dimensional, function. This "trust region" is chosen adaptively and the minimization is performed herein. MATLAB's lsqnonlin function, which was used in the example given below, implements all three procedures with a trust region-based method as the default. The iteration continues until a certain stopping criterion is matched, e.g., if the change in residual norm (or the relative parameter change) is smaller than a predefined value. Finally, postregression statistics such as the goodness of fit (GOF), parameter covariances and confidence limits are performed, as described below.

The success and robustness of any optimization procedure is intricately linked with the model dynamics and data complexity. For some data settings, e.g., oscillatory time series data, parameter estimating is particularly difficult due to many local minima in Eq. 7. In this case, a multiple shooting approach can greatly increase the convergence radius of the global minimum (6). Since all optimizers have intrinsic parameters which influence the success of the minimization we generally recommend to test different optimization settings as well as different optimizers.

In Table 2, we compare the performance of different minimization algorithms from MATLAB (lsqnonlin) and the Systems Biology Toolbox (www.sbtoolbox2.org). To this end, we used a model of the TGFβ signalling pathways which is described in detail below. We simulated time series data from the model including 14 observations with 20 data points each and fitted 16 model parameters. The table shows a clear result: only the gradient-based Trust-Region method of MATLAB's lsqnonlin leads to accurate parameter estimates and performs efficient and robust. However, this excellent performance requires the specification of the gradient of the residuals (Eq. 8). If this information is not given, the gradient is approximated by finite differences, which leads to a considerable increase in computation time (ODE integrations) and the number of convergent fits is reduced. We also tested the Levenberg-Marquardt procedure of MATLAB's lsqnonlin function, which performed very poorly. As the implementation cannot handle bounds on parameter values, the algorithm frequently ran into negative values, especially for the unidentifiable parameters. As a result, ODE integration failed and the optimization was canceled. The Trust-Region methods can easily handle parameter constraints and the ODE integration never failed in all 30 trails. In contrast to the previous methods, the simplex and simulated annealing algorithms provided by the Systems Biology Toolbox don't use gradient information. Table 2 shows that both perform poorly. The simplex method converged only in 4% (1 of 30 trials) to the true minimum as measured by the GOF. Simulated annealing converged to the true minimum in 17% (5 out of 30 trials). This is in stark contrast to the Trust-Region method which found the true minimum in 100% of all convergent fits (which are 97%). Interestingly, the number of identifiable parameters as measured by the coefficient of variation is low in case of simulated annealing. Nevertheless, the smaller number of identifiable parameters were fitted with very good accuracy (see parameter norm, last column). Note, that we used the standard settings provided for both methods and only changed those that control the total computation time (e.g. the total number of function evaluations) in order to allow sufficient time for each minimization. In summary, the gradient based Trust-Region method greatly outperforms the two other gradient free methods in computation time, robustness and accuracy.

Table 2
Performance of different optimization algorithms. All optimizers were run on the same data set containing all but the TGFβ receptor and the I-Smad mRNA states (20 data points and 10% measurement error). In total, 16 parameters where fitted. Results are averaged over 30 different parameter starting conditions and are given as median and interquartile ranges (IQR). The IQR is a robust measure of spread and indicates the width of the range containing 50% of the data. Column 1 states the optimization algorithm. Simplex is a gradient free, local minimization method while simulated annealing is a stochastic, global minimization method. All methods were run with the default settings except for those that restrict the total number of function evaluation. These were increased to 10^5 in order to allow for sufficiently long minimization times. Column 2 indicates whether the Jacobian of the residuals (Eq. 8) was used. Column 3: percentage of true convergent fits, as evaluated by the GOF probability $\Pr[\chi_{264}^2] \geq 0.1$. Column 4: average computation time in minutes. All minimizations were run on an Intel Xeon(R) CPU, 2.83 GHz. Column 5: optimal χ^2 value (bold font) and GOF probability (regular font). The χ^2 distribution has 264 degrees of freedom. Column 6: percent of identifiable parameters. Parameters are called identifiable if their coefficient of variation is smaller than 1. Column 7: norm of the relative parameter deviation defined as $\| \frac{p_{true} - p^*}{p_{true}} \|^2$. The norm is given for all (bold font) and only for the identifiable (regular font) parameter

Optimizer	Jacobian (Eq. 8) used	% True convergent fits	Computation time [min]	Goodness of fit: χ^2, $\Pr[\chi^2]$	% Identifiable parameter	Deviation of parameter estimates: all, identifiable
MATLAB lsqnonlin: Trust-region	Yes	97	2.3 (1.3)	**214.55 (1.25)**, 0.99 (0)	75 (0)	**1,513 (11,662)**, 0.42 (0.15)
	No	73	39.5 (12.2)	**187.84 (0.18)**, 0.99 (0)	75 (0)	**1,124 (15,958)**, 0.27 (0.1)
SB toolbox: simplex	No	3	7.8 (7.8)	**3,705.16 (2,758.20)**, 0 (0)	50 (23)	**551 (7,819)**, 0.95 (0.4)
SB toolbox: annealing	No	16	23.9 (6.7)	**1,677.12 (2,519.10)**, 0 (0)	38 (16)	**62,521 (89,370)**, 0 (0)

Average values are given in the following format: median (IQR)

3. Post-regression Diagnostics

After parameter fitting the quality of the fit should be evaluated. We start with an evaluation of the goodness-of fit followed by an estimation of the confidence intervals for the estimated parameters. These can be used to calculate confidence intervals for future predictions generated with the model. Finally, we evaluate the correlations between estimated parameters. Because of such correlations we may not be able to determine a unique set of parameters as best fits, but instead we may obtain families of parameter solutions.

3.1. Goodness of Fit

Since the measurement error is Gaussian distributed, the weighted residuals are also Gaussian distributed with unit variance. Therefore, the sum of squared residuals follows a χ^2 distribution.

$$\sum_{i=1}^{T}\sum_{j=1}^{M} R_{ij}(p)^2 \sim \chi^2(p). \tag{10}$$

Intuitively we expect from a good fit that the deviations of the model from the data should be of the same order as the measurement error, i.e., $R_{ij} \approx 1$, which means that the sum in Eq. 10 should be centered around $T \cdot M$. A much larger χ^2 value than $T \cdot M$ indicates some variation in the data which is not accounted for by the model. This fact can be used to evaluate the quality of the fit in a GOF test which gives the probability of observing an as large or larger value than the value of $\chi^2(p^*)$ at the minimum (5). However, since the parameters were adjusted in order to minimize Eq. 7 the degrees of freedom of the χ^2 distribution are $dof = T \cdot M - P$, where P is the number of parameters. Usually a cut-off value such as $\Pr[\chi^2_{\mathrm{dof}}(p^*)] < 0.05$ is used to reject the fit. Note that an underestimation of the measurements errors or non-normality of the errors also results in an exceptionally large $\chi^2_{\mathrm{dof}}(p^*)$ value, i.e., a small probability $\Pr[\chi^2_{\mathrm{dof}}(p^*)]$. The GOF test is not powerful in detecting overfitting. Overfitting results if a model, which is too complex, would also fit the particular realization of the measurement error and thus have a much smaller value of $\chi^2_{\mathrm{dof}}(p^*)$ than the expected value which is equal to *dof*. A more appropriate way to detect overfitting is the comparison with a simpler model through a likelihood ratio test (2).

3.2. Confidence Intervals

The complexity of nonlinear optimization precludes a straightforward way of calculating confidence limits for parameter estimates. However, we can employ an approximate result which is valid in the limit of infinitely many data and complete parameter identifiability. Specifically, one can relate the variance in the parameters to the curvature of the $\chi^2_{\mathrm{dof}}(p^*)$ function at its minimum in order to derive parameter covariances and asymptotic confidence intervals (5).

Intuitively, the curvature determines how well the minimum is confined and therefore how well parameter estimates can be defined. The curvature of $\chi^2_{\mathrm{dof}}(p^*)$ at the minimum is determined by the $dof \times P$ matrix of second derivatives, the so-called Fisher information matrix (FIM).

$$FIM = (\partial^2 R_{ij} / \partial p_l^2) \approx J^T J. \tag{11}$$

The approximation in Eq. 11 neglects the second derivative terms but is computationally inexpensive as $J_{ij}^{p_l} = (\partial R_{ij} / \partial p_l)$ is the calculated gradient matrix of the residuals during minimization. The covariance of the parameters C is related to the inverse of the FIM as

$$C = 2(FIM)^{-1}. \tag{12}$$

Asymptotic confidence intervals can be calculated by taking into account the distribution of the χ^2 values, which are approximately Gaussian for large degrees of freedom. The 95% confidence intervals are given by

$$p^* \pm 1.96\sqrt{diag(C)}. \tag{13}$$

Symmetric confidence intervals are problematic if a parameter estimate is close to the boundary of the admissible parameter space. Moreover, the approximation Eq. 13 gives misleading results whenever the above-stated assumptions are heavily violated, a fact which can be difficult to evaluate beforehand (7).

3.3. Bootstrap

This method offers an alternative to the asymptotic approximation of parameter uncertainties. It is a heuristic but nevertheless exact way of determining parameter uncertainties. Bootstrap methods construct an empirical distribution of the parameter estimates by a repeated data resampling and consecutive parameter estimation. Parameter uncertainties can be inferred from the shape of the empirical parameter distribution (8). The elegance and simplicity of the bootstrap comes at a price. As it involves many parameter estimations, it is computationally more expensive.

3.4. Parameter Correlation and Identifiability

Frequently, the optimization procedure does not yield a unique optimal parameter set, because there is no unique optimal $\chi^2(p^*)$ value given the available data. In this case the value of some or all parameters is nonidentifiable. Nonidentifiability is the result of a nonunique χ^2 minimum, which can be caused, e.g., by a very flat χ^2 landscape. The later implies a functional relation between parameters along which the χ^2 value is unaltered. Parameter estimates appear highly correlated if this functional relationship is linear. Hengl et al. (9) suggest an approach based on mutual information and bootstrap to detect parameter dependencies taking many, equally well-fitting, parameter estimates as an input. This heuristic approach is particularly easy to apply as it does not require any a priori knowledge on the model structure.

There are three common ways to deal with nonidentifiability. One solution is to fix some of the nonidentifiable parameter at educated values and only estimate the remaining parameters. These estimates are of course biased since their optimum is in a functional relation to the fixed parameters. Alternatively, subsequent analyses can be based on all admissible parameter sets and the parameter sets can then be clustered according to the predictions derived from them. A third approach might be to reduce the model such that it does not contain the nonidentifiable parameters, e.g., by phenomenological descriptions or timescale separation techniques (10). It is noteworthy that nonidentifiability of parameters does not imply a poor fit to the data, but that parameter values cannot be constrained to a unique value. The predictive power of the model will therefore be limited to model predictions that are not sensitive to nonidentifiable parameters.

4. An Application: TGFβ Signaling

We now apply the above-mentioned methods to a specific example, the TGFβ-signaling pathway. We will concentrate on a previously published TGFβ-signaling model (11) which is extended by a transcriptional negative feedback involving an inhibitory Smad. An outline of the TGFβ model is presented in Fig. 1. The model consists of 18 variables and 19 kinetic rate constants, most of which have been determined in previous studies (11; 12). A brief recapitulation of the system dynamics is given in the caption of Fig. 1. A practical guideline should offer means to easily recapitulate the main analysis steps. We therefore provide for the interested reader the model as well as all functions in MATLAB format to reiterate the results presented in this section by own computer simulations from the authors web page (http://www.bsse.ethz.ch/cobi). The experienced programmer can easily extend the code for own modeling projects.

4.1. Sensitivities

In order to better understand the functioning of the complex network, we will first turn our attention to the sensitivities of the model output with respect to the model parameters. Figure 2 summarizes our results. The accompanying MATLAB script script_sensitivity.m can be used to reproduce the results. Panel a shows time courses of the stimulation protocol and the output variable (nuclear Smad2* /Smad4 complex). We stimulate the model by an initial TGFβ pulse and apply an inhibitor of TGFβ-receptor auto-phosphorylation after 3h for a duration of one hour.

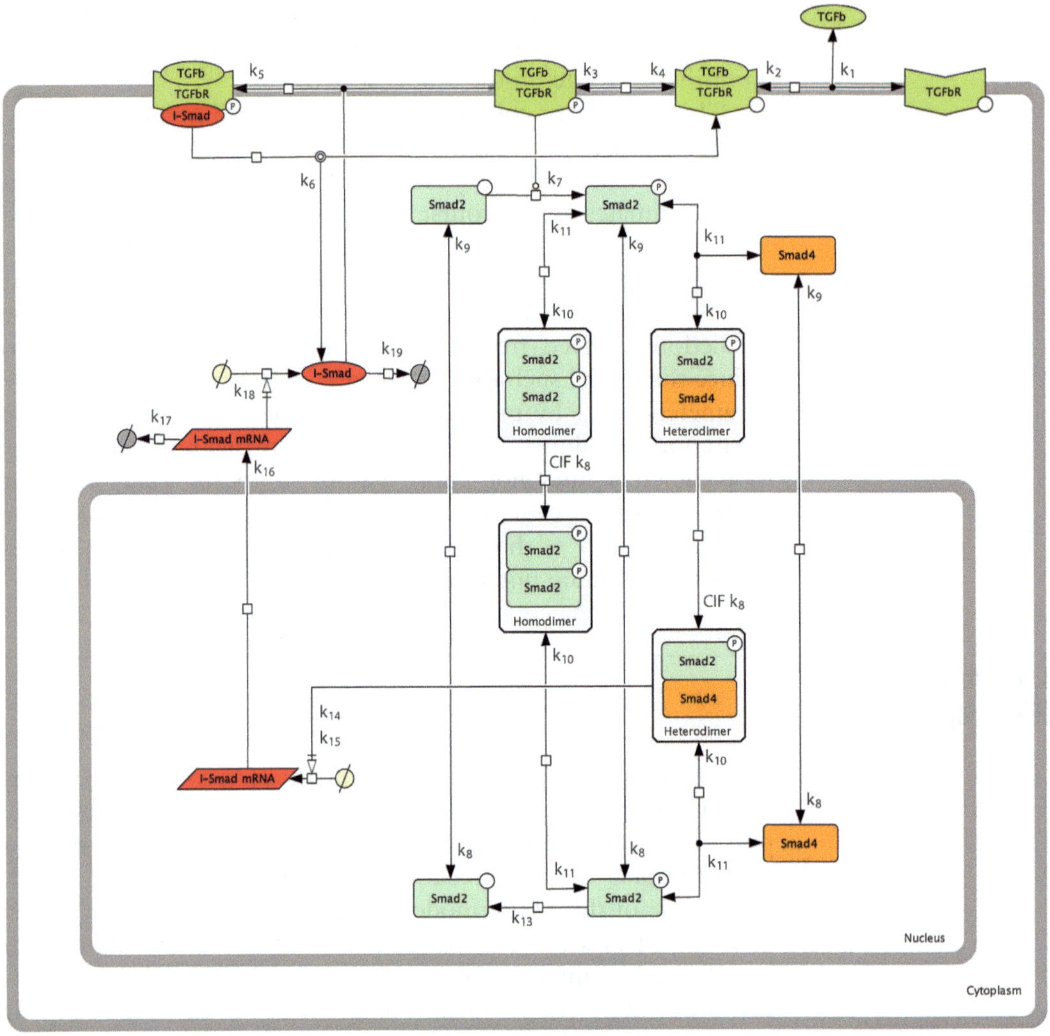

Fig. 1. Outline of the TGFβ model adopted from (11) including an inhibitory Smad. TGFβ binds with strong affinity to the receptor. The receptor complex gets autophosphorylated and signals by enhancing the phosphorylation of Smad2. The active form of Smad2 can either form Smad2 dimers or heterodimerize with Smad4. The Smads and their complexes can shuttle between the cytoplasm and the nucleus. The effect of a phosphatase (dephosphorylation of Smad2) is only considered in the nuclear compartment. The Smad2–Smad4 heterodimer serves as a transcription factor for the production of I-Smad mRNA. The mRNA needs to transfer to the cytoplasm to be translated into I-Smad protein. Finally the I-Smad can bind to the active ligand–receptor complex and decrease the total amount of the initial signal, thereby acting s as an inhibitor. The respective rate constants of each reaction are indicated as $k_1 - k_{19}$ and the complex import factor (CIF).

This stimulation protocol results in a complex response of the model. Note that the initial transient dynamics after stimulation is very fast due to the high affinity of TGFβ to its receptor. Panel **b** shows the time-resolved control coefficients of the five, most influential parameters controlling the dynamics of the nuclear Smad2*/Smad4 complex which is our model output of interest.

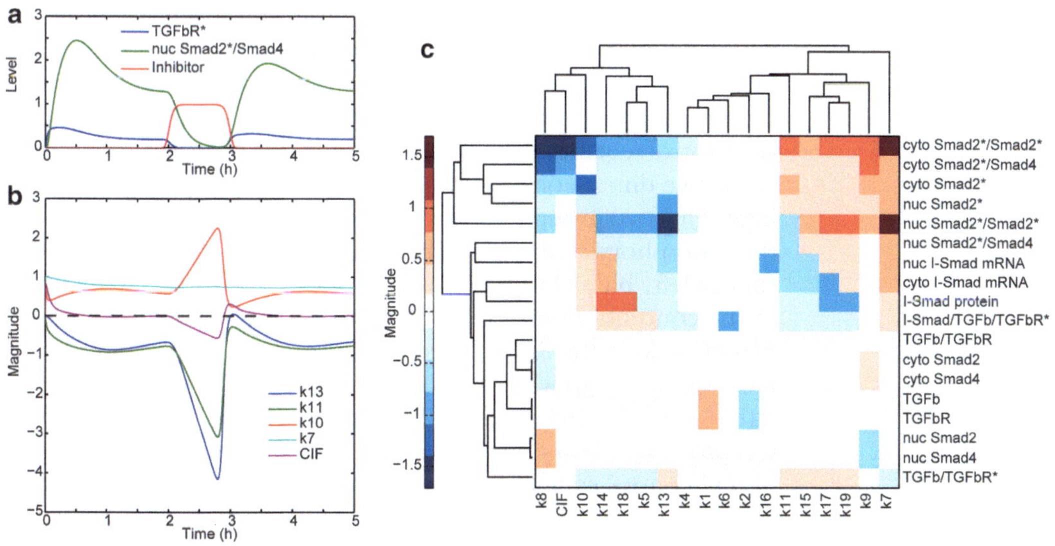

Fig. 2. Sensitivity analysis of the TGF-signaling model. (**a**) Time course of the stimulation protocol and the system output, nuclear Smad2 * /Smad4 complex. (**b**) Time-resolved control coefficients of parameters controlling the dynamics of the nuclear Smad2 * /Smad4 complex, i.e., the output signal of the signaling network. For details, see main text. (**c**) Clustergram of the steady-state control coefficients.

The time-dependent control coefficients are normalized sensitivities defined as

$$C_{p_j}^{m_i}(t) = \frac{p_j}{[m_i](t)} S_{p_j}^{m_i}(t), \tag{14}$$

where $[m_i](t)$ denotes the time-dependent concentration of the ith model output. The control coefficients of parameter k_7 (Smad2 phosphorylation rate) and k_{10} (Smad compex formation rate) are positive during the whole time course. Increasing these parameter will always lead to an elevation in the level of nuclear Smad2* /Smad4. On the contrary, k_{11} (complex dissociation rate) and k_{13} (Smad2 dephosphorylation rate) have mostly a negative effect with the strongest impact during the application of the TGFβ phosphorylation inhibitor. Note that the impact of a change in the complex import factor (CIF) can have a positive or a negative effect on the nuclear Smad2* /Smad4 concentration depending on the time point of a change. The remaining model parameters have a lower, but nevertheless nonzero impact on the nuclear Smad2* /Smad4 levels. In general, we expect parameters with small overall control coefficients to be difficult to identify. Parameters with strong correlations in the temporal profile of their control coefficients point to a strong underlying functional link and will also be correlated when estimated from data.

Next, we focus on the sensitivity of the steady-state levels with respect to all model parameters, again in terms of control coefficients. Figure 2c shows a clustergram of the steady-state control coefficients in which rows represent observables and columns represent parameters. Rows and columns of the clustergram are sorted such as to maximize the similarity between neighboring row- and column vectors. In this way, observables and parameters with similar control coefficients (and thus similar function) across their respective dimension are grouped closely to each other. Figure 2c highlights two groups of control coefficients which are all related to the phosphorylated observables. They show either a decrease (upper-left, blue cluster) or an increase (upper-right, red cluster) in the concentration of the phosphoforms. The latter set of parameters include the Smad phosporylation rate and the I-Smad degradation rate (k_{19}) while the former include the dephosphorylation rate of nuclear Smad and the I-Smad association rate to the active receptor (k_5). Some parameters (k_1, k_6, and k_{16}) have very low overall control coefficients, and we anticipate problems in estimating these parameters from steady-state data. In conclusion, sensitivity analysis can provide a first glimpse to the functioning of a complex model and allows to group reactions and parameters with similar impact on the system output into functional groups.

4.2. Parameter Estimation

We now turn to the inverse problem: the estimation of model parameters from experimental data. The interested reader can reiterate our results with the MATLAB script script_fit.m provided on the authors web site. To this end, we simulate data, including measurement errors, in order to estimate the original parameter values used for model simulation. This allows us to evaluate the performance of the fitting procedure as we know the true parameter values underlying our data. We apply the same stimulation protocol as in Fig. 2a and estimating all 19 model parameters given that all model species except the two mRNA species of I-Smad are observed. The simulated data includes 20 data points per species and 10% measurement error. Figure 3a shows a fit of the active receptor and nuclear Smad2* /Smad4 trajectories to a simulated data set ($\chi^2_{342} = 365.2$, $\Pr[X \geq \chi^2_{342}] = 0.19$). The color-shaded area underlying each trajectory is the approximated uncertainty of a trajectory based on the uncertainty in the parameter estimates. It is calculated by error propagation as

$$\mathrm{Cov}[y(t)] = S_p^y(t)CS_p^y(t)^T, \qquad (15)$$

where S_p^y is the matrix of the sensitivities of the observations defined by Eq. 1 and C is the parameter covariance matrix determined by Eq. 12. Correlations between the parameter estimates are shown in panel **b**. It is apparent that all parameter related with nuclear import/export and complex formation/dissociation are highly correlated. Additionally, parameters related to the TGFβ ligand/receptor interaction show a large positive correlation. Since this interaction happens on a fast timescale due to a high TGFβ receptor affinity the parameters cannot be well identified with the given temporal resolution of the data. Additionally, all parameters related to the I-Smad expression are highly correlated. This is in part expected as the I-Smad mRNA states are not observed and therefore parameters related to the mRNA dynamics cannot be inferred from the data. Note that a strong correlation does not necessarily imply a bad identification as the coefficient of variation for the single parameters can still be very small. We will focus on these intrinsic variations in the next paragraph.

Instead of determining correlations by a single fit as in Fig. 3a, we next investigated parameter uncertainties and correlations arising from many fits to data sets that only differ in the noise

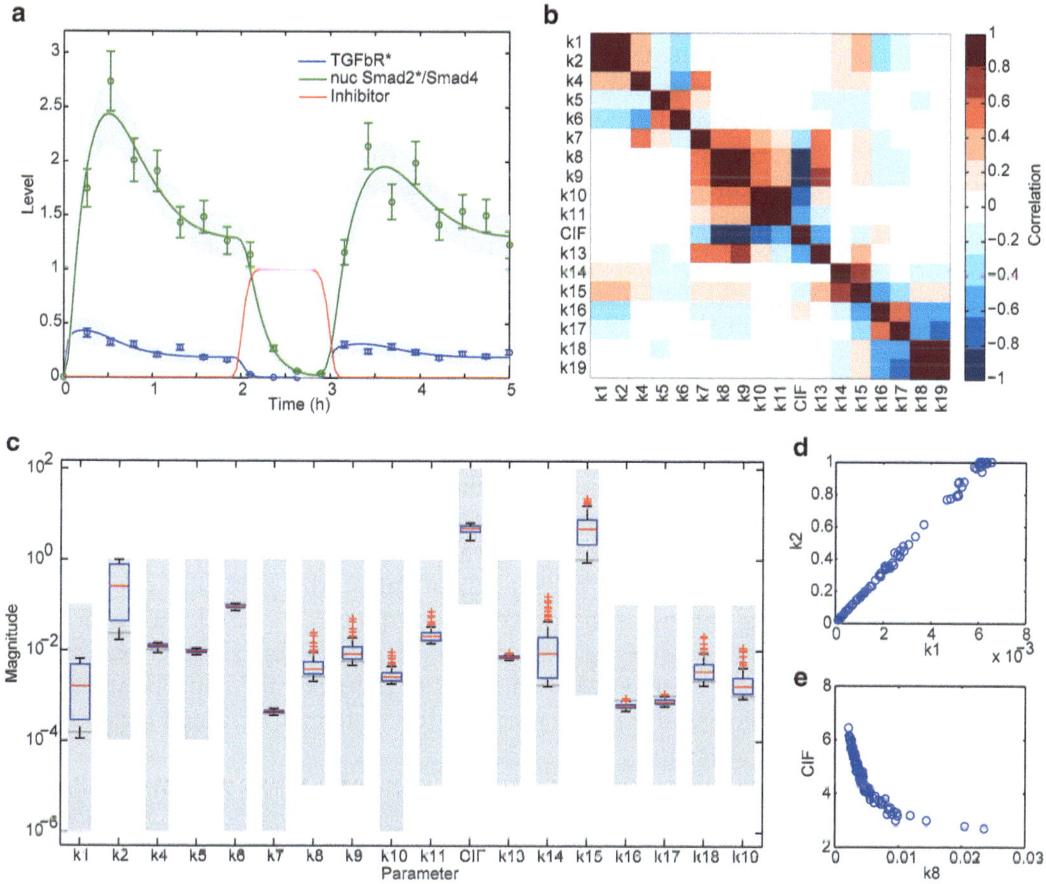

Fig. 3. Parameter estimation in the TGFβ model. The model was fitted to 16 observations (all model species except the two mRNA species of I-Smad; 20 data points each, 10% measurement error) optimizing 19 parameters. (**a–b**) Fit to one representative data set. (**a**) Data and time courses of phosphorylated TGF receptor complex and nuclear, phosphorylated Smad2/Smad4. Error bars denote data mean and standard deviation. *Gray-shaded areas* highlight error intervals for the trajectories calculated by error propagation (Eq. 15). (**b**) Correlation matrix of parameter estimates from fit in (**a**). (**c–e**) Evaluation of the expected variance in parameter estimates. Parameters were fitted to 100 data sets, each with a different realization of the measurement noise. (**c**) Box plots of estimated parameter sets (median=*red line*, 25% and 75% quantiles=*blue lines*, 5% and 95% quantiles=*black lines*, *red asterisk*=outliers). Parameter-fitting ranges are indicated in *light gray*, true parameter values are given in *dark gray*. (**d**,**e**) Scatter plots revealing a functional relation between the estimates for k_1 and k_2 (**d**) and k_8 and CIF (**e**).

realization. The corresponding MATLAB script is script_multi_fit.m. This strategy is similar to a bootstrap in case of a sufficiently high number of replicates per measurement point and reflects more accurately the expected parameter uncertainties than a single fit does. It also allows to clarify the expected accuracy in parameter estimates given the experimental design and measurement error. Figure 3c shows boxplots of the optimized parameter estimates from 100 data sets. Parameter limits used for the estimation are indicated as light gray bars. Some parameter estimates have a small variance and nicely fit the true parameter values indicated as dark gray bars. Some parameters can be sufficiently well identified despite having considerable correlations, e.g., k_7 and k_{13}. However, most parameter estimates have a large variance and do not match the true parameter on average. In fact, a closer look at the parameter distributions (e.g., by means of (9)) reveals a strong functional dependence with other parameters, meaning that they cannot be identified under the given experimental setup. Figure 3d, e shows two representative scatter plots of pairs of parameter estimates which are functionally related. The hyperbolic relation between the estimates of k_8 and CIF is apparent from the model formulation, as CIF is only a scaling factor for the import rate of the complexes. For other functional relations, the underlying mechanism is less clear. Most parameters related to I-Smad expression, which is only observed on the protein level, cannot be well identified. Generally, unobserved processes will corrupt the identifiablity of the model parameters. It should be noted that nonidentifiability due to a lack of sensitivity, which is sometimes interpreted as systems robustness, does not imply a lack in functionality of the respective parameters or processes. Sensitivities are always dependent on the particular way of defining and measuring a system output. A focussed, and necessarily limited, investigation cannot in general assess functionality in a larger context. This highlights the need for appropriate experiments and mathematical models which are designed for a particular question.

5. Conclusion

Parameter estimation from experimental data is a central part of modeling and analyzing biological signaling networks. It is, in general, an iterative process, and robust and efficient algorithms are key to obtain good estimates with reasonable computational effort. This chapter summarizes the main steps to attain these goals. The optimization problem is formulated in terms of a likelihood function and a gradient-based minimization algorithm is suggested to determine the parameter set that maximizes the likelihood. It is important to subsequently analyze the GOF and to derive confidence intervals for the parameter estimates. Approximate symmetric confidence intervals can be formulated in terms of the variance in the data and parameter sensitivities. More accurate estimates can be obtained with computationally more demanding Bootstrap methods. Before and after the regression the identifiability of parameters should be analyzed and the number of parameters in the model should either be reduced accordingly or parameters should be excluded from the estimation. The further analysis of the model should, in any case, consider all admissible parameter sets and cluster these according to the predictions of the model.

Acknowledgments

We thank Andreas Raue and Martin Peifer for discussions and members of the Iber group for the critical reading of the manuscript. The work was partially funded through a SystemsX iPhD scholarship to G.F. and a SystemsX grant as part of the RTD InfectX.

References

1. Jaqaman K, Danuser G (2006) Linking data to models: data regression. Nat Rev Mol Cell Biol 7(11):813–819
2. Kreutz C, Timmer J (2009) Systems biology: experimental design. FEBS J 276(4):923–942
3. Kreutz C, Bartolome Rodriguez MM, Maiwald T, Seidl M, Blum HE, Mohr L, Timmer J (2007) An error model for protein quantification. Bioinformatics 23(20):2747–2753
4. Silvey SD (1970) Statistical inference. Penguin Books, Baltimore
5. Press WH (2007) Numerical recipes: The art of scientific computing, 3rd edn. Cambridge University Press, New York
6. Peifer M, Timmer J (2007) Parameter estimation in ordinary differential equations for biochemical processes using the method of multiple shooting. IET Syst Biol 1(2):78–88
7. Raue A, Kreutz C, Maiwald T, Bachmann J, Schilling M, Klingmuller U, Timmer J (2009) Structural and practical identifiability analysis of partially observed dynamical models by exploiting the profile likelihood. Bioinformatics 25(15):1923–1929
8. Efron B, Tibshirani R (1986) Bootstrap methods for standard errors, confidence intervals, and other measures of statistical accuracy. Statist Sci 1(1):54–75
9. Hengl S, Kreutz C, Timmer J, and Maiwald T (2007) Data-based identifiability analysis of non-linear dynamical models. Bioinformatics 23(19):2612–2618
10. Murray JD (2002) Mathematical biology: I. an introduction, 3rd edn. Springer, Berlin
11. Schmierer B, Tournier AL, Bates PA, Hill CS (2008) Mathematical modeling identifies smad nucleocytoplasmic shuttling as a dynamic signal-interpreting system. Proc Natl Acad Sci USA 105(18):6608–6613
12. Clarke DC, Liu X (2008) Decoding the quantitative nature of tgf-beta/smad signaling. Trends Cell Biol 18(9):430–442

Chapter 3

A Tutorial on Mathematical Modeling of Biological Signaling Pathways

Zhike Zi

Abstract

Mathematical models have been widely used in the studies of biological signaling pathways. Among these studies, two systems biology approaches have been applied: top-down and bottom-up systems biology. The former approach focuses on X-omics researches involving the measurement of experimental data in a large scale, for example proteomics, metabolomics, or fluxomics and transcriptomics. In contrast, the bottom-up approach studies the interaction of the network components and employs mathematical models to gain some insights about the mechanisms and dynamics of biological systems. This chapter introduces how to use the bottom-up approach to establish mathematical models for cell signaling studies.

Key words: Mathematical modeling, Signaling pathway, Reaction kinetics, Systems biology

1. Introduction

Traditional cell signaling studies have focused on understanding the functions of individual signaling components. It is now realized that cell behaviors are not only determined by the characteristics of individual system components, but also by the interactions of such components acting together as a system. Conventional biochemistry and cell biology approaches are useful to describe and discover the components of cellular signaling pathways. However, most of the descriptions are carton-type hypotheses that give us the static and qualitative information about the system. The nonlinear interactions of the components might exhibit some emergent properties, such as bistability, oscillation, and robustness, which are difficult to be discovered by experimental analyses but are easier to be investigated by kinetic analysis of mathematical models.

On the other hand, mathematical modeling of signaling pathways requires experimental data and prior knowledge of the pathways. Therefore, systems biology research could not be

Xuedong Liu and Meredith D. Betterton (eds.), *Computational Modeling of Signaling Networks*, Methods in Molecular Biology, vol. 880, DOI 10.1007/978-1-61779-833-7_3,

done without the contribution of traditional biochemical and cell biology. However, classical cell signaling studies face the limitations, such as inaccuracy, inability to deal with emergent properties, and so on (1). The bottom-up systems biology approaches could fetch up these limitations by merging model simulations and quantitative experimental analyses. Therefore, systems biology approach with mathematical models has been emerged as a powerful tool in studying signaling pathways because it provides a way to study the signaling dynamics and the underlying principles.

Many sources of uncertainty, including errors, noise, incomplete information, and poor experimental design, impose limitations on our confidence of the experimental data. Some experimental data might lead to a misunderstanding of the true property of biological systems due to the limited number of experimental data. For example, we might get a time course data for protein X as shown in Fig. 1a. The amount of protein X oscillates with the increase of time. The experimentalists may judge the number of periods by counting the number of peaks appearing in the data by connecting the discrete data. In this case, the number of peaks will be counted as 3 (Fig. 1b). However, if we perform model simulation for this system, we will find that the number of the oscillations during the investigated time might be 7, rather than 3 (Fig. 1c). Inspired by the model prediction, we could design a new experiment to verify the actual number of oscillations by taking more samplings at proper time points.

2. Materials

The material requirement for mathematical modeling of cellular signaling pathways is not demanding. However, getting high quality of quantitative experimental data sets is still one of the bottlenecks for the development of the mathematical models. Normally, computer and software packages are required to perform model simulations. The specific configuration of the computers and software depends on the modelers' preference. Till now, more than 200 software tools have been developed with the support of the Systems Biology Markup Language (SBML) format (2). A summary of these software packages is available in the official SBML Web site (http://sbml.org).

3. Method

The mathematical format for modeling signaling pathways depends on the properties of the studied system and the specific questions that are going to be answered. The most commonly used mathematical forms are ordinary and partial differential equation systems, which can be in either deterministic or stochastic format (3).

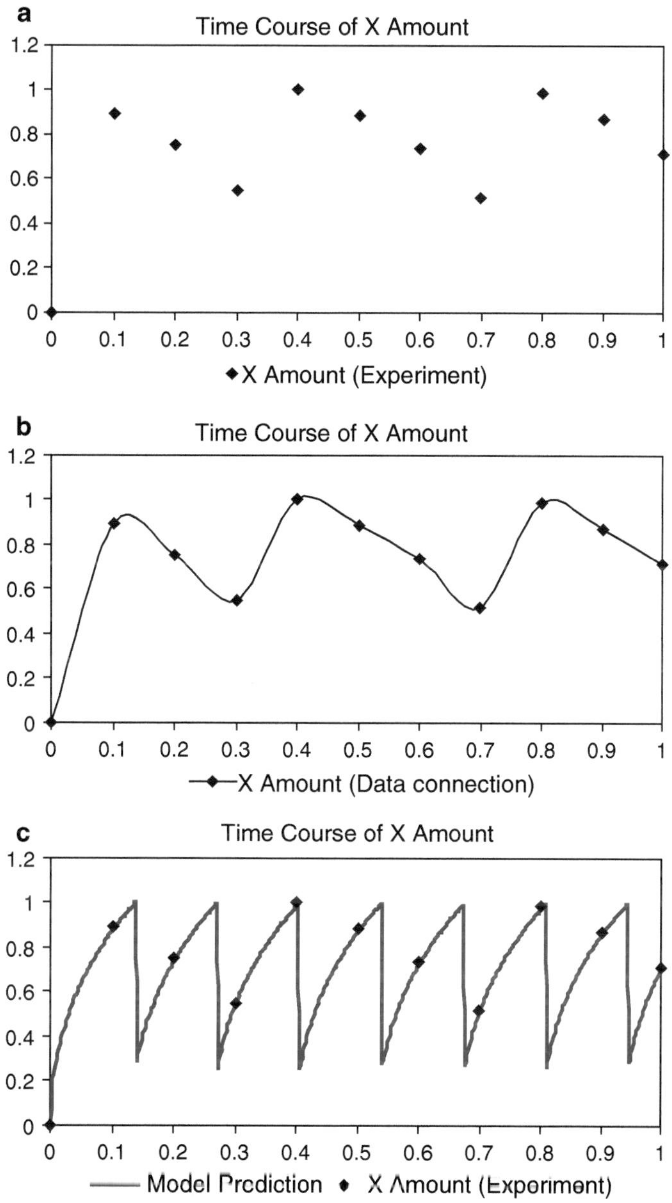

Fig. 1. Illustration of the role of model prediction. (**a**) Time course data for the amount of protein X. (**b**) The intuitive data connection predicts three oscillation periods. (**c**) Computational model simulation of the underlying network suggests that protein X has oscillated seven times.

Stochastic equations are useful to investigate the properties of the system arising from random fluctuations or noise. Here, we introduce how to establish mathematical models for the signaling pathways with a system of ordinary differential equations (ODE). The ODE modeling approach can be applied to the signaling studies when the components of the signaling pathway can be assumed to homogenously distribute in the cell and the stochastic effects could be ignored.

3.1. The Choice of Reaction Kinetics

There are different ways to model the reaction kinetics in signaling pathways. Typical reaction kinetics include the law of mass action, Hill equation, and Michaelis–Menten kinetics. Here, we use the example of ligand receptor interaction to illustrate how to use these kinetics to describe signaling reactions.

(1) *The law of mass action*: In the nineteenth century, Guldberg and Waage introduced the law of mass action to describe the biochemical kinetics (4). The law of mass action states that the reaction rate is proportional to the probability of the collision of the reactants. This probability is also proportional to the concentration of reactants to the power of their molecularity, the number of them entering the specific reaction (5).

The ligand and receptor interaction is a signaling event that happens in most signaling pathways. With the binding of the ligand (L), the receptor (R) forms a ligand–receptor complex (LRC) with a characteristic rate constant k_{on}. On the other hand, the LRC can dissociate to ligand and receptor with a rate constant k_{off}. The ligand–receptor interaction can be described with the following reaction:

$$\mathrm{L} + \mathrm{R} \underset{k_{off}}{\overset{k_{on}}{\Leftrightarrow}} \mathrm{LRC}. \tag{1}$$

By the definition of mass action law, we can derive the concentration change over time of ligand, receptor, and LRC by the following ODE system:

$$\frac{\mathrm{d}[L]}{\mathrm{d}t} = k_{off}[\mathrm{LRC}] - k_{on}[L][R], \tag{2}$$

$$\frac{\mathrm{d}[R]}{\mathrm{d}t} = k_{off}[\mathrm{LRC}] - k_{on}[L][R], \tag{3}$$

$$\frac{\mathrm{d}[\mathrm{LRC}]}{\mathrm{d}t} = k_{on}[L][R] - k_{off}[\mathrm{LRC}]. \tag{4}$$

(2) *Hill equation*: The signaling receptors might exist not only as monomers, but also form dimers or oligomers with two or more identical binding sites for the ligand. The bound subunit has a cooperative effect on the later binding subunits, which means that the affinity of the receptor to the later bound ligand is significantly increased by the already bound ligand. A typical example that has such a property is the binding of oxygen to hemoglobin (Hb). The early experimental research found that the fractional saturation of Hb with oxygen had a sigmoid response to the oxygen partial pressure. Hill explained the interaction between the oxygen

binding sites of Hb subunits by the following fractional saturation (6):

$$\Upsilon = \frac{[L_n R]}{[R_t]} = \frac{[L]^n}{(K_a)^n + [L]^n}, \tag{5}$$

where n is known as Hill coefficient and K_a corresponds to the ligand concentration occupying half of the receptor-binding sites.

In general, for a receptor with n subunits, the reaction rate of the receptor to the ligand can be approximated as:

$$v = V_{\max}\Upsilon = \frac{V_{\max}[L]^n}{(K_a)^n + [L]^n}. \tag{6}$$

The Hill coefficient (n) can be calculated by

$$n = \frac{\log(81)}{\log(S_{90}/S_{10})}. \tag{7}$$

Here, S_{10} and S_{90} correspond to the ligand (stimulus) levels that are required to achieve 10% and 90% of the maximum activation, respectively (7).

One can use Hill equation kinetics with a larger value of Hill coefficient (n) to approximately model some switch-like processes, for example the turn on of ion channels.

(3) *Michaelis–Menten kinetics*: When an enzyme (kinase or phosphatase) catalyzes a biochemical reaction, it is not consumed or produced by this reaction, but forms a temporary complex with the substance in the reaction. For such a reaction, we can use Michaelis–Menten kinetics to describe the reaction rate under the key assumption of quasi-steady-state approximation, which is valid when the enzyme concentration is much lower than the substrate concentrations and when the enzyme is not allosteric. Michaelis–Menten kinetics is named after Leonor Michaelis and Maud Menten and has the following formulation:

$$v = \frac{V_{\max} S}{K_m + S}, \tag{8}$$

where K_m is the Michaelis constant and is equal to the substance concentration that causes the half-maximal reaction rate $V_{\max}$.

3.2. How to Construct the Systems of Ordinary Differential Equations?

The ODE modeling approach describes the concentration change of a component over time by a system of ordinary differential equations. For a certain component's concentration $[C_i]$ evolved over time, we calculate both the sum of the reaction rates producing C_i (the reactions coming towards component C_i, v^+) and the sum of the rates consuming C_i (the reaction going outwards component

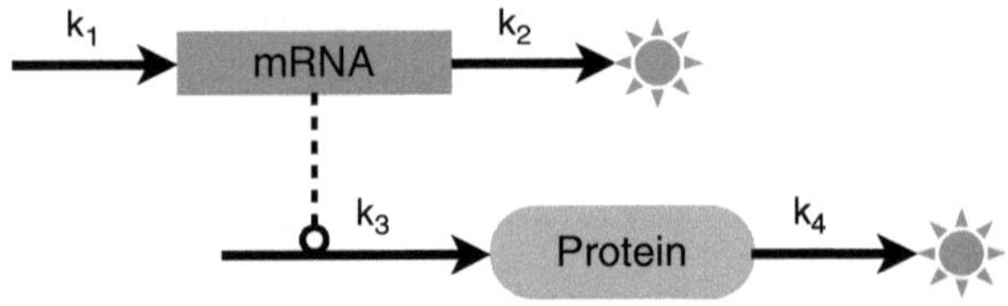

Fig. 2. A simple network for the production and degradation of mRNA and protein.

C_i, v^-). The system of ODEs is determined by the subtraction of the former to the latter item as the following equation:

$$\frac{d[C_i]}{dt} = \sum v^+ - \sum v^-. \tag{9}$$

More examples about how to construct the ODE system will be illustrated in the following section.

3.3. How to Model Different Signal Transduction Steps?

Some signal transduction steps are shared in different signaling pathways. In order to establish a complete model for a certain signal pathway, it would be useful to understand how to model representative signal transduction steps (8).

3.3.1. Production and Degradation of mRNA and Protein

The production and degradation of mRNA and protein are typical processes involved in signaling pathways (Fig. 2). Here, we can assume that there is a constitutive mRNA synthesis with a production rate constant k_1. The degradation of mRNA follows first-order mass-action kinetics with degradation rate constant k_2. The protein is produced with a rate constant k_3 and is proportional to the concentration of mRNA. Similarly, the protein is degraded with rate constant k_4.

For this simple network, we can use the following system of ODEs to describe its dynamics.

$$\frac{d[\text{mRNA}]}{dt} = k_1 - k_2[\text{mRNA}], \tag{10}$$

$$\frac{d[\text{Protein}]}{dt} = k_3[\text{mRNA}] - k_4[\text{Protein}]. \tag{11}$$

It is worth noting that mRNA has appeared in Eq. 11 for the protein production, but it does not mean that mRNA is a reactant or substrate for protein production. Therefore, parameter k_3 should not appear in Eq. 10. mRNA is used as a template during protein translation and mRNA itself is not converted to protein. In SBML, such molecules are defined as "modifiers."

For Eq. 10, we can derive an analytic solution to calculate the mRNA amount at time t with the following equation:

$$[\text{mRNA}]_t = \frac{k_1}{k_2} - \left(\frac{k_1}{k_2} - [\text{mRNA}]_0\right)e^{-k_2 t}, \tag{12}$$

where $[\text{mRNA}]_0$ is the concentration of mRNA at initial time 0.

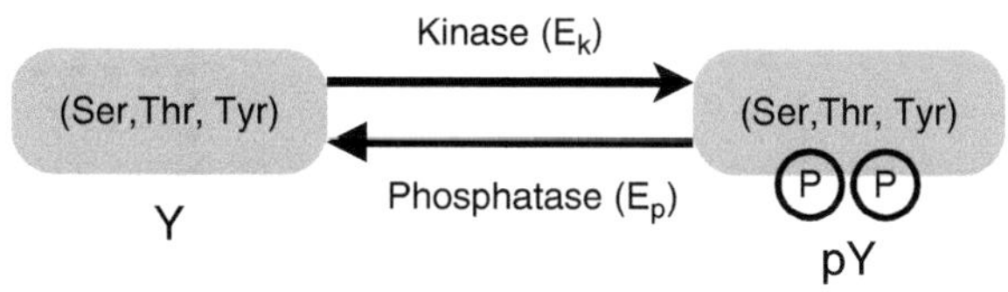

Fig. 3. Scheme of protein phosphorylation and dephosphorylation.

3.3.2. Phosphorylation and Dephosphorylation

Protein phosphorylation and dephosphorylation take place very frequently in cellular signaling transduction. In Fig. 3, the protein $\varUpsilon$ is phosphorylated to p$\varUpsilon$ by a kinase and then the phosphorylated p$\varUpsilon$ is dephosphorylated to $\varUpsilon$ by a phosphatase. Here, we assume that the total amount of $\varUpsilon$ and p$\varUpsilon$ is constant and equals to $\varUpsilon_t$.

For this process, there are two different ways to model it. We can choose mass-action kinetics or Michaelis–Menten kinetics depending on the context of the signaling pathway. If the concentrations of kinase (E_k) and phosphatase (E_p) are much lower than their substrates, we can model this simple system with Michaelis–Menten kinetics:

$$\frac{d[\varUpsilon]}{dt} = -\frac{V_{\max 1}[\varUpsilon]}{K_{m1} + [\varUpsilon]} + \frac{V_{\max 2}(\varUpsilon_t - [\varUpsilon])}{K_{m2} + (\varUpsilon_t - [\varUpsilon])}, \tag{13}$$

$$[pY] = \varUpsilon_t - [\varUpsilon]. \tag{14}$$

In some cases, the kinase or phosphatase can be one component of the signaling pathway and its concentration is in a similar range as the substrate. The assumption for Michaelis–Menten kinetics is not valid anymore. In this case, mass-action kinetics might be better to model the system. It might be necessary to include the intermediate steps. For simplicity, we can model this system with mass-action kinetics in the following way.

$$\frac{d[\varUpsilon]}{dt} = -k_1[\varUpsilon] + k_2(\varUpsilon_t - [\varUpsilon]), \tag{15}$$

$$[pY] = \varUpsilon_t - [\varUpsilon]. \tag{16}$$

3.3.3. Feedbacks

Feedback regulations are important strategies for cells to control the signaling pathways. For example, positive feedback may allow signaling pathways to read continuously graded signal input and generate digital signaling output (9). Moreover, cells can adapt to some types of signals by using negative-feedback regulation (10). In some cases, a signaling pathway with delayed negative feedback could generate an oscillatory response of the signaling output.

Figure 4 shows a simple example of feedback network, in which the activation of molecule A leads to the activation of molecule B and activated B (B *) has a positive or negative feedback on the activation of A. We can use mass-action kinetics to model the activation of A and

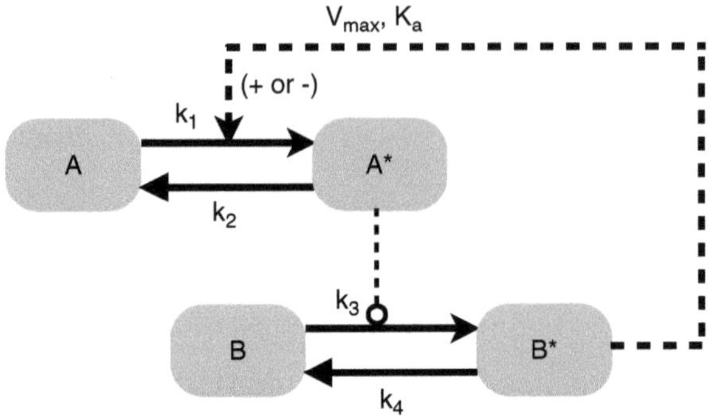

Fig. 4. Scheme of network feedback.

B. For the feedback of activated B on the activation of A, we can choose Hill equation to model it. Here, the total amount of A and A* is A_t, and the total amount of B and B* is B_t. Therefore, for the network shown in Fig. 4, it is possible to model it with the following system of ODE:

$$\frac{\mathrm{d}[A^*]}{\mathrm{d}t} = k_1(A_t - [A^*]) - k_2[A^*] + \frac{V_{\max}(B^*)^n}{(K_a)^n + (B^*)^n}(A_t - [A^*]), \tag{17}$$

$$\frac{\mathrm{d}[B^*]}{\mathrm{d}t} = k_3[A^*](B_t - [B^*]) - k_4[B^*]. \tag{18}$$

In Eq. 17, parameter $V_{\max}$ denotes how strong the feedback is. Parameter K_a and Hill coefficient n indicate how sensitive the feedback is to the activated B. If n is negative, then the feedback is negative regulation. Otherwise, it represents a positive feedback.

3.3.4. How to Model Signaling Inputs?

A variety of signaling sources trigger different types of signaling pathway, which includes environment stimuli and signaling molecules. According to the specific signaling input profile, different mathematical approximations can be used to model the signaling input change over time. Two typical approximations are employed to model the step change and the depletion of signaling input (S) change.

For step change of signal, the signal profile is constant, for example the osmotic change for the stress signaling pathways in bacterial and yeast cells. In this case, the ODE for the signal input is defined as:

$$\frac{\mathrm{d}[S]}{\mathrm{d}t} = 0. \tag{19}$$

Therefore,

$$[S_t] = [S_0]. \tag{20}$$

In mammalian cells, the signaling molecules (or ligands) can be internalized into the cells and then degraded. The signaling molecules in the extracellular medium will be depleted over time. In this case, for a certain dose of signaling molecules, we can assume that the degradation of the signaling molecules is proportional to their concentration, which reads as

$$\frac{d[S]}{dt} = -\lambda[S]. \tag{21}$$

The signaling molecules' time course profile is a decaying function.

$$[S_t] = [S_0]e^{-\lambda t} \tag{22}$$

Here, λ characterizes the speed of ligand depletion.

It is worth noting that the value of parameter λ may depend on the ligand dose and cell density (11). For a very high dose of ligand, λ is approximate to 0. Therefore, the depletion of the signaling molecules can be ignored and it is approximately constant. However, when the ligand dose is relatively low, the depletion of signaling molecules plays an important role and λ cannot be ignored. When the signaling responses to different ligand doses are modeled, it would be better to develop a more complicated mathematical model, including the ligand–receptor interaction, and internalization and degradation of the ligand.

3.4. How to Run Simulation for the Mathematical Model?

Two pieces of important information are essential for implementing simulation for the ODE mathematical model, which are (a) the initial condition of the variables and (b) the values for the reaction rate constants.

The initial condition of the ODE system corresponds to the abundance of the proteins involved in the signaling pathway. Normally, it is difficult to know the complete information about the absolute number of molecules per cell and it is still a challenge to directly measure the rate constants by in vivo experiments. Some model parameter values have to be estimated based on quantitative experimental data. The estimation of the model parameter values is called parameter estimation. The goal of parameter estimation for the ODE models is to find the possible parameter sets that minimize the difference between the model simulation data and experimental data, which is formalized as:

$$F_{\text{obj}} = \sum_{i=1}^{n} \left(\frac{y_i^{\text{sim}} - y_i^{\text{data}}}{\sigma_i} \right)^2. \tag{23}$$

Here, F_{obj} is called the objective function (cost function) or sum of squared errors. y_i^{sim} is computer simulation data and y_i^{data} is the corresponding experimental data. σ_i represents the noise or standard deviation of the experimental data. Different optimization algorithms can be used for parameter estimation. The performance of the algorithms usually depends on the specific models (12). Different tools are available for parameter estimations, for example Copasi (13), SBML-PET-MPI (14, 15), PottersWheel (16), and so on.

When we know the initial conditions and parameter values for the ODE model, it is possible to run computer simulation for the model with software tools. Model simulation is usually implemented with some ODE solver algorithms. The user needs to specify the ODE solver with the initial condition, parameter values, and ODE system. Some SBML software tools are able to automatically generate the ODE equations if the modeler can define all reactions involved in the pathway and set the reaction kinetics. Copasi (18) and CellDesigner (17) are two popular model construction and simulation tools. Details about the software tools are available in the user guide of these tools or other resources.

3.5. A General Workflow for Modeling of Cellular Signaling Pathways

The procedure of building a mathematical model for biological signaling pathways can be summarized as the following steps.

(a) Draw a complete reaction interaction map of the signaling pathway based on the existing biological knowledge. Occasionally, the signaling pathway might be disconnected; new reactions can be added to bridge the disconnected subsystems together. Such a map is a schematic representation of the mathematical model. The modeler can define the reactions with some software tools in different ways, for example write the reaction in text format (for example, Copasi) or draw the graphic reaction map (for example, CellDesigner).

(b) Assign appropriate reaction kinetics for all the reactions that are involved in the signaling pathway.

(c) Generate the ODE equations for the signaling pathway. For a simple system, it is possible to write it down manually. However, for large signaling network, it is more convenient to generate the ODE system by SBML software packages, which avoid typo mistakes as well.

(d) Set the initial concentration (or initial amount) of the molecules in the signaling pathway. The initial concentrations of the molecules might be unknown and need to be estimated with quantitative experimental data sets.

(e) Set the values of the parameter appeared in the reaction kinetics. If some parameter values are unknown, perform parameter estimation based on quantitative experimental data sets.

(f) Run simulations for the mathematical model. With the mathematical model, one can implement additional kinetic analysis of the signaling pathway, for example steady-state analysis, sensitivity analysis, and other perturbation analysis such as over-expression, down-regulation, knockout, and so on.

Acknowledgments

This work was supported by the Excellence Initiative of the German Federal and State Governments (EXC 294).

References

1. Boogerd FC, Bruggeman FJ, Hofmeyr J-HS, Westerhoff HV (2007) Systems biology: philosophical foundations, 1st edn. Elsevier, Amsterdam
2. Hucka M, Finney A, Sauro HM, Bolouri H, Doyle JC, Kitano H et al (2003) The systems biology markup language (SBML): a medium for representation and exchange of biochemical network models. Bioinformatics 19:524–531
3. Aldridge BB, Burke JM, Lauffenburger DA, Sorger PK (2006) Physicochemical modelling of cell signalling pathways. Nat Cell Biol 8:1195–1203
4. Guldberg CM, Waage P (1879) Concerning chemical affinity. Erdmann's J Pract Chem 127:69–114
5. Klipp E, Herwig R, Kowald A, Wierling C, Lehrach H (2005) Systems biology in practice: concepts, implementation and application, 1st edn. Wiley, Berlin
6. Hill AV (1910) The possible effects of the aggregation of the molecules of haemoglobin on its dissociation curves. J Physiol 40:4–7
7. Legewie S, Bluthgen N, Herzel H (2005) Quantitative analysis of ultrasensitive responses. FEBS J 272:4071–4079
8. Szallasi Z, Stelling J, Periwal V (2006) System modeling in cellular biology: from concepts to nuts and bolts, 1st edn. The MIT, Boston, MA
9. Ferrell JE Jr (2008) Feedback regulation of opposing enzymes generates robust, all-or-none bistable responses. Curr Biol 18:R244–R245
10. Ma W, Trusina A, El-Samad H, Lim WA, Tang C (2009) Defining network topologies that can achieve biochemical adaptation. Cell 138:760–773
11. Zi Z, Klipp E (2007) Cellular signaling is potentially regulated by cell density in receptor trafficking networks. FEBS Lett 581:4589–4595
12. Moles CG, Mendes P, Banga JR (2003) Parameter estimation in biochemical pathways: a comparison of global optimization methods. Genome Res 13:2467–2474
13. Hoops S, Sahle S, Gauges R, Lee C, Pahle J, Simus N et al (2006) COPASI—a COmplex PAthway SImulator. Bioinformatics 22:3067–3074
14. Zi Z (2011) SBML-PET-MPI: a parallel parameter estimation tool for Systems Biology Markup Language based models. Bioinformatics 27:1028–1029
15. Zi Z, Klipp E (2006) SBML-PET: a Systems Biology Markup Language-based parameter estimation tool. Bioinformatics 22:2704–2705
16. Maiwald T, Timmer J (2008) Dynamical modeling and multi-experiment fitting with PottersWheel. Bioinformatics 24:2037–2043
17. Tanimura AFMMHKN (2003) CellDesigner: a process diagram editor for gene-regulatory and biochemical networks. Biosilico 1:159–162
18. Mendes P, Hoops S, Sahle S, Gauges R, Dada J, Kummer U (2009) Computational modeling of biochemical networks using COPASI. Methods Mol Biol 500:17–59

Chapter 4

Bistability in One Equation or Fewer

Graham A. Anderson, Xuedong Liu, and James E. Ferrell, Jr.

Abstract

When several genes or proteins modulate one another's activity as part of a network, they sometimes produce behaviors that no protein could accomplish on its own. Intuition for these emergent behaviors often cannot be obtained simply by tracing causality through the network in discreet steps. Specifically, when a network contains a feedback loop, biologists need specialized tools to understand the network's behaviors and their necessary conditions. This analysis is grounded in the mathematics of ordinary differential equations. We, however, will demonstrate the use of purely graphical methods to determine, for experimental data, the plausibility of two network behaviors, bistability and irreversibility. We use the *Xenopus laevis* oocyte maturation network as our example, and we make special use of iterative stability analysis, a graphical tool for determining stability in two dimensions.

Key words: Graphical method, Bistability, Irreversibility, Iterative stability analysis, Steady state, Ultrasensitivity, *Xenopus laevis*, Oocyte maturation

1. Introduction

Let us begin by considering the point in describing a biological circuit at which natural language fails. We will trace a chain of causalitiy through a schematic of the *Xenopus* oocyte maturation machinery (Fig. 1a).

Xenopus oocytes are arrested in Meiosis I early in a female frog's development. In adulthood, oocytes are triggered to mature, an all-or-nothing and irreversible (1) process of cell growth and progression to metaphase of meiosis II, at which point the oocyte again arrests awaiting fertilization. Maturation is initiated by progesterone. Imagine an oocyte is exposed to a small amount of progesterone, which leads to a small amount of phosphorylated and activated Mos. This kinase in turn phosphorylates and activates the kinase MEK, which does the same to the kinase p42 MAPK. But then p42 MAPK phosphorylates and activates a little more Mos, which activates a little more MEK, and so on. In the absence of any

Xuedong Liu and Meredith D. Betterton (eds.), *Computational Modeling of Signaling Networks*, Methods in Molecular Biology, vol. 880, DOI 10.1007/978-1-61779-833-7_4,

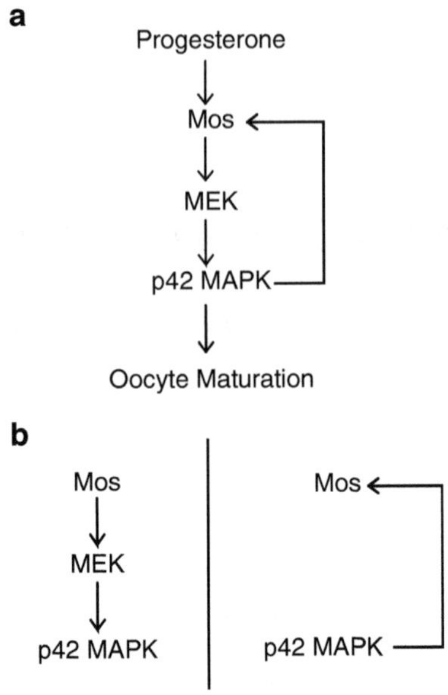

Fig. 1. A schematic of the *Xenopus laevis* MAP kinase pathway (**a**) as connected *in vivo* and (**b**) separated into a cascade and a positive feedback, for purposes of theoretical analysis.

phosphatases, this positive feedback loop (2) would clearly continue on in this fashion, leading to the total activation of all the available components of the cascade. In the presence of phosphatases, however, things get trickier.

We could imagine that there is enough phosphatase activity acting throughout the loop to balance the collective activity of the kinases. In this case, the positive feedback loop will not explode to total activation, but will come to rest at some intermediate level. Now what happens if we add just a little more progesterone to a loop at rest at this intermediate position? Will the loop then go on to explode up to total activation? Or will the loop instead remain at rest, but with a little more kinase activity?

Furthermore, what if not all the kinases in the loop pass on activation in the same way? Specifically, what if some kinases respond with a nonlinear response that most of us know as *cooperativity*? If this were true, then, for example, MEK may not phosphorylate p42 MAPK very much when just a small fraction of MEK is phosphorylated, but when phosphorylated MEK reaches higher levels, it may begin to become active and phosphorylate p42 MAPK with great alacrity. If you think imaginitively and with a bit of forsight, you might even imagine that in this case of a cooperative MEK, the feedback loop could be balanced by phosphatase activity twice—once when MEK activity is still not responding well to phosphorylation and again when MEK kicks into gear.

As you can see, analyzing a feedback loop presents an inherent challenge. We understand the world in terms of cause and effect, but in a feedback loop, cause and effect are all mixed up. The preceding exploration of the oocyte maturation network's potential behaviors was reaching the limit of what natural language can describe precisely. Mathematics is a good choice for the language of our conversation, but we will keep our dialogue focused on the cognates that natural language and math share in common: curves and graphs. In doing so, we will explore some of the more biologically relevant concepts in *dynamical systems theory*, the branch of mathematics used to describe the behavior of sets of nonlinear differential equations.

2. Steady State and Stability

Let's begin by obtaining a clear definition of two useful concepts: steady state and stability.

Imagine a valley with very high and steep sides. At the bottom of the valley floor is a small ridge, which effectively divides the valley in two. Additionally, the valley is smooth, uniform, and symmetric—like a half-pipe—so it is enough to consider the cross section of the valley, as we do in Fig. 2. Now, drop a perfectly spherical rock into the valley. Where will it come to rest? Intuition tells us that as long as there is some friction to slow things down, the rock will come to rest at the bottom on one side or the other of the small ridge. Now what if the rock is centered exactly on the top of the ridge? In an idealized situation, the rock will stay perched atop the ridge forever.

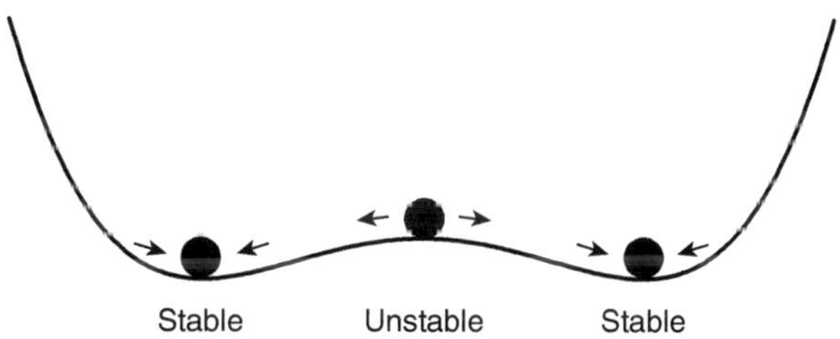

Fig. 2. Cross section of a landscape containing a perfectly spherical rock. The rock can come to rest at three positions: the two valley floors, and at the very top of the small hill. If the rock is perched at the hilltop, and it is pushed, it will clearly accelerate in the direction it was pushed and roll down the hillside. Thus, the hilltop position is unstable. A rock perched at a valley bottom, however, if pushed, will stop rolling, then reverse direction and eventually settle again at the valley bottom. The two valley bottom positions are stable. You can map the mathematical concept of stability onto an intuitive understanding of stability you already possess.

Every location where the rock is able to come to rest is a *steady state*. There are three steady states in total: the two bottoms of the mini-valleys on either side of the ridge and the top of the ridge itself. Only the two valley bottoms, however, can be occupied by the rock in anything less than an idealized world. If the rock should fall perfectly atop the ridge, the slightest breeze or tremor would knock it to one side or the other. The valley top is, it follows, an *unstable* steady state. When the rock is at the valley bottom, on the other hand, breezes, tremors, even hard shoves may push the rock up the valley wall, but it will always roll back down and come to rest at its original location. These are *stable* steady states.

Hopefully, the parable of the spherical rock and the oddly shaped valley has allowed you to begin mapping the mathematical definition of stability onto the common-sense idea of stability. Explicitly, dynamical stability occurs when small perturbations from a steady state shrink over time. This is clearly true of the rock at a valley bottom—a shove up the wall is reversed due to gravity. You've probably already guessed that instability occurs when small perturbations grow over time. This happens when a tiny breeze pushes the ridge-top rock to one side, and the displacement quickly grows as the rock rolls down the ridge.

These definitions of steady states and stability will serve us well in analyzing biochemical circuits, but we still need a few more tools in our kit.

3. Open Loops

We seek a method of understanding a feedback loop that is both graphical in nature and closely related to experiments. In the case of the MAPK system governing *Xenopus* oocyte maturation, we can divide the system into two interacting pieces: (1) The cascade from Mos to p42 MAPK, and (2) the positive feedback from p42 MAPK to Mos (Fig. 1b). We can then imagine measuring a dose–response curve for each piece in isolation. We will call this curve an *open-loop dose–response curve*. The open-loop dose–response curve from Mos to p42 MAPK has been experimentally measured (1), a fact we will return to. For now, let's start our analysis by assuming both open-loop dose–response curves for the MAPK system are saturating Michaelis–Menten shaped curves.

Each curve represents a series of steady states that would exist for p42 MAPK and Mos when half of the feedback loop is operating in isolation. So, if we can find a pair of values for p42 MAPK and Mos that are steady states in both open-loop dose–response curves, that pair of values are also a steady state for the entire MAPK system. In other words, any states that are steady for all the isolated

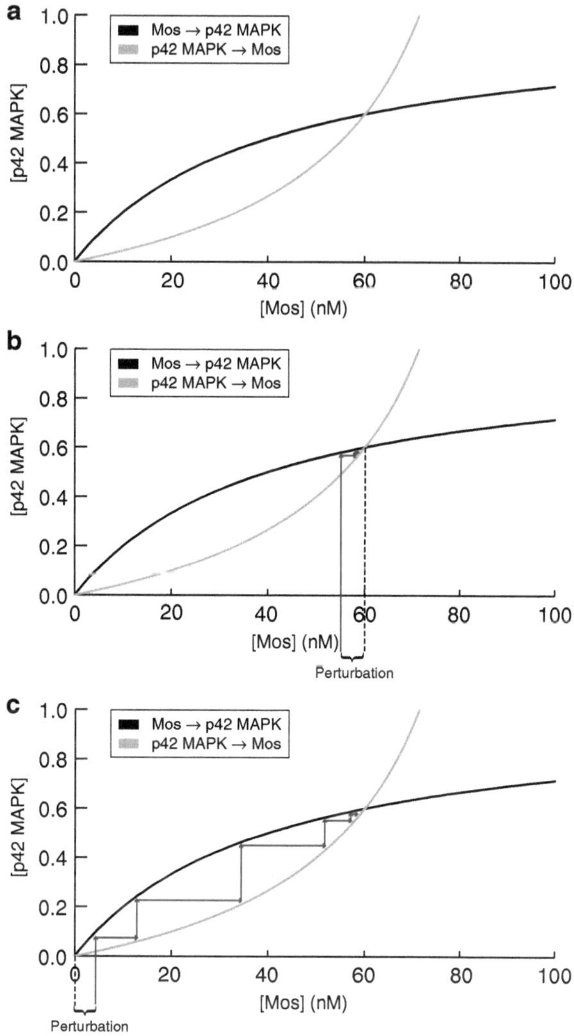

Fig. 3. Open-loop dose–response curves for the separated legs of the MAPK system, assuming Michaelis–Menten-shaped curves. Note the two intersection points (**a**), indicating two steady states. Suppose the system comes to rest at the nonzero steady state (**b**). Following a small perturbation with iterative stability analysis, we end up back at the steady-state; the nonzero steady-state is stable. If the system is at rest at the graph's origin (**c**), then a small perturbation will grow with each iteration until the system comes to rest at the stable steady state. Thus, the steady state at the origin is unstable.

pieces of a system must also be steady for the integrated system. Graphically, this statement has compelling simplicity: steady states for a positive feedback loop occur at intersections of the component open-loop dose–response curves. To see this, we will graph open-loop curves for both separated pieces of the MAPK system on the same set of axes (Fig. 3a). The Mos → p42 MAPK curve is plotted as a saturating Michaelis–Menten shaped curve, and the MAPK → Mos curve, with the same shape, is reflected and rotated in order to be displayed on the same set of axes.

Now, we just have to find the intersection of these two curves to find the steady states of the entire positive feedback loop. But in order to analyze the potential for interesting behaviors and phenotypes, we need to be able to determine the stability of any steady state we come across. To see if our rock was in a stable state, we simply pushed it and observed what happened. We'll now give the MAPK feedback loop a shove and watch what happens.

Because we have separated the MAPK signaling system into two parts, each with its own steady-state dose–response curve, our version of "watching what happens" will exist within this piecewise framework. In real life, the consequences of a small perturbation to a steady state of MAPK signaling simultaneously involve both the cascade from Mos to MAPK and the feedback from MAPK to Mos. But for small enough perturbations, we can get away with examining each piece in turn. If we begin at a steady state, then "push" the amount of active MEK away from the steady state, we can follow the causality through each piece of the MAPK system in series, then determine if our perturbation has grown, indicating an unstable steady state, or has shrunk back toward the fixed point, indicating a stable steady state.

For the first iteration of our piecewise method, start at the systemwide steady-state value of MAPK on the *x*-axis of Fig. 3b, then move a bit to the right or left to create a small perturbation. In this case, a small enough perturbation is one over which the slopes of the steady-state curves do not change sign. Now, trace a line up to the cascade steady-state curve (black) and read off the y-value of MAPK activity. For the second step, begin with that same y-value of MAPK activity, and trace a line over to the feedback steady-state curve (gray). You should now arrive at a value on the *x*-axis that you can compare to the starting point. Is the new value of MEK activity closer to the fixed point or farther away? If it is closer, then we know that the MAPK circuit has decreased the size of the perturbation after one round of our iterative analysis, and it will do so again on the next iteration. This fits our definition of a stable steady state. On the other hand, if the perturbation has grown, and one round of iterative analysis has brought MEK signaling away from the fixed point, we know that the steady state is unstable.

4. Bistability

In Fig. 3b, 3c, we apply iterative stability analysis to two different steady states. We find that the steady state at the origin is unstable. Any small perturbation climbs toward the second steady state, which is stable. The biochemical interpretation is that any small amount of activation in the MAPK system will lead to activation at the level of the stable steady state. In the presence of any perturbation, whether biological noise or progesterone, the loop will turn on and stay on. If we were to add progesterone, we could push this steady state to a higher degree of activation (not pictured), and if we took progesterone away, we could lower the steady state. But without a second stable steady state, the system is irreversibly on. This constitutes a trivial case of irreversibility, one in which there is no "off" state. In fact, we wouldn't really consider this to be an irreversible cell-fate decision. We would just consider it to be a constitutively activated signaling pathway.

What could create a second stable steady state in the MAPK positive feedback loop? We made some guesses about nonlinear activation in the introduction, but let's keep our thinking constrained for now by the system's open-loop dose–response curves. One way to make more steady states might be to bend one of these curves, such that they intersect at more points. This makes sense—more bends in a curve provide more chances for intersecting a straight curve.

It turns out that the experimentally measured Mos → p42 MAPK curve does contain such a bend. In Fig. 4a, we reprint data from (1). On a log horizontal axis, even a Michealis-Menten-shaped curve appears in an "S"-shape, but you can see that the experimental Mos → p42 MAPK curve is significantly sharper than a Michaelis–Menten curve. We plot the same data on a linear axis for further comparison in Fig. 4b.

After adding a similar bend to our idealized representation of the Mos → p42 MAPK curve, the two open loop dose–response curves now intersect at three points (Fig 5a). Furthermore, using our iterative stability analysis, we find that a small perturbation in Mos from the modified first steady state now translates into a smaller perturbation, not a larger one, after a trip through the feedback loop (Fig. 5b). Before bending the Mos → MAPK curve, we obtained the opposite result, that perturbations in Mos are magnified. Really, the only thing that has changed here at the origin is the relative slopes of the steady-state curves. The Mos → p42 MAPK curve (black) previously exited the origin above the p42 MAPK → Mos curve (gray). But now it exits below, and the stability of the steady state has switched from unstable to stable.

The newly created middle steady state has the property that the origin used to have: The Mos → p42 MAPK curve (black) exits above and to the right of the p42 MAPK → Mos curve. If all that matters for stability is the relative slopes of the curves at a steady state, we would predict this state to be unstable, just like the origin was in Fig. 3. Iterative stability analysis shows that this is indeed the case (Fig. 5c), as perturbations from the middle steady state inevitably arrive at the first or the third state.

What does this mean for the MAPK feedback loop? There are now clearly two stable steady states that the loop may come to rest in, one low and one high. More importantly, there are no stable

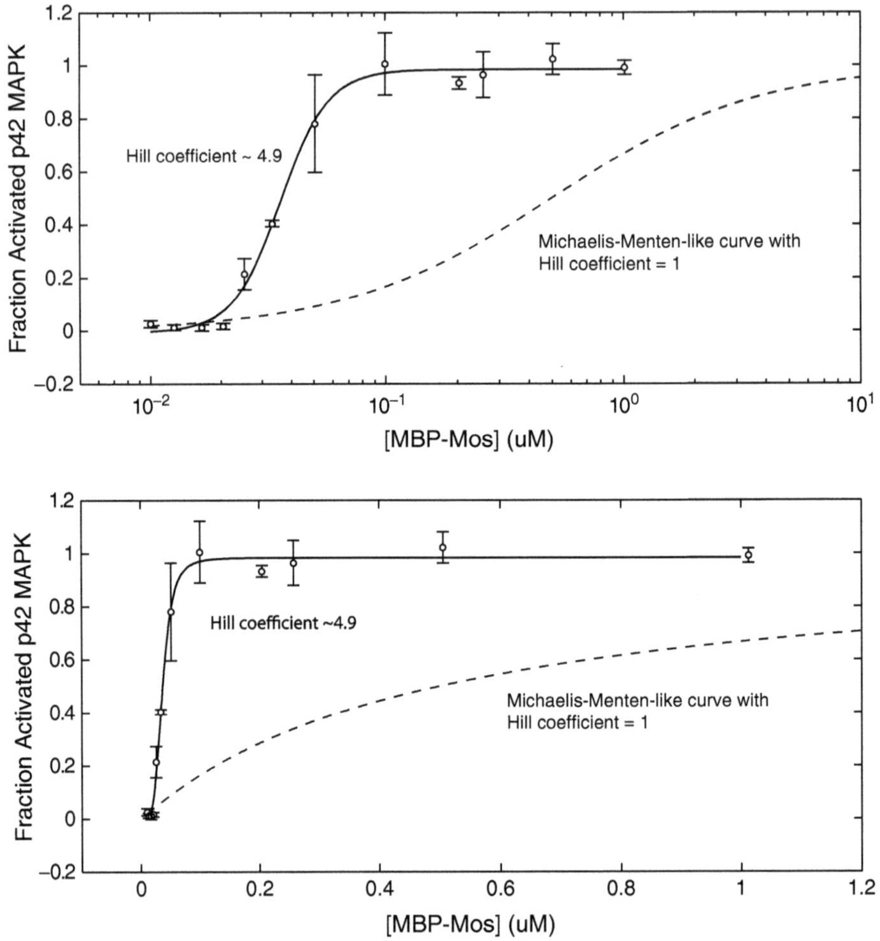

Fig. 4. The in vivo p42 MAPK → Mos curve displays a sigmoidal, not a Michaelis–Menten shape. A hill coefficient of 4.9 causes p42 MAPK activity to markedly accelerate with respect to Mos concentration before saturating. For reference, a Michaelis–Menten-shaped curve contains no such acceleration. The difference in shape can be seen on both semi-log (**a**) and linear (**b**) axes. Data reprinted from (1).

steady states connecting the "off" and "on" states. Just by bending one of the open-loop dose–response curves in our positive feedback loops, we have created *bistability*.

5. Ultrasensitivity

What actually happened when we bent the Mos → p42 MAPK curve? What is the biological meaning of such a change? It turns out that the exact shape we bent the curve into is not as important as how steep the curve was in the bent region. A steeper curve is more likely to intersect another curve at more than one point.

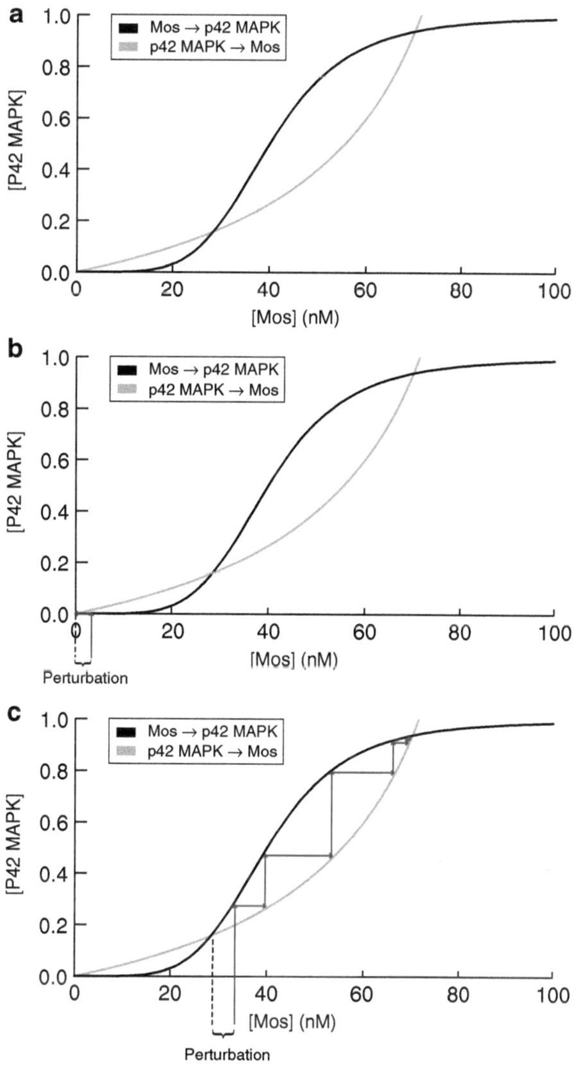

Fig. 5. We induce some sigmoidal curvature in the Mos → p42 MAPK open-loop curve (**a**). Note there are now three intersections of the two open-loop curves and three steady-states. A perturbation to a system at rest at the origin now results in a return to the origin—the origin has become a stable steady state (**b**). Perturbations from the middle steady state will grow, coming to rest at either the origin or the upper steady state (**c**). In total, there are now two stable steady states and one unstable.

Indeed, steep *ultrasensitive* dose–response curves are often present in networks that display the hallmakrs of bistability (3; 4; 5). But what sort of biological mechanisms could affect the steepness of a dose–response curve?

There are in fact too many steepening mechanisms to list here (6; 7; 8; 9; 10; 11), but one that many biochemists and cell biologists are familiar with is cooperativity. Cooperativity occurs whenever some process is required to be completed more than once in order for an activation to occur. It shows up when more than one transcription factor is required to bind to a promoter for full gene expression, and when one of the subunits of hemoglobin must be occupied with oxygen in order for the other subunits to display an increased affinity. The *Xenopus* p42 MAPK cascade may exhibit cooperativity due to the distributive phosphorylations required to activate each kinase (12).

But there are other noncooperative ways to bend a dose–response curve (13; 14). For example, ultrasensitivity may arise when enzymes in the MAPK cascade are nearly saturated with substrate. When this happens, an opposing kinase and phosphatase engage in a tug-of-war for modifying their shared substrate. Any excess of the kinase's activity over the phosphatase's, for example, results in nearly all of the substrate becoming phosphorylated. The transition in this tug-of-war is sharp, leading to a sigmoidal dose–response curve, though there was never a cooperative mechanism at work. We will call the general property of a nonlinear, bent, sigmoidal curve ultrasensitivity. In practice, we quantify the degree of ultrasensitivity present in experimental data by fitting a Hill equation $Y = min + (max - min)\,\frac{X^{n_H}}{K^{n_H} + X^{n_H}}$ where Y is any biological output, X is any biological input, max and min are the bounds of the output, K is the level of X at which Y is halfway activated between min and max, and n_{H} is the Hill coefficient, which controls ultrasensitivity. Equations fit to more bent, sigmoidal graphs will have higher Hill coefficients. You may also notice that this is a Michaelis–Menten equation with renamed constants and the addition of the Hill coefficient. In fact, a Michaelis– Menten equation is a Hill equation with $n_{\mathrm{H}} = 1$.

Using a standard nonlinear curve-fitting algorithm in MATLAB we can fit the Hill equation to experimentally measured data to find that the Mos → p42 MAPK dose–response-curve has an nH value of 4.9 (1). In fact, we have based our example Mos → MAPK open-loop dose–response curve on precisely these data. The curve from MAPK → Mos has not been measured, so we will for now stick with our assumption that this dose–response curve is a saturating Michaelis–Menten-shaped curve.

6. Reversibility

What do we do with these steady states now that we've found them and characterized their stability? We already know that the existence of more than one stable steady state is bistability. But we probably want to know if bistability exists over a broad range of the values of some parameter, or if bistability was more of an ephemeral property of the system. And what happens when bistability appears and disappears as we vary a parameter?

Let's choose that varied parameter to be progesterone concentration. In the *Xenopus* oocyte, progesterone initiates maturation and leads to self-sustaining MAPK signaling. The specific way in which progesterone interacts with the MAPK feedback loop is poorly understood, so we will simply assume for now that it leads to an amount of activated p42 MAPK proportional to progesterone concentration. This progesterone-derived p42 MAPK signaling is added to the activation already brought about by Mos, and it shifts the Mos → MAPK open-loop dose–response curve vertically by an amount proportional to progesterone concentration. The open-loop response curve accounts for MAPK activation coming from all sources. MAPK activation coming from progesterone is constant at all levels of Mos, so it shifts the curve vertically without altering its shape.

Now, let's simply add a series of concentrations of progesterone, vertically shifting the Mos → MAPK curve (Fig. 6a), keeping track of where the steady states are found (Fig. 6b). (In these figures, we've adjusted the saturating values of the two open-loop dose–response curves to make the following points more clear.) We notice right away that adding progesterone moves the first stable steady state (lower black line in Fig. 6b) and the unstable steady state (gray line in Fig. 6b) closer together. Eventually, the two points meet, and with another addition of progesterone, disappear. At this high level of progesterone, the two open-loop curves intersect at only one point (top black line in Fig. 6b).

What happens to the MAPK system if it occupied the stable steady state that has now disappeared? This problem at first seems difficult, since we haven't encountered a previous case in which the system was not already at a steady state. But we can perform iterative stability analysis and get very close to the answer we would arrive at with more sophisticated treatments. We find that after the lower stable steady state disappears, the feedback loop shoots up to the upper stable-steady state, undergoing a rapid activation of MAPK signaling.

This type of behavior exactly matches descriptions of the all or nothing nature of oocyte maturation (1). Furthermore, it is an example of how a bistable system moves between its two stable steady states. The system never comes to rest in any state that is intermediate between the two stable steady states. From the point of view of an experiment conducted at steady state, the transition is absolute and discontinuous. This change in the number and type of steady states is termed *bifurcation*. In the bifurcation we just examined, the lower stable steady state and the unstable steady state merge and annihilate. If a system occupies a steady state that is annihilated in a bifurcation, the system will be forced to move to a remaining stable steady state, creating an all-or-nothing transition.

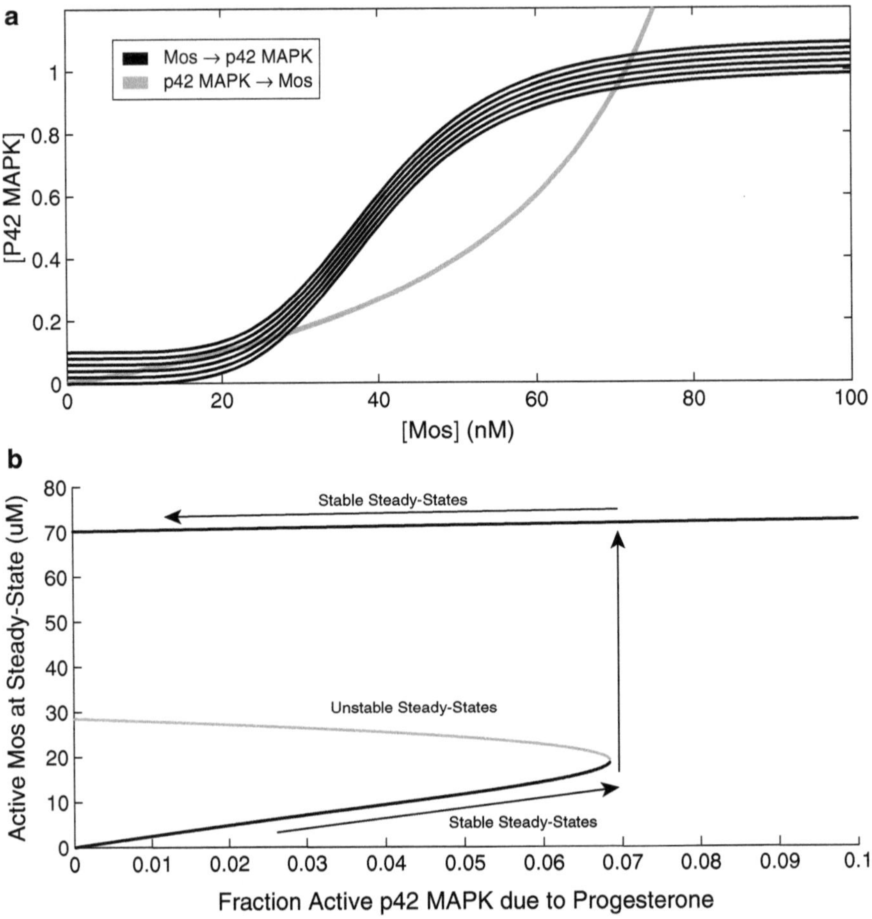

Fig. 6. We incrementally increase the basal activity of p42 MAPK, which represents the effect of progesterone in vivo. This shifts the Mos → p42 MAPK open-loop curve in the vertical axis (**a**). Each shift has the effect of changing the positions of the three steady states in the Mos axis. Now, we can plot the location of these steady states while varying basal p42 MAPK activity over a wide range (**b**). As basal p42 MAPK activity increases, the two stable steady states move up the Mos axis (now displayed as the vertical axis), while the unstable steady state moves down the Mos axis. At a high-enough basal p42 MAPK activity, the lower stable steady state and the unstable steady state meet and annihilate, leaving only the upper stable steady state. A system at rest at the lower stable steady state, when experiencing this annihilation event, will explosively jump from a low Mos concentration to a high Mos concentration without coming to rest at any intermediate levels. Even if the basal p42 MAPK level is lowered after this explosive event, the system will forever remain at the higher of the two stable steady states. The transition from low to high Mos was irreversible.

Of course, lowering the concentration of progesterone will now allow the lower stable steady state and the middle unstable steady state to reappear. The bifurcation will occur again, but in reverse. Will the MAPK loop then decrease its activity to reoccupy the newly formed lower stable steady state? No, it will not. Because the upper stable steady state is still present, the MAPK system will remain in its high-activity state, even though the lower-expression option exists (15).

Importantly, these persistent states and the all-or-nothing behaviors they produce are properties of the network of kinases itself, rather than of any individual enzyme. If the connections between kinases were altered, the number and type of steady states might change. Conversely, individual kinases can be degraded and resynthesized while the network remains in a persistent steady state (16; 17).

Through this simple graphical method, we have uncovered something quite profound. The behavior of our model of the MAPK feedback loop depends on its history. In the bistable region, the loop will occupy the lower stable steady state if it was previously in this state, and it will occupy the upper stable steady state if it was instead previously in that state. This history dependence of a system's behavior is *hysteresis*. You are already familiar with hysteresis in blowing up a rubber balloon. The first time you inflate a balloon, it requires quite a bit of pressure. But this extra pressure goes into stretching the rubber inelastically, and it never returns to its original state after being stretched. The next time you blow up the balloon, it requires much less pressure to inflate the balloon to the same degree.

Will the MAPK feedback loop ever return to its original state once it is activated? In Fig. 6a, we continue reducing the concentration of progesterone bit-by-bit, but we find that the upper stable steady state never disappears. We record this fact in Fig. 6b as an upper steady state (top black line) that intersects the *y*-axis at 0 nM progesterone. Like a stretching balloon, the MAPK system undergoes an irreversible transition the first time it is activated. Unlike our first encounter with this property in Fig. 3c, there is now a stable "off" state that can be initially occupied.

Our model now displays an "off" state, an "on" state, and an irreversible all-or-nothing transition from "off" to "on". With just graphs and one equation, we've recapitulated some of the most biologically significant behaviors of cell-fate determination. We can even make a simple prediction. Because the all-or-nothing nature of this decision is a consequence of bistability, which emerges from the interaction of the cascade from Mos → p42 MAPK and the feedback from p42 MAPK → Mos, we should be able to abolish irreversibility and the all-or-nothing character by inhibiting one of the legs of the feedback loop. Graphically, if we remove the p42 MAPK → Mos curve (gray in Figs. 3 and 5), then every point along the remaining Mos → p42 MAPK curve will be a steady state. The output of the cascade will depend continuously on the input, with no irreversibility. Indeed, these predictions that abolishing positive feedback in the *Xenopus* MAPK cascade also converts the normally all-or-nothing and irreversible response of the cascade to progesterone into a continuous, reversible response have been born out experimentally (18; 19).

7. Conclusion

The concepts of bistability, bifurcations, and hysteresis have made their way into mainstream cell biological and biochemical literature, though many working scientists did not have access to these concepts as a part of their training. Fortunately, only basic calculus and graphical methods are required for a working knowledge of these dynamical phenomena. By separating a nonlinear system such as a feedback loop into at least two chunks, one can graph the open-loop dose–response curves, or estimates thereof, and then, by applying iterative stability analysis, arrive at the number and location of stable and unstable steady states. Whether achieved using graphical techniques, as we have here, or using analytical techniques, as you may go on to master, identifying and characterizing steady states is the foundation of dynamical systems analysis.

Once identified and characterized, one can observe how the steady states change as parameters of a biological system are varied. Here, we chose to vary progesterone concentration, though we could have chosen to vary the sensitivity of the cascade instead, and we would have obtained the graph of steady-state position as a function of sensitivity. Bistability is simply the property that two stable steady states are present at a single parameter value. Bifurcations occur whenever the number of steady states (stable or unstable) changes as the parameter varies, and hysteresis, always present in the case of bistability, refers to the history dependence a system displays as it moves along the graph of stable steady states as the parameter is varied.

Bistability, when coupled with some degree of cell-to-cell variability, often presents itself in a population as two peaks in the histogram of some single-cell measurement. To be sure, this bimodality is a hallmark of bistability, but not a sufficient condition (20). Hysteresis can be thought of as a more reliable hallmark of bistability (21).

If this was your first intellectual jaunt into the methods of dynamical systems, we recommend adding one additional graphical method to your toolbox, rate–balance analysis (22). It is the graphical method of choice for systems that can be written in a single variable, and characterizing steady states in rate–balance analysis has a more obvious link to the mathematical definition of stability.

Finally, we can recommend Stephen Strogatz's textbook Nonlinear Dynamics and Chaos (23) as a lucid and eminently readable next step in your dynamical systems education.

References

1. Huang CY, Ferrell JE Jr (1996) Ultrasensitivity in the mitogen-activated protein kinase cascade. Proc Natl Acad Sci USA 93(19):10078–10083
2. Matten WT, Copeland TD, Ahn NG, Vande Woude GF (1996) Positive feedback between MAP kinase and Mos during Xenopus oocyte maturation. Dev Biol 179:485–492
3. Bagowski CP, Besser J, Frey CR, Ferrell JE Jr (2003) The JNK cascade as a biochemical switch in mammalian cells: ultrasensitive and all-or-none responses. Curr Biol 13(4):315–320
4. Bagowski CP, Ferrell JE Jr. Bistability in the JNK cascade. Curr. Biol. 2001 11, 1176–1182.
5. Kim SY, Ferrell JE Jr (2007) Substrate competition as a source of ultrasensitivity in the inactivation of Wee1. Cell 128(6):1133–1145
6. Ferrell JE Jr (1996) Tripping the switch fantastic: how a protein kinase cascade can convert graded inputs into switch-like outputs. Trends Biochem Sci 21(12):460–466
7. Burack WR, Sturgill TW (1997) The activating dual phosphorylation of MAPK by MEK is nonprocessive. Biochemistry 36:5929–5933
8. Chen M, Li D, Krebs EG, Cooper JA (1997) The casein kinase II beta subunit binds to Mos and inhibits Mos activity. Mol Cell Biol 17:1904–1912
9. Ferrell JE Jr (1997) How responses get more switch-like as you move down a protein kinase cascade. Trends Biochem Sci 22:288–289
10. Ferrell JE Jr (1999) Building a cellular switch: more lessons from a good egg. Bioessays 21 (10):866–870
11. Ferrell JE Jr (2008) Feedback regulation of opposing enzymes generates robust, all-or-none bistable responses. Curr Biol 18(6): R244–R245
12. Markevich NI, Hoek JB, Kholodenko BN (2004) Signalling switches and bistability arising from multisite phosphorylation in protein kinase cascades. J Cell Biol 164:353–359
13. Goldbeter A, Koshland DE (1981) An amplified sensitivity arising from covalent modification in biological systems. Proc Natl Acad Sci USA 78(11): 6840–6844.
14. Goldbeter A, Koshland DE Jr (1982) Sensitivity amplification in biochemical systems. Q Rev Biophys 15:555–591
15. Ferrell JE Jr (2002) Self-perpetuating states in signal transduction: positive feedback, double-negative feedback and bistability. Curr Opin Cell Biol 14:140–148
16. Lisman JE (1985) A mechanism for memory storage insensitive to molecular turnover: a bistable autophosphorylating kinase. Proc Natl Acad Sci USA 82:3055–3057
17. Thomas R, Kaufman M (2001) Multistationarity, the basis of cell differentiation and memory. I. Structural conditions of multistationarity and other nontrivial behavior. Chaos 11:170–179
18. Ferrell JE Jr, Machleder EM (1998) The biochemical basis of an all or none cell fate switch in *Xenopus* oocytes. Science 280 (5365):895–898
19. Xiong W, Ferrell JE Jr (2003) A positive-feedback-based 'memory module' that governs a cell fate decision. Nature 426(6965):460–465
20. To TL, Maheshri N (2010) Noise can induce bimodality in positive transcriptional feedback loops without bistability. Science 327 (5969):1142–1145
21. Pomerening JR, Sontag ED, Ferrell JE Jr (2003) Building a cell cycle oscillator: hysteresis and bistability in the activation of Cdc2. Nat Cell Biol 5:346–351
22. Ferrell JE Jr, Xiong W (2010) Bistability in cell signalling: how to make continuous processes discontinuous, and reversible processes rreversible. Chaos 11:227–236
23. S. Strogatz, (2001) "Nonlinear dynamics and chaos," Westview Press; 1 edition

Chapter 5

Mathematical Investigation of How Oncogenic Ras Mutants Promote Ras Signaling

Edward C. Stites and Kodi S. Ravichandran

Abstract

We have used a mathematical model of the Ras signaling network to link observable biochemical properties with cellular levels of RasGTP. Although there is abundant data characterizing Ras biochemistry, attributing specific changes in biochemical properties to observed phenotypes has been hindered by the scope and complexity of Ras regulation. A mathematical model of the Ras signaling module, therefore, appeared to be of value for this problem. The model described the core architecture shared by pathways that signal through Ras. Mass action kinetics and ordinary differential equations were used to describe network reactions. Needed parameters were largely available in the published literature and resulted in a model with good agreement to experimental data. Computational analysis of the model resulted in several unanticipated predictions and suggested experiments that subsequently validated some of these predictions.

Key words: Ras, Cancer, Oncogene, ODE model, Simulation, Module, Network

1. Introduction

A theme long present in the Ras literature is that the key to understanding how oncogenic Ras mutants promote cancer will be best attained through defining the biochemical differences between wild-type and mutant Ras proteins. There is a large literature characterizing the many Ras reactions for wild-type and mutant Ras proteins that have certainly helped us understand the biochemistry of Ras proteins (1). Perhaps the most striking outcome of these studies is that rather than a single property differing between wild-type and mutant-containing networks there are multiple properties that differ (2–4). It is not trivial to attribute each biochemical difference directly with its proportional contribution to the increased cellular levels of guanosine-triphosphate (GTP-bound) Ras (RasGTP) as the Ras signaling module is complex and nonlinear. To overcome the experimental limitations in "isolating"

Xuedong Liu and Meredith D. Betterton (eds.), *Computational Modeling of Signaling Networks*,
Methods in Molecular Biology, vol. 880, DOI 10.1007/978-1-61779-833-7_5,

the individual properties, we developed a mathematical model that had the potential to provide a link between protein biochemistry and observed levels of RasGTP in a cancer cell (5). By quantitatively describing the individual reactions and their dependence upon the measured rate constants, it should in principle be possible for a mathematical model to serve as a bridge between biochemical data and observed patterns of RasGTP. This approach should, therefore, be capable of providing a more rigorous link between the numerous reactions and their properties with cellular levels of RasGTP.

1.1. The Ras Signaling Module

As a GTPase, Ras reversibly binds the guanine nucleotides GTP and guanosine diphosphate (GDP) and has the ability to hydrolyze bound GTP to GDP (Fig. 1). Under basal, unstimulated conditions, wild-type Ras (RasWT) is primarily bound to GDP; after stimulation, such as through the platelet-derived growth factor receptor (PDGFR), RasWT has an increased fraction bound to GTP (6). Oncogenic mutants have impaired GTPase activity and are found with a higher percentage bound to GTP than wild-type Ras (6). Ras proteins adopt slightly different structural conformations in a manner dependent upon the nucleotide bound (7). Many proteins in the cell, termed Ras effectors, have domains that allow them to specifically interact with the GTP-bound Ras (RasGTP). These interactions promote downstream signaling from RasGTP (8).

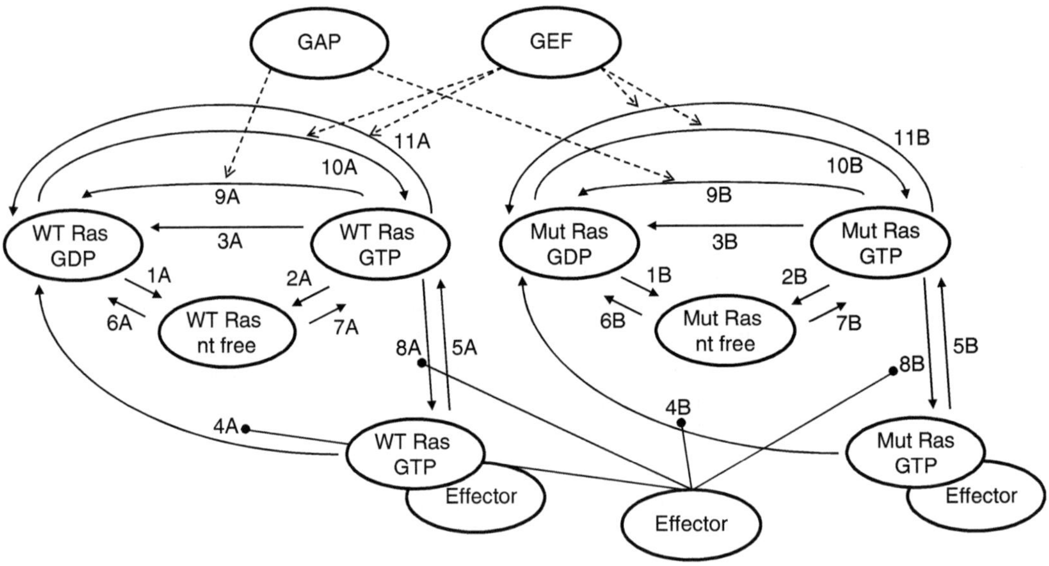

Fig. 1. Schematic diagram of the reactions regulating RasGTP levels for a network that contains both wild-type and mutant Ras proteins. Each *arrow* indicates a biochemical reaction, and the label corresponds to a reaction rate equation specified in Tables 1 and 2. *Dashed arrows* indicate that a reaction is catalyzed by the species at the *head of the arrow*. For example, the *dashed arrow* from GAP pointing to the RasGTP to RasGDP reaction labeled 9A is the GAP reaction catalyzing the conversion of RasGTP to RasGDP. For bimolecular interactions involving effector proteins, the *solid line* with a *closed circle* indicates that it participates in the nearby reactions. GDP and GTP levels are assumed to be constant and these species are not included in the diagram for the relevant reactions (1A, 1B, 2A, 2B, 6A, 6B, 7A, and 7B).

GTPase-activating proteins (GAPs) bind Ras and catalyze the GTPase reaction. GAPs are believed to be primarily responsible for maintaining Ras in a GDP-bound state (9). The oncogenic Ras mutants commonly found in cancer are unable to undergo GAP-mediated GTP hydrolysis (10). Guanine nucleotide exchange factors (GEFs) are another class of proteins that interact with Ras (9). GEFs facilitate nucleotide dissociation and association. As there is more GTP than GDP in the cell, interactions between Ras and one of its GEFs tends to promote GTP association/loading and Ras signal activation. Changes in GEF activity on oncogenic Ras mutants are much less well understood and are not believed to play a role in oncogenic Ras mutant-mediated carcinogenesis (11).

Increased RasGTP is believed to promote cancer as a result of the signals transmitted through its interactions with downstream effector proteins (1). RasGTP interactions with Raf participate in the activation of the Raf proteins, which leads to activation of the extracellular signal-regulated kinase, mitogen-activated protein kinase pathway (ERK/MAPK) that promotes cellular proliferation (12). RasGTP also interacts with phosphoinositide 3-Kinase (PI3K), which promotes signaling through the PTEN/Akt pathway that promotes cellular survival (13). Self-sufficiency in growth signals and pro-survival/anti-apoptotic signals are two of the hallmarks of cancer believed to be shared by all cancers (14). This linkage presents a basic framework for understanding how oncogenic Ras mutants promote cancer as a result of their increased levels of RasGTP.

2. Model Development

2.1. Considerations Prior to Constructing a Model

The model building process requires careful consideration of the biology and mathematics in order to build a model with a maximal chance of serving as a valuable tool for investigating biological questions. Before proceeding with actual mathematical model construction, we considered several factors.

1. *Question(s) to investigate.* This necessarily influences many of the other factors considered. We wished to understand how changes in rate constants to the different reactions that regulate Ras signals result in the pathological increase of RasGTP found in cells with oncogenic Ras mutants. For this, we needed to include both the "fast" reactions (e.g. reactions catalyzed by GEFs and GAPs) and "slow" reactions (e.g. the uncatalyzed counterparts). The slow reactions have typically been ignored in studies of receptor tyrosine kinase/Ras/MAPK dynamics (e.g., refs. 15, 16); previously, analysis of oncogenic mutants has not considered the fast, enzymatic reactions (10) (see Note 1).

2. *Model boundaries.* One important decision in model development was determining how far upstream and downstream within the Ras signaling pathway to include in the model. All biomolecules are part of one large interaction network. Most mathematical models necessarily focus on a subset of this entire network with the idea that the behavior of a subset of the network may still be observable in the larger network. This is not necessarily true; for example, the input/output function of the MAPK module in isolation appears to differ from the input/output function of the entire EGFR/MAPK network (17).

   Working upstream from Ras, the amount of RasGTP in the cell reflects the activity of GEFs and GAPs (9). The activity of GEFs and of GAPs in turn reflects the activity of their upstream regulatory proteins. Biologically, there are several GEFs and GAPs that regulate RasGTP signals, and the amount of basal control exerted by each is not fully understood. Additionally, the regulation of GEFs and GAPs is not well understood compared to the regulation of RasGTP itself. There is no mathematical reason why the model could not extend further upstream than GEFs or GAPs, and many mathematical models have included simplifications of GEF and GAP activation (15, 16, 18). These models, however, focused on how receptor tyrosine kinase activation resulted in MAPK activation, and the inclusion of such steps was necessary for the questions being investigated. For our investigation of RasGTP levels, it seemed that the most important consideration is the amount of GEF or GAP acting on Ras, and not how those GEFs or GAPs came to be active. Due to the relative absence of data regarding regulation and roles of specific GEFs and GAPs compared to the amount of data regarding the regulation of Ras, we decided to focus on basally active levels of GEF or GAP and work toward determining the steady-state levels of basally active GEFs and GAPs. More recently, additional details of the activation of the Ras GEF Sos have been experimentally elucidated allowing more physiological models of Ras GEF activation to be developed (19).

   Working downstream from Ras, a RasGTP–effector complex can shield the bound RasGTP from being acted on by GAPs and GEFs. The amount of RasGTP–effector, therefore, influences the amount of RasGTP in the cell and the RasGTP–effector interaction must be included (20, 21). As with GEFs and GAPs, there are multiple Ras effectors in the cell. There is some uncertainty, however, as to which Ras family proteins some of the putative Ras effectors physiologically interact. Although the activation of some effector proteins like Raf continues to be elucidated (22), it is accepted that interaction with RasGTP is an important process. As with GEFs and GAPs, there is no mathematical reason that we could not extend our model further downstream of Ras and many models have taken this step (15–17, 19, 23). But as our model output of interest was RasGTP levels rather than other downstream activation states, we chose to set the downstream boundary of the model at the level of the effectors that may interact with Ras.

3. *Processing.* Posttranslational modifications are required for the proper membrane localization of Ras proteins (24). This could also be modeled (25). However, since we were focusing on the steady-state at the membrane Ras signals, we chose the assumption that Ras concentrations are constant in time. We are not aware of large differences in processing for different mutants.
4. *Isoforms and proteins with similar function.* There are three Ras isoforms: H-Ras, N-Ras, and K-Ras. Although isoform-specific signaling differences have been proposed (26), no biochemical properties differ largely between the isoforms, their interactions with GEFs, GAPs, or effectors. All three isoforms have been linked to different forms of cancer. We, therefore, choose to group them all together as a single "Ras" species that includes all Ras isoforms. Similarly, there are many different Ras GEFs and Ras GAPs. Which of these contribute to basal Ras activation is unclear for GEFs. The Ras GAP neurofibromin clearly has a role in the basal regulation of RasGTP levels (27), but the tissue-specific effects of genetic loss of neurofibromin seem to suggest that other Ras GAPs may also play a role. As we are also not modeling the regulation of GEFs or GAPs, we group all basally active GEFs together and also group all basally active GAPs together. We similarly group all Ras effectors as we are interested only in how they influence RasGTP levels and not specific downstream signaling outcomes.
5. *Membrane localization.* Which membrane Ras signals from, e.g., the plasma membrane or endocytic vesicle membrane, is believed to be a component of the Ras signal (28). Similarly, the signaling of Ras within different microdomains of a membrane is also believed to be important for Ras signals (29). A mechanistic understanding is not presently available and a mathematical model of microdomain signaling had to impose microdomain signaling (23). Moreover, a difference in membrane localization or microdomain localization for oncogenic mutants is not evident; we, therefore, decided to proceed without accounting for localization to various, specific, membranes. However, we recognized that membrane localization might have large effects (i.e., isolating proteins from different interacting partners, etc.), and this decision was made with the understanding that it may need to be reconsidered if large deviations were observed between model predictions and experimental data.

6. *Feedback*. Feedback loops are believed to modulate the kinetics of Ras/MAPK signals. Both positive and negative feedback loops are known to play a role in physiological Ras/MAPK pathway signals (16, 30). Negative feedback loops (31) have also been described for the persistent RasGTP levels that would occur with an oncogenic Ras mutant. It is not well understood if feedback loops significantly influence steady-state RasGTP levels after the acquisition of a Ras point mutant. Like membrane localization, we decided to model the Ras signaling module without feedback loops with the understanding that it may need to be reconsidered if large deviations were observed between model predictions and experimental data. Should a large difference be observed, mathematical models would be extremely well suited for investigating the effects of multiple, and competing, feedback loops.
7. *Parameters*. Another key decision that needed to be made was how to obtain parameters for the model. There is an abundant literature of in vitro rate constants for the Ras network, and the goal of this analysis was to determine if mathematical methods could add any additional insights from these data. We, therefore, chose to incorporate these values. One potential problem is that rate constants are typically measured in solution, while physiologically many of these reactions are restricted to occur on a 2D cellular membrane. We, therefore, corrected for membrane localization using a method used previously to study Ras signaling (18).
8. *Method*. There are multiple methods available for studying a signaling network. We chose to use ordinary differential equations (ODEs) as this method seemed most consistent with the prevailing ideas used by experimental biologists as they consider Ras signals.

2.2. Model Building

1. *Describing the reactions individually.* For the Ras signaling module (Fig. 1), we described the individual reactions with the same mathematical descriptions that biochemists use to study them in isolation (Tables 1 and 2). First-order mass action kinetics were used to describe GTPase hydrolysis, Ras-GTP dissociation, Ras-GDP dissociation, and Ras-effector dissociation. Second-order mass action kinetics were used to describe Ras-GTP association, Ras-GDP association, and Ras-effector association. The individual steps of the GAP reaction could be described with mass action kinetics (18). However, the full GAP reaction may involve additional steps (32). The traditional Michaelis–Menten equation is a simplification of either of these cases, and it also has the advantage that rate constants k_{cat} and K_m are commonly measured experimentally. The individual steps of the GEF reaction could also be

Table 1
Reaction rate equations for the reactions acting on wild-type Ras (WT Ras)

Reaction #	Description	Reaction rate equation
1A	Dissociation of GDP from WT RasGDP	$k_{d,GDP,WT}[WTRasGDP]$
2A	Dissociation of GTP from WT RasGTP	$k_{d,GTP,WT}[WTRasGTP]$
3A	GTPase activity of WT RasGTP	$k_{GTPase,WT}[WTRasGTP]$
4A	GTPase of WT RasGTP–effector complex	$k_{GTPase,WT}[WTRasGTP \cdot Effector]$
5A	Dissociation of WT RasGTP–effector complex	$k_{d,Effector,WT}[WTRasGTP \cdot Effector]$
6A	Association of GDP to nucleotide-free WT Ras	$k_{a,GDP,WT}[WTRasntfree][GDP]$
7A	Association of GTP to nucleotide-free WT Ras	$k_{a,GTP,WT}[WTRasntfree][GTP]$
8A	Association of effector to WT RasGTP	$k_{a,Effector,WT}[WTRasGTP][Effector]$
9A	GAP-mediated GTP hydrolysis of WT RasGTP	$\frac{[GAP]k_{cat,WT}[WTRasGTP]}{K_{M,WT}(1+(MutRasGTP)/(K_{M,mut}))+[WTRasGTP]}$
10A	GEF-facilitated exchange of GDP to GDP on WT RasGDP	$\frac{[GEF]((k_{cat,GDP,WT})/(K_{M,GDP,WT}))[WTRasGDP]}{1+([WTRasGDP]/K_{M,GDP,WT})+([WTRasGTP]/K_{M,GTP,WT})+([MutRasGDP]/K_{M,GDP,mut})+([MutRasGTP]/K_{M,GTP,mut})}$
11A	GEF-facilitated exchange of GTP to GDP on WT RasGTP	$\frac{[GEF](k_{cat,GTP,WT}/(K_{M,GTP,WT})[WTRasGTP]}{1+([WTRasGDP]/K_{M,GDP,WT})+([WTRasGTP]/K_{M,GTP,WT})+([MutRasGDP]/K_{M,GDP,mut})+([MutRasGTP]/K_{M,GTP,mut})}$

The reaction number corresponds to the labeled reaction in Fig. 1. The reaction rate equation is derived from simple first-order mass action kinetics (1A–5A), second-order mass action kinetics (6A–8A), Michaelis–Menten enzyme kinetics (9A), or reversible Michaelis–Menten kinetics (10A, 11A)

described with mass action kinetics. Alternatively, we used the reversible Michaelis–Menten equation. Wild-type and mutant Ras proteins undergo identical reactions but with different rate constants (2). Therefore, the mathematical equation used for these reactions is identical in form, but the values of the rate constants that go in the model differ when considering wild-type or mutant protein (i.e., the reactions in Table 2 are

Table 2
Reaction rate equations for the reactions acting on mutant Ras (Mut Ras)

Reaction #	Description	Reaction rate equation
1B	Dissociation of GDP from Mut RasGDP	$k_{d,GDP,mut}[MutRasGDP]$
2B	Dissociation of GTP from Mut RasGTP	$k_{d,GTP,mut}[MutRasGTP]$
3B	GTPase activity of Mut RasGTP	$k_{GTPase,mut}[MutRasGTP]$
4B	GTPase of Mut RasGTP–effector complex	$k_{GTPase,mut}[MutRasGTP \cdot Effector]$
5B	Dissociation of Mut RasGTP–effector complex	$k_{d,Effector,mut}[MutRasGTP \cdot Effector]$
6B	Association of GDP to nucleotide-free Mut Ras	$k_{a,GDP,mut}[MutRasntfree][GDP]$
7B	Association of GTP to nucleotide-free Mut Ras	$k_{a,GTP,mut}[MutRasntfree][GTP]$
8B	Association of effector to Mut RasGTP	$k_{a,Effector,mut}[MutRasGTP][Effector]$
9B	GAP-mediated GTP hydrolysis of Mut RasGTP	$\frac{[GAP]k_{cat,mut}[MutRasGTP]}{K_{M,mut}(1 + (WTRasGTP/K_{M,WT})) + [MutRasGTP]}$
10B	GEF-facilitated exchange of GDP to GDP on Mut RasGDP	$\frac{[GEF](k_{cat,GDP,mut}/K_{M,GDP,mut})[MutRasGDP]}{1 + ([WTRasGDP]/K_{M,GDP,WT}) + ([WTRasGTP]/K_{M,GTP,WT}) + ([MutRasGDP]/K_{M,GDP,mut}) + ([MutRasGTP]/K_{M,GTP,mut})}$
11B	GEF-facilitated exchange of GTP to GDP on Mut RasGTP	$\frac{[GEF](k_{cat,GTP,mut}/K_{M,GTP,mut})[MutRasGTP]}{1 + ([WTRasGDP]/K_{M,GDP,WT}) + ([WTRasGTP]/K_{M,GTP,WT}) + ([MutRasGDP]/K_{M,GDP,mut}) + ([MutRasGTP]/K_{M,GTP,mut})}$

The reaction number corresponds to the labeled reaction in Fig. 1. The reaction rate equation is derived from simple first-order mass action kinetics (1B–5B), second-order mass action kinetics (6B–8B), Michaelis–Menten enzyme kinetics (9B), or reversible Michaelis–Menten enzyme kinetics (10B, 11B)

equivalent to those in Table 1, but the values of the parameters for each may be quite different).

2. *Describing reactions collectively.* Each of the reactions in Tables 1 and 2 gives the rate of a reaction at a given moment of time. The concentration of an entity being considered in the model (e.g., the amount of RasWT bound to GTP) could be influenced by several different reactions. This can be expressed mathematically by creating a differential equation that gives

Table 3
Complete set of ordinary differential equations for the modeled components of the Ras network

$\frac{d[\text{WT RasGDP}]}{dt} = -1A + 3A + 4A + 6A + 9A - 10A + 11A$
$\frac{d[\text{WT RasGTP}]}{dt} = -2A - 3A + 5A + 7A - 8A - 9A + 10A - 11A$
$\frac{d[\text{WT Ras nt free}]}{dt} = 1A + 2A - 6A - 7A$
$\frac{d[\text{WT RasGTP} \cdot \text{Effector}]}{dt} = -4A - 5A + 8A$
$\frac{d[\text{Effector}]}{dt} = 4A + 5A - 8A + 4B + 5B - 8B$
$\frac{d[\text{Mut RasGDP}]}{dt} = -1B + 3B + 4B + 6B + 9B - 10B + 11B$
$\frac{d[\text{Mut RasGTP}]}{dt} = -2B - 3B + 5B + 7B - 8B - 9B + 10B - 11B$
$\frac{d[\text{Mut Ras nt free}]}{dt} = 1B + 2B - 6B - 7B$
$\frac{d[\text{Mut RasGTP} \cdot \text{Effector}]}{dt} = -4B - 5B + 8B$

The numbers refer to the reaction rate equations specified by this reaction number in Tables 1 and 2. The positive or negative sign reflects whether the individual reaction results in an increase or decrease of the concentration of the protein (or complex) described by the differential equation

the rate of change in the amount of the entity being considered (i.e., as RasGTP) as a function of the different, individual biochemical reactions. This was done for all of the species being considered (Table 3). This resulted in a system of ODEs.

3. *Parameters for WT Ras.* Most of the needed rate constants and other needed parameters have been experimentally measured and are available in the peer-reviewed literature (see ref. 5 for a complete list of parameters and their values). We tried to find values obtained with rigorous, accurate, experimental methods. When more than one publication with data from high-quality experiments was available, we tried to use the more recently published values, simply because the experimental approaches for measuring some of these values have improved over the years (see Note 2).
4. *Parameters for mutant Ras.* Several Ras mutants, particularly RasG12V and RasG12D, have been well characterized (2). As with RasWT, the published, peer-reviewed literature was

used to determine the properties of oncogenic mutants. As the studies comparing an oncogenic mutant to the wild-type protein might not have been the same studies that provided the wild-type parameters, we adjusted the wild-type parameter value by the relative change in the mutant parameter to the wild-type parameter found in the same study.

5. *Program into MATLAB.* These systems of ODEs are not generally solvable analytically. They can be simulated to numerically solve the ODEs. The simulation essentially lets the biochemical reactions "occur" in the computer, where concentrations, reaction rates, and other properties that would be of interest to a biologist can easily be "measured." We used commercially available MATLAB software for our simulations and used its built-in numerical integration regimens to solve the equations. In the context of cancer, we are interested in the steady-state levels of RasGTP; MATLAB simulations were run until steady-state solutions were achieved.

6. *Parameter fitting.* Despite Ras biochemistry being very well described, there were some parameters that could not be found in the literature. Specifically, concentrations of the total amount of basally active GEFs, GAPs, and effectors have not been described. As multiple proteins contribute to each of these cases and only a fraction of total GEF and total GAP are basally active, this concentration is not easily determined experimentally. Additionally, it is well known experimentally that the amount of RasGTP in the cell depends strongly on the amount of GEF and GAP in the cell.

   We, therefore, fit the amount of basally active GEF and GAP in our model. There were two different sets of experiments in the published literature that helped us determine the amount of basally active GEF and GAP. Interestingly, levels of RasGTP seem to be relatively constant between different cell lines at about 2% of total Ras bound to GTP at basal, unstimulated conditions. The rate of nucleotide exchange in the absence of an activation signal has also been determined. Different levels of GEF and GAP were considered for different levels of Ras and effector concentrations (as it is also known that concentrations of proteins can vary largely between different cell types) and simulations helped determine the fraction of basal RasGTP and nucleotide exchange. The amount of basally active GEF and GAP was chosen to minimize the normalized root mean square error.

7. *Algebraic solutions.* Alternatively, the steady-state solutions of the ODE model can be found by setting the ODEs equal to zero and solving algebraically. To do this, we also used MATHEMATICA software to determine steady-state solutions for given sets of parameters. We were, therefore, able to confirm many results with both simulations in MATLAB and analytical, algebraic manipulations in MATHEMATICA.

3. Model Validation

As discussed earlier, the development of this mathematical model (or any mathematical model) requires many simplifications. It is possible that processes left out of the model for the sake of simplification could be essential to making a meaningful prediction. It is also possible that information that went into the model could be erroneous, for example a rate constant measured in vitro might not apply to the same reaction in vivo. Model predictions ultimately need to be compared to experimental data to be confident that the numerous sources of potential "error" (e.g., simplifications and biological data) do not confound model predictions.

1. *Predictions*: To test the ability of the model to predict levels of RasGTP, we considered several cases that could be compared to experimental data. One condition was the loss of Ras GAP activity so that predictions could be compared with measurements from neurofibromin-deficient cells (27). Experiments have previously been performed, where mutant RasG12V or RasG12D had been transfected into cells and total RasGTP or mutant RasGTP measured (6, 33, 34). This was modeled with a network with the full amount of WT Ras (as the endogenous Ras in the cells from the experiments was WT Ras) and then with an additional amount of mutant Ras added to the network. These experiments had all been performed previously, and model predictions matched well with experimental data.
2. *Sensitivity analysis*: One important consideration is the sensitivity of the model to the specific values of the parameters used. By running simulations with different values of the rate constants (either individually or in various combinations), it is possible to investigate the robustness of the model. Sensitivity analysis showed that for many of the parameters predictions would be largely unaffected by a change in the value used. However, there were several parameters, where a change would have a large effect on RasGTP levels. Each of these cases, however, corresponded to a known biological mechanism that resulted in increased RasGTP levels (5). This further suggested that the model was reproducing the features of interest for the Ras biological system.
3. *Decision points*: At various points, it is important to evaluate if the model is worth proceeding with or if discrepancies between model predictions and experimental data suggest that the model needs to be adjusted. Up to this point, no new experiments had been necessary as the Ras literature has published so many useful quantitative studies. If model performance seems

poor or if clear areas for improvement are evident, it may be better to develop a "better" model rather than proceed with new experiments (see Note 3).

4. *Novel predictions and experiments*: The model appeared to be able to reproduce known features from the Ras literature. Further confidence in the model would come from making additional predictions that could subsequently be experimentally validated (see Note 4). We considered the case of varying expression levels of Ras mutants RasF28L and RasG12V. In addition to contributing to model validation, this also allowed for an investigation into a biologically interesting question: If both GAP-insensitive Ras mutants (of which RasG12V is a member) and fast-cycling Ras mutants (of which RasF28L is a member) have cancer-promoting properties when studied in vitro, why are only GAP-insensitive Ras mutants found in human cancer?

4. Model-Based Investigation

The biochemical reactions of the Ras signaling module have been well characterized for both WT and oncogenic Ras mutants. There is clear consensus over which properties change significantly after mutation and which properties seem much less affected. Two properties that change for all members of the class of Ras mutants commonly found in cancer are (a) impaired intrinsic GTPase activity (2) and (b) insensitivity to GAP-mediated GTP hydrolysis (4). For the commonly found Ras mutant with a codon 12 glycine to valine nonsense mutation (Ras G12V), (c) an increased affinity for RasGTP with downstream effector Raf has also been described (3).

Germline loss of the Ras GAP neurofibromin shows that GAPs play an important role in maintaining wild-type Ras in a predominantly GDP-bound state in basal, non-signaling conditions (27), suggesting that GAP activity regulates basal RasGDP levels along with intrinsic GTPase activity. It is notable that the kcat for the GAP-catalyzed reaction is approximately 10,000 times larger than the rate constant for the non-catalyzed GTPase reaction (35). The slower GTPase reaction is reduced by a factor of ten or less for most oncogenic Ras mutants (10, 36). But for most oncogenic Ras mutants, the GAPs appear to have no effect in accelerating the intrinsic GTPase activity of Ras, terming these mutants "GAP insensitive." As the changes to a mutant are a tenfold change to the slow GTPase reaction and effectively complete elimination of the physiologically important and 10,000 times faster GAP reaction, the popular belief is that GAP insensitivity alone contributes to the increased RasGTP levels in cells containing oncogenic Ras mutants.

This hypothesis has not been experimentally tested. Since the behaviors of even simple nonlinear systems can be nonintuitive, simple order-of-magnitude arguments could be misleading.

With the goal of investigating the relative contribution of the different altered Ras biochemical properties to establishing RasGTP levels and with a mathematical model that appeared to be in good agreement with experimental data, we proceeded as follows.

1. *Define test conditions*: We needed to systematically include or exclude each of the altered properties. To exclude GAP insensitivity, the modeled "mutant" received the parameters of wild-type Ras for the Ras-GAP interactions. To exclude GTPase deficiency, the modeled "mutant" had the wild-type Ras GTPase rate constant. To exclude competitive inhibition, the competitive inhibition term $(1 + [I]/K_I)$ was removed from the Michaelis–Menten GAP equation.
2. *Generate conditions for simulations*: A collection of simulated Ras "mutants" with properties mixed from RasWT and the oncogenic Ras mutant being investigated (here, RasG12D) were generated. This was done so that each property would be excluded or included in all possible combinations (Table 4) (see Note 5).
3. *Simulations*: For each set of parameters, simulations were used to determine the levels of RasGTP for wild-type and "mutant" Ras (Table 4).
4. *Extend to other Ras mutants*: Repeat 2 and 3 with parameters for other Ras mutants (i.e., RasG12V) to investigate whether results are robust to which GAP-insensitive Ras mutant is modeled.
5. *Sensitivity analysis of predictions.* To evaluate if interesting observations were robust, we performed simulations for different concentrations of network proteins, different levels of total Ras mutated, and different changes to model parameters.
6. *Experimental tests*: To test our model-based predictions, we focused on testing the prediction that competitive inhibition of GAP by mutant Ras proteins (thereby preventing GAP binding to wild-type Ras proteins) resulted in an increased fraction of RasWT bound to GTP. The basis of our experimental test was that there are three Ras isoforms (H-Ras, N-Ras, and K-Ras) and we measured the amount of RasGTP for each of the isoforms in cell lines with mutations to one of these Ras isoforms or with no Ras mutation. We found increased levels of RasGTP to the non-mutated isoforms, consistent with the predictions of the model (5). Similarly, another group has also published the observation that increased levels of wild-type N-Ras-bound GTP are found in cells that harbor a K-Ras mutation (37) (see Note 6).

Table 4
Parameters for simulations and results

	Parameters				Results		
Modeled Ras	k_{cat}	k_{GTPase}	CI	Nuc exch	Total Ras % RasGTP	Mut Ras % RasGTP	WT Ras % RasGTP
WT	WT	WT	+	WT	2	N/A	2
G12D	G12D	G12D	+	G12D	37	66	17
G12D, no GI	WT	G12D	+	G12D	2	2	2
G12D, no GD	G12D	WT	+	G12D	29	51	15
G12D, no CI	G12D	G12D	–	G12D	27	67	1
G12D, no GI, GD	WT	WT	+	G12D	2	2	2
G12D, no GI, CI	WT	G12D	–	G12D	1	1	1
G12D, no GD, CI	G12D	WT	–	G12D	21	51	1
G12D, no GI, GD, CI	WT	WT	–	G12D	1	1	1

The table shows the values used to simulate different conditions, including wild-type Ras, mutant RasG12D, and simulated mutants with biochemical properties between these two cases. For each set of parameters, the percentage of total Ras bound to GTP, mutant Ras bound to GTP, and WT Ras bound to GTP is also specified. For all simulations, 60% of total Ras was considered wild type and 40% of total Ras was considered mutant. *GI* GAP insensitivity. *GD* GTPase deficiency, *CI* competitive inhibition. To leave out competitive inhibition, the competitive inhibition term in equations 9A and 9B of the form $(1 + [I]/K_I)$ is removed. k_{cat}, k_{GTPase}, and Nuc exch, respectively, are the rate constant for the catalytic step of GAP-mediated RasGTP hydrolysis, the rate constant for Ras intrinsic GTP hydrolysis, and the rate constants for nucleotide exchange; *WT* indicates that it has the parameter value of WT Ras; and *G12D* indicates that it has the parameter value of RasG12D

5. Discussion

The previously held belief based upon order-of-magnitude arguments that GAP insensitivity is the only significant contributor to Ras activation after an oncogenic mutant was based on correct assumptions. The order-of-magnitude argument, however, failed to account for additional factors, such as the relative concentrations of enzymes and the nonlinearities in the system. The advantage of our mathematical model is that it is not biased by preconceived ideas and accounts for rate constants, concentrations, and nonlinearities. Our model with biologically meaningful parameters allowed us to characterize these processes through computer modeling. Also, years of mental consideration regarding the many Ras-regulatory reactions individually, but not as a “network,” resulted in the oversight of the contribution of competitive inhibition to Ras activation. Our mathematical model finds behaviors that result from system-level interactions as evidenced by the observation of competitive inhibition of Ras GAPs by GAP-insensitive Ras mutants leading to increased WT Ras bound to GTP. Both of these findings highlight the importance of integrating mathematical models into efforts to understand the molecular basis of cancer.

6. Notes

1. Interestingly, one analysis of oncogenic mutants that ignored GEF, GAP, and effector binding achieved a remarkably good correlation between biochemical property changes and the increase in RasGTP after expression of this mutant (10). As large changes in GEF properties are not believed to contribute to oncogenic Ras activation, as Ras mutants are insensitive to GAPs, and assuming effector affinity changes are also minimal, this approach may capture many of the essential processes and be able to make good predictions. This choice of model boundaries, however, limits the questions that can be asked of the model and prevents system-level effects from being observed.
2. Sensitivity analysis demonstrated that results would be largely similar regardless of what parameters were selected from the literature (5).
3. One early, common complaint of modeling advanced by experimental biologists was that biological measurements are not precise enough for mathematical models. Accordingly, the high level of agreement between experimental data and model predictions as is found in physical science models should not be expected. A model that matches the biological system qualitatively and quantitatively sufficiently well can be a very useful tool for biological investigations.
4. A consequence of simplifications made in the model building process is that the model could make predictions that are not easily tested experimentally. Also, for a very well-studied system like Ras signaling, it may seem that all quantities that could reasonably be tested experimentally have already been measured.
5. The model-based approach allowed for a systematic consideration of each property individually and in various combinations, analogous to a well-controlled experiment. To do this same analysis experimentally would require a collection of Ras mutants with all of these intermediate properties; such a collection of Ras mutants has not been described.

6. Simplifications made in the model building process should be reconsidered when determining what experimental tests of the model to pursue. For example, our model did not include isoforms, but the behavior of isoforms could easily be inferred. By considering our prediction that oncogenic mutants should result in increased wild-type RasGTP through competitive inhibition, we easily inferred that a K-Ras oncogenic mutant should result in increased H-RasGTP and increased N-RasGTP. This allowed us to perform experiments that utilized Ras isoform-specific antibodies to measure changes in RasGTP levels. Such an experiment could easily be overlooked if one did not reconsider the simplifications.

References

1. Malumbres M, Barbacid M (2003) RAS oncogenes: the first 30 years. Nat Rev Cancer 3 (6):459–465
2. Eccleston JF, Moore KJ, Brownbridge GG, Webb MR, Lowe PN (1991) Fluorescence approaches to the study of the p21ras GTPase mechanism. Biochem Soc Trans 19(2):432–437
3. Chuang E, Barnard D, Hettich L, Zhang XF, Avruch J, Marshall MS (1994) Critical binding and regulatory interactions between Ras and Raf occur through a small, stable N-terminal domain of Raf and specific Ras effector residues. Mol Cell Biol 14(8):5318–5325
4. Ahmadian MR, Hoffmann U, Goody RS, Wittinghofer A (1997) Individual rate constants for the interaction of Ras proteins with GTPase-activating proteins determined by fluorescence spectroscopy. Biochemistry 36 (15):4535–4541
5. Stites EC, Trampont PC, Ma Z, Ravichandran KS (2007) Network analysis of oncogenic Ras activation in cancer. Science 318 (5849):463–467
6. Gibbs JB, Marshall MS, Scolnick EM, Dixon RA, Vogel US (1990) Modulation of guanine nucleotides bound to Ras in NIH3T3 cells by oncogenes, growth factors, and the GTPase activating protein (GAP). J Biol Chem 265 (33):20437–20442
7. Milburn MV, Tong L, de Vos AM et al (1990) Molecular switch for signal transduction: structural differences between active and inactive forms of protooncogenic Ras proteins. Science 247(4945):939–945
8. Macara IG, Lounsbury KM, Richards SA, McKiernan C, Bar-Sagi D (1996) The Ras superfamily of GTPases. FASEB J 10(5):625–630
9. Bos JL, Rehmann H, Wittinghofer A (2007) GEFs and GAPs: critical elements in the control of small G proteins. Cell 129(5):865–877
10. Donovan S, Shannon KM, Bollag G (2002) GTPase activating proteins: critical regulators of intracellular signaling. Biochim Biophys Acta 1602(1):23–45
11. Mistou MY, Jacquet E, Poullet P et al (1992) Mutations of Ha-Ras p21 that define important regions for the molecular mechanism of the SDC25 C-domain, a guanine nucleotide dissociation stimulator. EMBO J 11 (7):2391–2397
12. Sebolt-Leopold JS, Herrera R (2004) Targeting the mitogen-activated protein kinase cascade to treat cancer. Nat Rev Cancer 4(12):937–947
13. Vivanco I, Sawyers CL (2002) The phosphatidylinositol 3-Kinase AKT pathway in human cancer. Nat Rev Cancer 2(7):489–501
14. Hanahan D, Weinberg RA (2000) The hallmarks of cancer. Cell 100(1):57–70
15. Schoeberl B, Eichler-Jonsson C, Gilles ED, Muller G (2002) Computational modeling of the dynamics of the MAP kinase cascade activated by surface and internalized EGF receptors. Nat Biotechnol 20(4):370–375
16. Bhalla US, Ram PT, Iyengar R (2002) MAP kinase phosphatase as a locus of flexibility in a mitogen-activated protein kinase signaling network. Science 297(5583):1018–1023
17. Chen WW, Schoeberl B, Jasper PJ et al (2009) Input-output behavior of ErbB signaling pathways as revealed by a mass action model trained against dynamic data. Mol Syst Biol 5:239
18. Markevich NI, Moehren G, Demin OV, Kiyatkin A, Hoek JB, Kholodenko BN (2004) Signal processing at the Ras circuit: what shapes Ras activation patterns? Syst Biol (Stevenage) 1(1):104–113
19. Das J, Ho M, Zikherman J et al (2009) Digital signaling and hysteresis characterize Ras activation in lymphoid cells. Cell 136(2):337–351

20. Warne PH, Viciana PR, Downward J (1993) Direct interaction of Ras and the amino-terminal region of Raf-1 in vitro. Nature 364 (6435):352–355
21. Zhang XF, Settleman J, Kyriakis JM et al (1993) Normal and oncogenic p21ras proteins bind to the amino-terminal regulatory domain of c-Raf-1. Nature 364(6435):308–313
22. Rajakulendran T, Sahmi M, Lefrancois M, Sicheri F, Therrien M (2009) A dimerization-dependent mechanism drives RAF catalytic activation. Nature 461(7263):542–545
23. Tian T, Harding A, Inder K, Plowman S, Parton RG, Hancock JF (2007) Plasma membrane nanoswitches generate high-fidelity Ras signal transduction. Nat Cell Biol 9(8):905–914
24. Konstantinopoulos PA, Karamouzis MV, Papavassiliou AG (2007) Post-translational modifications and regulation of the RAS superfamily of GTPases as anticancer targets. Nat Rev Drug Discov 6(7):541–555
25. Legewie S, Sers C, Herzel H (2009) Kinetic mechanisms for overexpression insensitivity and oncogene cooperation. FEBS Lett 583(1):93–96
26. Haigis KM, Kendall KR, Wang Y et al (2008) Differential effects of oncogenic K-Ras and N-Ras on proliferation, differentiation and tumor progression in the colon. Nat Genet 40 (5):600–608
27. Basu TN, Gutmann DH, Fletcher JA, Glover TW, Collins FS, Downward J (1992) Aberrant regulation of Ras proteins in malignant tumour cells from type 1 neurofibromatosis patients. Nature 356(6371):713–715
28. Chiu VK, Bivona T, Hach A et al (2002) Ras signalling on the endoplasmic reticulum and the Golgi. Nat Cell Biol 4(5):343–350
29. Hancock JF, Parton RG (2005) Ras plasma membrane signalling platforms. Biochem J 389(Pt 1):1–11
30. Shin SY, Rath O, Choo SM et al (2009) Positive- and negative-feedback regulations coordinate the dynamic behavior of the Ras-Raf-MEK-ERK signal transduction pathway. J Cell Sci 122(Pt 3):425–435
31. Courtois-Cox S, Genther Williams SM, Reczek EE et al (2006) A negative feedback signaling network underlies oncogene-induced senescence. Cancer Cell 10(6):459–472
32. Kotting C, Kallenbach A, Suveyzdis Y, Wittinghofer A, Gerwert K (2008) The GAP arginine finger movement into the catalytic site of Ras increases the activation entropy. Proc Natl Acad Sci USA 105(17):6260–6265
33. Bollag G, Adler F, el Masry N et al (1996) Biochemical characterization of a novel KRAS insertion mutation from a human leukemia. J Biol Chem 271(51):32491–32494
34. Boykevisch S, Zhao C, Sondermann H et al (2006) Regulation of Ras signaling dynamics by Sos-mediated positive feedback. Curr Biol 16(21):2173–2179
35. Wittinghofer A, Scheffzek K, Ahmadian MR (1997) The interaction of Ras with GTPase-activating proteins. FEBS Lett 410(1):63–67
36. Der CJ, Finkel T, Cooper GM (1986) Biological and biochemical properties of human rasH genes mutated at codon 61. Cell 44(1):167–176
37. Keller JW, Haigis KM, Franklin JL, Whitehead RH, Jacks T, Coffey RJ (2007) Oncogenic K-RAS subverts the antiapoptotic role of N-RAS and alters modulation of the N-RAS: gelsolin complex. Oncogene 26 (21):3051–3059

Chapter 6

Modeling miRNA Regulation in Cancer Signaling Systems: miR-34a Regulation of the p53/Sirt1 Signaling Module

Xin Lai, Olaf Wolkenhauer*, and Julio Vera*

Abstract

MicroRNAs (miRNAs) are a family of small regulatory RNAs whose function is to regulate the activity and stability of specific messenger RNA targets through posttranscriptional regulatory mechanisms. Most of the times signaling systems involving miRNA modulation are not linear pathways in which a certain transcription factor activate the expression of miRNAs that posttranscriptionally represses targeting proteins, but complex regulatory structures involving a variety of feedback-loop architectures.

In this book chapter, we define, discuss, and apply a Systems Biology approach to investigate dynamical features of miRNA regulation, based on the integration of experimental evidences, hypotheses, and quantitative data through mathematical modeling. We further illustrate the approach using as case study the signaling module composed by the proteins p53, Sirt1, and the regulatory miRNA miR-34a. The model was used not only to investigate different possible designs of the silencing mechanism exerted by miR-34a on Sirt1 but also to simulate the dynamics of the system under conditions of (pathological) deregulation of its compounds.

Key words: Systems biology, p53, miR-34a, Sirt1, DBC1, ODE model, Silencing, posttranscriptional regulation, Mathematical modeling, Power-law model

1. Background

Any attempt to update the paradigm of the genetic information flow represented by the Central Dogma of Molecular Biology (DNA → mRNA → protein) should include at least the role of two new key players intimately involved in transcriptional and posttranscriptional regulation: transcription factors (TFs) and microRNAs (miRNAs) (see Fig. 1). TFs, a key element in gene expression networks, are a class of proteins able to promote and regulate the expression of given genes. Experimental evidences suggest that TFs are involved

*Olaf Wolkenhauer and Julio Vera contributed equally.

Xuedong Liu and Meredith D. Betterton (eds.), *Computational Modeling of Signaling Networks*, Methods in Molecular Biology, vol. 880, DOI 10.1007/978-1-61779-833-7_6,

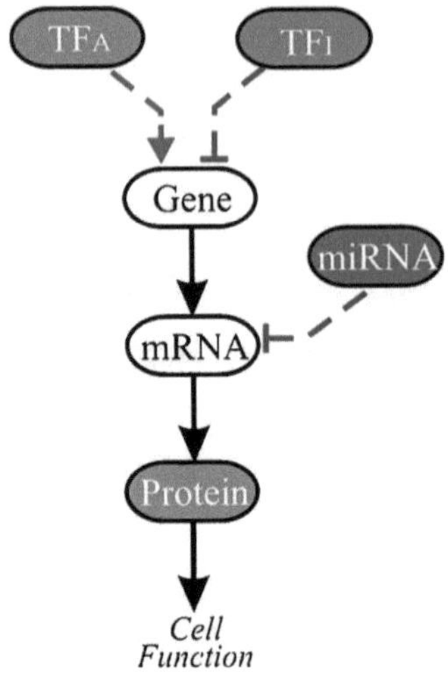

Fig. 1. Sketch of the classical conception of genetic information flow (the so-called Central Dogma of Molecular Biology) updated with new players in transcriptional and posttranscriptional regulation. TFs can promote (TF_A) or block (TF_I) the transcription of genes.

in a variety of feedback loop signaling systems that give to the processing of the genetic information the necessary reliability, robustness, and responsiveness regarding changes in internal and external signals (1).

In the last decade it has been experimentally established that the existence of small regulatory RNAs, called miRNAs, whose function is to regulate the activity and stability of messenger RNA (mRNA) targets, in a sort of posttranscriptional regulation mechanism (2, 3). Longer precursor transcripts with hairpin structures are synthesized by RNA polymerase II. After processing of the precursors by Drosha and Dicer ribonucleases, mature miRNAs are generated, with length of typically ~22 nucleotides in humans. Depending on the degree of complementarity to target sequences, miRNAs induce translational repression or cleavage of mRNAs. A single miRNA can target hundreds to thousands of mRNAs, rendering it challenging to attribute distinct functions to specific miRNAs (4). miRNAs are supposed to account for 1–5% of animal genes (5, 6), making them one of the most abundant gene product regulators. Recent experiments, using in vitro translation systems, have indicated that miRNAs can repress translation, at least in part, by inhibiting translation initiation (7). However, a clear picture of the mechanism does not exist and a number of possible reaction schemes, in which this miRNA-mediated silencing occurs, have been described in the literature and proved for given miRNA systems (8, 9). Moreover, very little is known about the regulation of miRNA biogenesis and processing to this day. miRNAs have been implicated in tumor progression in several cancer types. miRNAs have target genes which play a role in tumor development and progression. Examples are *RAS*, *c-Myc*, *BCL2*,

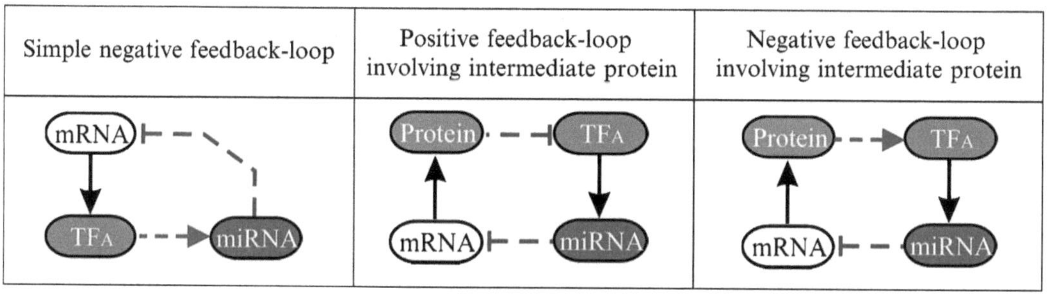

Fig. 2. Different signaling regulatory structures involving miRNA posttranscriptional regulation.

and cyclin-dependent kinases (*CDKs*) (10, 11). Depending on the targets, they can act as tumor suppressors when they downregulate oncogenes (12–14), or as oncogenes by negatively regulating the expression of tumor suppressor genes (12, 15).

Interestingly enough, a number of designs in the architecture of regulatory networks involving miRNAs, TFs, and other signaling proteins have been already described in the literature (16). Instead of linear signaling pathways in which given TFs activate the transcription of miRNAs, very often more complex regulatory schemes emerge involving a variety of feedback-loop architectures (see Fig. 2). miRNAs may directly inhibit the expression of its own TFs in a kind of direct homeostatic loop (see Fig. 2, left), but also they may posttranscriptionally regulate third signaling proteins involved in the positive (see Fig. 2, middle) or negative (see Fig. 2, right) regulation of the miRNAs' TFs, with remarkable dynamical consequences in the system (2, 16, 17). When one or more of these regulatory structures appear in a biochemical system, intuition is no more a valuable tool to design experiments and construct hypotheses concerning the behavior and structure of the biochemical network. Thus, the integration of experimental evidences, hypotheses and quantitative data through mathematical modeling (the elements of the Systems Biology approach) becomes necessary (18).

In our case study, tumor suppressor p53 is considered one of the key regulators of the transient or permanent cell cycle arrest and/or the programmed cell death in response to the DNA damage. The mutations of p53 has been found in approximately half of human cancers, and most of the remaining malignancies directly or indirectly target the p53 pathway by upregulating its inhibitors, downregulating its activators or inactivating its downstream targets (19, 20). Global miRNA expression analyses identified a cohort of miRNAs that exhibit p53-dependent expression following DNA damage. One of these miRNAs, miR-34a, is commonly deleted in human cancers and has been proved to act as modulator in fine-tuning the gene expression program initiated by p53 (21). In addition, Yamakuchi and his colleagues have found that miR-34a

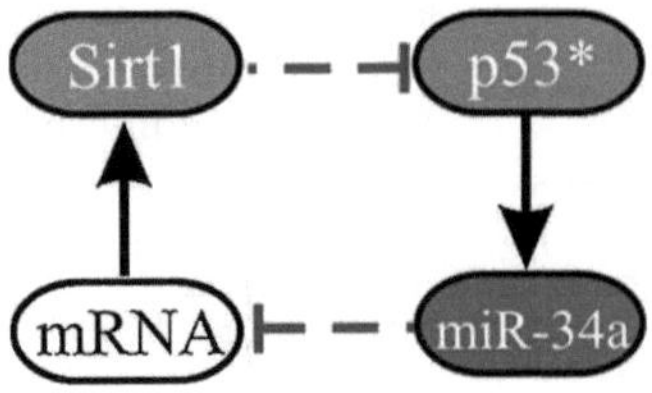

Fig. 3. Sketch of the regulatory structure of signaling module composed by p53, miR-34a, and Sirt1. Upon activation, active p53 (p53*) promotes the transcription of miR-34a, which is able to downregulate the expression of Sirt1 through interacting with the mRNA of Sirt1. Sirt1 can repress the production of p53* through deacetylation. As a result of this, we have signaling module displaying mutual inhibition, a dynamic motif which provokes the emergence of a positive feedback loop.

establishes a positive feedback regulatory loop with p53. In this regulatory structure miR-34a inhibits the expression of sirtuin1 (Sirt1), an NAD-dependent deacetylase which inhibits p53 acetylation and in turn provokes an increase in the fraction of acetylated/ active p53 (22) (see Fig. 3). Other than miR-34a, DBC1 (deleted in breast cancer 1) has been proved to be another native inhibitor of Sirt1 in human cells, repressing Sirt1 and therefore increasing the level of acetylated p53 and subsequently upregulating p53-mediated pathways (23).

In this book chapter, we define, discuss, and apply a Systems Biology approach to investigate the signaling module composed by the proteins p53 and Sirt1 and the regulatory miRNA miR-34a. Particularly, we derived, characterized, and analyzed a mathematical model in ordinary differential equations (ODEs), which was used to analyze different hypothesis concerning the repression of Sirt1 by miR-34a. Moreover, the nonlinear dynamical features of the feedback-loop-controlled signaling system integrated by p53, miR-34a, and Sirt1 was also investigated with the help of the mathematical model.

2. Materials and Methods

Mathematical modeling. For the investigation of the dynamic features of the signaling pathway p53/Sirt1/miR-34a, we derived, expanded, and calibrated a model in ordinary differential equations (ODEs). The procedure that we followed to set up the mathematical model includes an iterative cycle with the following steps (for further description, see ref. 18): (1) the list of relevant proteins, miRNA, and basic interactions as well as the structure of the mathematical model in mathematical terms are determined by using information extracted from biomedical literature and biological databases; (2) suitable quantitative data, specifically retrieved from the literature, are used in combination with optimization algorithms to calibrate the model (24, 25); (3) the quality of the resulting model is tested by computational analysis and additional quantitative experiments, and refined if required by the iteration of the previous steps; and (4) predictive simulations and other methodologies are used to identify key biochemical processes and unravel dynamical features of the signaling pathway.

For the mathematical model we chose rate equations based on power-law terms. Power-law models are a kind of kinetic equations, in which kinetic orders have noninteger values and even negative values when inhibition is considered (see Box 1 and ref. 26 for further information). There are biophysical (27, 28) and practical reasons (26) to use power-law representations for modeling signaling pathways. Power-law equations have been widely used to model different kinds of biochemical systems, including metabolic systems (29, 30), cell signaling pathways (26, 31, 32), and gene networks (33).

Box 1
The Definition of Power-Law Models

For our model we used the power-law formulation with the following structure:

$$\frac{d}{dt}X_i = \sum_j c_{ij} \cdot \gamma_j \cdot \prod_{k=1}^{p} X_k^{g_{jk}} \quad i = 1, \ldots, n_d.$$

Here, X_i represents any of the n_d time-dependent variables of the model (protein, mRNA, or miRNA expression levels). The biochemical rate j is expanded as a product of a rate constant (γ_j) and the p variables of the system with kinetic orders (g_{jk}), while c_{ij} are the stoichiometric coefficients. In our model formulation, kinetic orders can have noninteger values accounting for model aggregation of several biochemical processes into simplified representations. Negative values for the kinetic order represent inhibition, while 0 indicates that the variable does not affect the described process. In case of positive kinetic orders, values between 0 and 1 represent a saturation-like behavior, and higher ones represent cooperativity in the modeled process (see below).

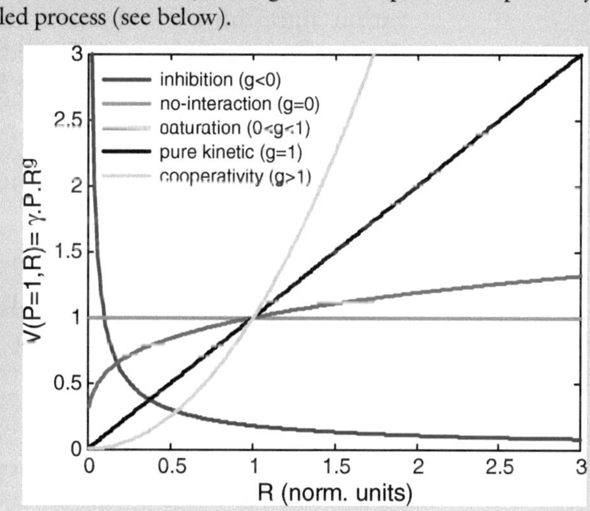

(continued)

Box 1
(continued)

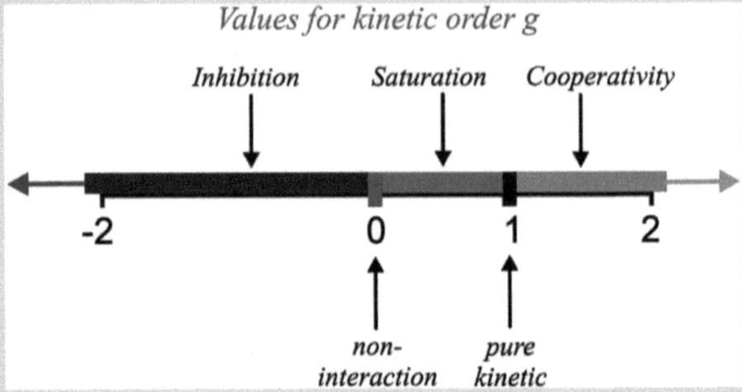

Parameter estimation. We characterized the model parameters with the following procedure: (1) existing data concerning protein half-life were used to assign values to the degradation rates of some molecules in the model (Mdm2, K_{dmd}, miR-34a, K_{dmi}, Sirt1, K_{dsi}, and DBC1, K_{ddb}); (2) the MATLAB's build-in interpolation function was used to characterize the dynamics of the exogenous DNA damage, which was used as input signal in the parameter estimation procedure (see Appendix Fig. A1); (3) suitable quantitative experimental data concerning the dynamics of the system under experimental conditions similar to the one analyzed were retrieved, processed, and used to estimate the rest parameter values (22, 34, 35). Parameter estimation was performed using standard parameter estimation functions implemented in SBtoolbox 2 (36), a toolbox developed under MATLAB. The computing time was in the order of minutes for each parameter estimation round. In brief, the parameter estimation uses the implemented local or global optimization algorithms to minimize the cost function defined by the distance between the model predictions and the data for given experimental conditions (see Box 2).

Box 2
Parameter Estimation

In the parameter estimation procedure optimization algorithms are used to search for optimal parameter set that minimizes the distance between model simulations and experimental data, which is described with the cost function:

$$F_{\mathrm{error}} = \frac{1}{n_{\mathrm{exp}} \cdot n_{\mathrm{var}} \cdot n_{\mathrm{tp}}} \sum_{k=1}^{n_{\mathrm{exp}}} \sum_{j=1}^{n_{\mathrm{var}}} \sum_{i=1}^{n_{\mathrm{tp}}} \left(X_{k,j}(t_i) - X_{k,j}^{\mathrm{exp}}(t_i)\right)^2,$$

where n_{exp} is the number of experiments, n_{var} is the number of measured quantities (observables), n_{tp} is the number of time points where each observable was measured, $X_{k,j}(t_i)$ denotes the value of the *j*th observable

(continued)

Box 2
(continued)

at the *i*th time point obtained after numerical integration of the solution for the *k*th experiment. $X_{k,j}^{exp}(t_i)$ is the value of the *j*th observable at the *i*th time point measured in the *k*th experiment. In the most advanced setting for this process, global optimization algorithms based on probability approach (e.g., particle swarm and evolutionary algorithms) are implemented and used to find the optimal parameter set (see 24 for further explanations).

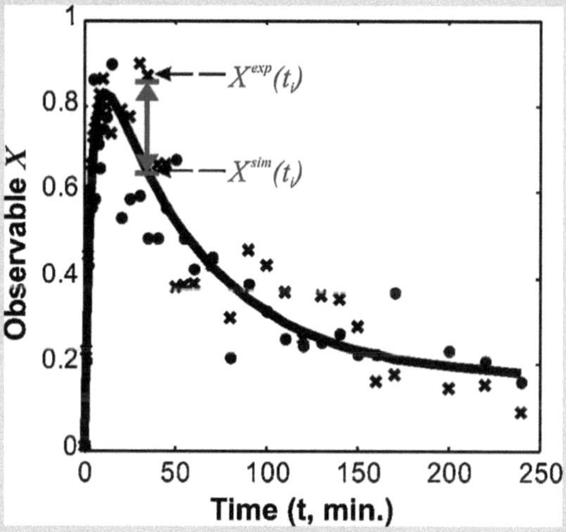

Illustration of the parameter estimation process. Optimization algorithms are used to estimate the optimal parameter values that minimize the distance (red arrow) between the model predictions (X^{sim}) and the experimental data (X^{exp}) for the considered experimental conditions.

Predictive simulations. After calibration and refinement of the mathematical model, we used it to perform computational simulations focusing on (a) the model-based investigation of the hypotheses concerning mechanisms of Sirt1 repression by miR-34a and (b) the predictions of the systems's dynamics under certain biological conditions, which are relevant for the designed system but are not tested by the experiments. The predictive simulations were manually designed and implemented, and performed using SBtoolbox 2 (36).

3. Results

3.1. Mathematical Model

Following the described procedure, we set up a power-law model accounting for the dynamics of miR-34a regulation in p53/Sirt1 signaling pathway (see Box 3). The model is composed of four main parts: (1) the activation of p53 in response to DNA damage-mediated signals (37–39); (2) the positive feedback loop integrated by p53, miR-34a, and Sirt1 (22, 40, 41); (3) the DNA damage induced enhancement in DBC1 activity and its inhibitory effect on Sirt1 (4, 42–45); and (4) the negative feedback loop between Mdm2 and active p53 (31, 46).

The model includes differential equations for the following model variables: expression level of the p53-inhibitor MDM2 (*Mdm2*, Eq. 1); fraction of non-active (deacetylated) p53 (*p53*, Eq. 6.2); fraction of active (acetylated) p53 (*p53**, Eq. 3); expression level of the miR-34a (*miR-34a*, Eq. 6.4); expression level of the p53 acetylation inhibitor Sirt1 (*Sirt1*, Eq. 5); and expression level of the Sirt1-activity inhibitor DBC1 (*DBC1*, Eq. 6).

For p53, the model contains rate equations accounting for: (1) basal protein synthesis; DNA damage-mediated synthesis of p53; (2) p53 acetylation and activation; (3) Sirt1-mediated p53 deacetylation; and (4) Mdm2-mediated p53 degradation. Mdm2 dynamics is modeled with rate equations describing the basal protein synthesis, the p53-mediated synthesis and the protein degradation. In case of the miR-34a dynamics, the rate equations consider p53-mediated synthesis and miRNA degradation. For Sirt1, the model accounts for miR-34a inhibited protein synthesis and protein degradation. Finally, the model includes rate equations describing the basal and DNA-damage-mediated synthesis of DBC1, as well as the protein degradation. In addition, an algebraic equation describes the total amount of p53 as the sum of the non-active and active fractions (Eq. 7).

Box 3
The Model Equations and Scheme

$$\frac{d}{dt}Mdm2 = K_{smd1} + K_{smd2} \cdot p53^*(t - \tau_1) - K_{dmd} \cdot Mdm2 \quad (1)$$

$$\frac{d}{dt}p53 = K_{sp53i1} + K_{sp53i2} \cdot Dd - K_{tp53i} \cdot p53 \cdot Dd + K_{tp53p} \cdot p53^* \cdot Sirt1 \cdot DBC1^{g_1} - K_{dp53i} \cdot p53 \cdot Mdm2 \quad (2)$$

$$\frac{d}{dt}p53^* = K_{tp53i1} \cdot p53 \cdot Dd - K_{tp53p} \cdot p53^* \cdot Sirt1 \cdot DBC1^{g_1} - K_{dp53p} \cdot p53^* \cdot Mdm2 \quad (3)$$

$$\frac{d}{dt}miR34 = K_{smi1} + K_{smi2} \cdot p53^*(t - \tau_2)^{g_2} - K_{dmi} \cdot miR34 \quad (4)$$

$$\frac{d}{dt}Sirt1 = K_{ssi} \cdot miR34(t - \tau_3)^{g_3} - K_{dsi} \cdot Sirt1 \quad (5)$$

(continued)

Box 3
(continued)

$$\frac{d}{dt}DBC1 = K_{sdb1} + K_{sdb2} \cdot Dd - K_{ddb} \cdot DBC1 \tag{6}$$

$$p53_{\text{total}} = p53 + p53^{*} \tag{7}$$

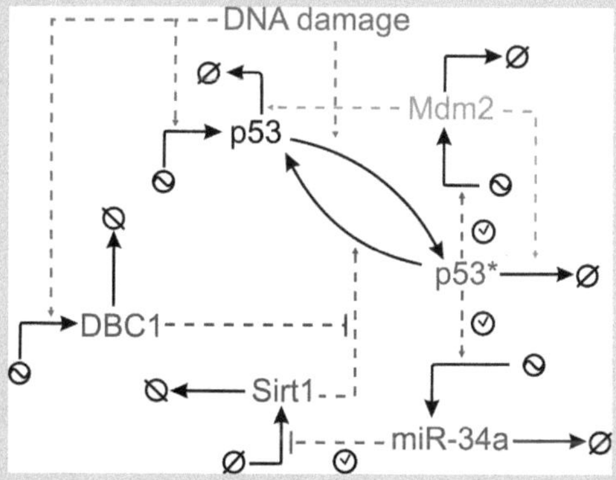

The model equations (top) and the graphic representation (bottom) of the signalling module p53/Sirt1 regulated by miR-34a. In the figure, the dashed lines with arrow mean that the species from where the arrow emanates activates the process where the arrow terminates, the dashed lines with hammer head means that the species inhibit the process where the hammer points to; the solid lines represent rate processes; the symbols ⊗ and Ø refer to synthesis and degradation processes, respectively; the "clock" symbol ⊘ refers to a time delay in the considered process.

Our model was constructed using power-law rate equations following the procedure discussed in Subheading 2. Actually, our simplified power-law model (26) differs from a pure kinetic model only in the description of inhibitory processes (DBC1-mediated inhibition of Sirt1 deacetylase activity and miR-34a-mediated repression of Sirt1 expression), for which we assigned a negative kinetic order ($g_1, g_3 = -1$). Besides, we also introduced three biologically proved time delays in the model: (1) time delay associated with the *p53**-mediated activation of Mdm2 synthesis (τ_1); (2) time delay attributed with the *p53**-mediated transcriptional activation of miR-34a expression (τ_2); and (3) time delay associated with the repression of Sirt1 by miR-34a (τ_3).

3.2. Parameter Estimation

The procedure described in Subheading 2 was used to characterize the model parameter values. Before the parameter estimation, each parameter value is constrained in an interval, which is physically plausible. The estimated parameter values for the derived model as

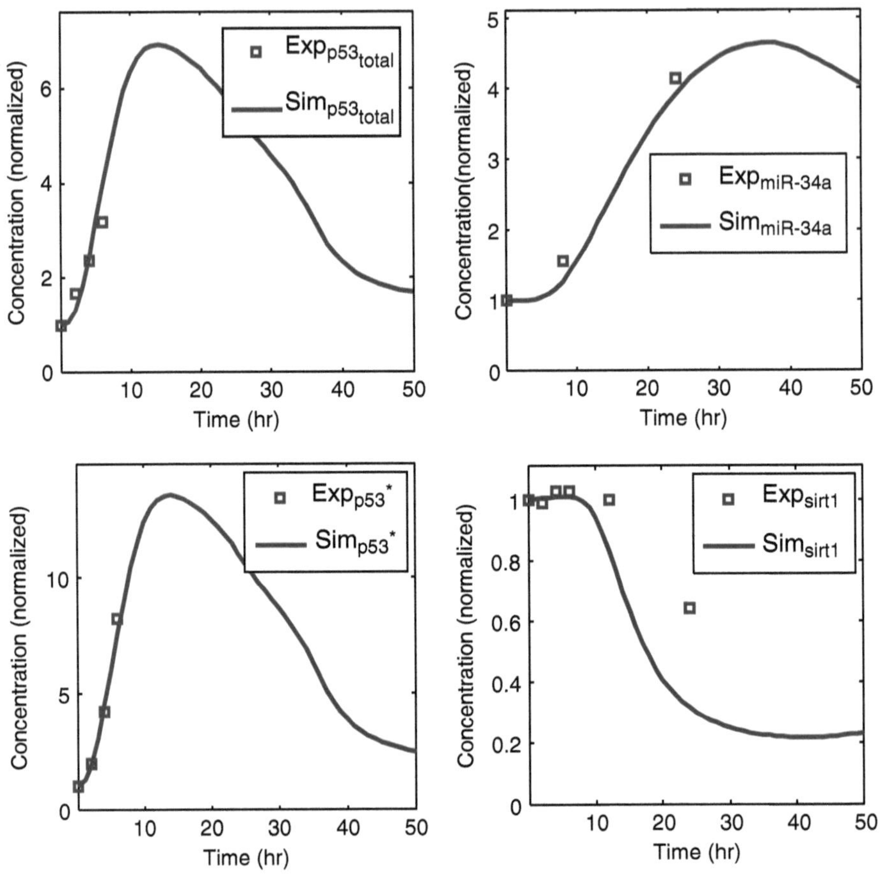

Fig. 4. Comparison between the model trajectories obtained for the selected model variables and the experimental time-series data used for model calibration: total amount of p53 ($p53_{total}$), active p53 ($p53^*$), miRNA-34a (*miR-34a*), and Sirt1 (*Sirt1*). *Solid lines* represent model simulations and *square markers* denote experimental data.

well as their initial values used for parameter estimation are summarized in Table A1 included in Appendix. The model trajectories obtained for the selected variables are depicted in Fig. 4, which compare the time-series experimental data with model simulations. In general, there is a good agreement between the model predictions and experimental data. More specifically, in response to DNA damage the total amount ($p53_{total}$) and the active fraction of p53 ($p53^*$) increase and reach their peaks several hours later after the stimulation. Upon accumulation of $p53^*$, the upregulation of miR-34a is observed with several hours of latency due to the necessary time (τ_2) needed for transcriptional activation by $p53^*$. Subsequently, with the increasing level of miR-34a, the concentration of Sirt1 starts to decline with several hours delay (τ_3). Because of the persistence inhibitory effect of miR-34a, the concentration of Sirt1 stays at low level until the end of the simulation.

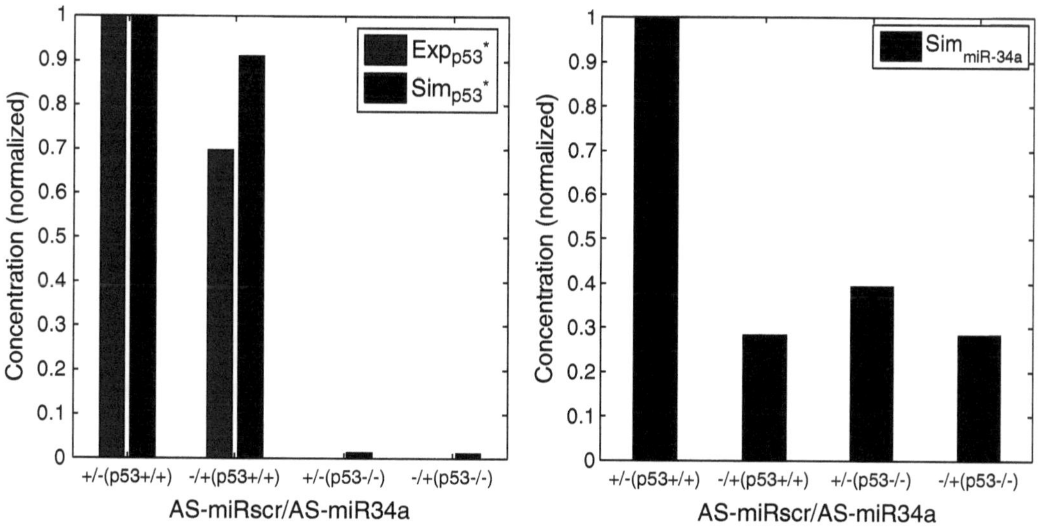

Fig. 5. *Left*: Effects of knockdown of endogenous miR-34a on concentration of acetylated p53. The bar plot represents the comparison between the model simulations (*left*) and the experimental data (*right*). In Yamakuchi et al. (22), the protein activation profiles were measured after 16-h treatment with 5-FU (Fluorouracil) to induce p53 in two different cell lines (shown in the *bracket*): (1) wild-type HCT116 cell lines, indicated by (p53+/+) and (2) p53-mutated HCT116 cell lines (p53−/−). Both cell lines were transfected with antisense oligonucleotides (AS-miR34a) or scrambled oligonucleotides (AS-miR34scr), respectively ("+" and "−" represent transfect and nontransfect of the specified oligonucleotides). *Right*: Model simulations for the levels of miR-34a in the same biological conditions described before.

To further validate the predictive ability of the model, we compared the model simulations with the experimental data showed in the work by Yamakuchi and coauthors (22) concerning the protein-expression profiles when wild-type and p53-mutated HCT116 cells were transfected with antisense oligonucleotides to miR-34 (AS-miR34a) to knock down the endogenous miR-34a or scrambled oligonucleotides (AS-miRscr). As we see in Fig. 5, model predictions are in a good agreement with the experimental data. Thus, we can conclude that the characterized model is capable of making reliable predictions for the defined biological system.

3.3. Predictive Simulations

Two groups of predictive simulations were performed. Firstly, we analyzed different hypothese concerning the mechanisms by which miR-34a silences Sirt1. Secondly, we simulated and analyzed the nonlinear dynamical features of the feedback-loop signaling system integrated by p53, miR-34a, and Sirt1 under nonexperimentally tested conditions.

Analysing the mechanism of miR-34a posttranscriptional repression of Sirt1. In the literature, there are several mechanisms by which small interference RNA (siRNA) and miRNA can cause gene silencing including: inhibition of the target mRNA by siRNA-mediated

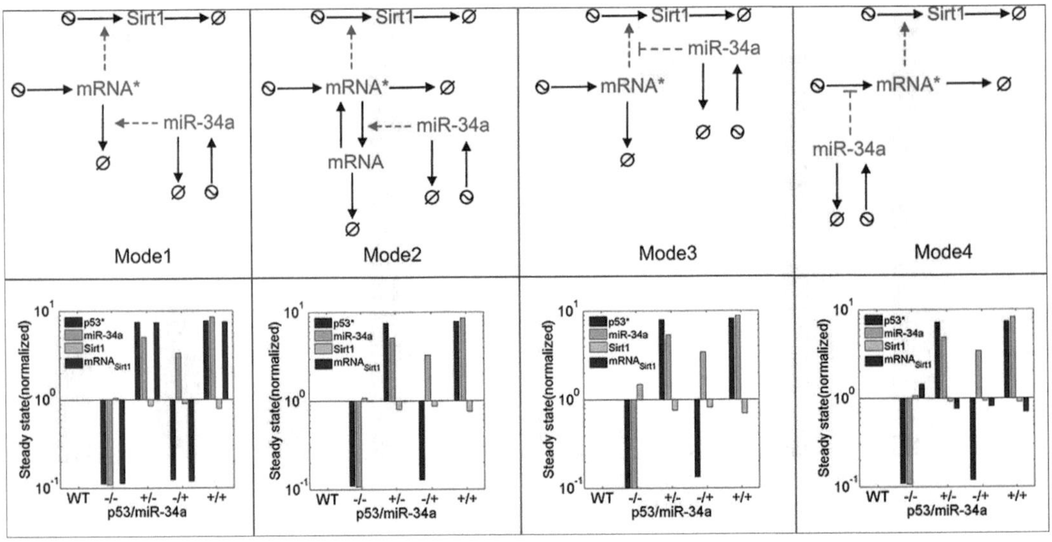

Fig. 6. The schemes of four hypothetical mechanisms by which miR-34a silences Sirt1. *Dashed lines* with arrow or hammer heads represent activation or inhibition, respectively. *Solid lines* represent synthesis (when starting with a *circle* and *curved line* ⦸ or degradation (when finishing in the symbol Ø). *Mode-1* mRNA cleavage, *Mode-2* posttranscriptional repression, *Mode-3* translation repression, *Mode-4* pretranscriptional repression. Each group of bars are arranged in order of active p53 (p53*), miR-34a (miR-34a), the mRNA ($mRNA_{Sirt1}$) and protein (Sirt1) of Sirt1 (from *left* to *right*).

methylation (transcriptional silencing); miRNA can specify cleavage if the target mRNA has sufficient complementarity to miRNA (mRNA cleavage), and destabilization or deactivation of target mRNA by non-perfect complementary binding (translational repression) (47).

Here we tested whether the combination of available quantitative data and model simulation allows discriminating the precise silencing mechanism exerted by miR-34a on Sirt1. Towards this end, we formulated four hypothetical mechanisms of miR-34a-mediated Sirt1 repression, namely: (1) miR-34 enhances degradation of the mRNA of Sirt1 (mRNA cleavage, mode 1, Fig. 6, left); (2) miR-34a promotes a reversible deactivation of the Sirt1 mRNA (posttranscriptional repression, mode 2, Fig. 6, center-left); (3) miR-34a inhibits the translation of mRNA and therefore inhibits the production of Sirt1 (translational repression, mode 3, Fig. 6, center-right); and (4) miR-34a directly inhibits the synthesis of the mRNA (pretranscriptional repression, mode 4, Fig. 6, right).

We expanded our model according to these hypotheses (see Appendix Table A2) and run some predictive simulations describing the behavior of the systems under different experimental conditions concerning p53 and miR-34a expression levels (see Fig. 6, bottom). Precisely, we simulated the expression levels for the components of the system (p53*, miR-34a, $mRNA_{Sirt1}$, Sirt1) for five different experimental conditions: wild-type (WT), combined downregulation of p53 and miR-34a (−/−), upregulation of p53 and downregulation of miR-34a (+/−), downregulation of p53 and upregulation of miR-34a (−/+), and upregulation of both p53 and mi-34a (+/+). The aim was to predict whether an experiment with such design could help discriminating between the different regulatory mechanisms. In order to allow easy comparison, the expression levels in Fig. 6 are displayed in a logarithmic scale normalized for the WT expression levels. Interestingly, we found that all the formulated mechanisms show the same expression pattern for p53 and mir-34a, but differ in the expression levels of Sirt1 protein and mRNA. The simulations indicate that the proposed experimental design could discriminate between mechanisms accounting for mRNA cleavage (mode 1), pretranscriptional repression (mode 4) and the other two models (modes 2 and 3), but it would fail to distinguish between posttranscriptional repression (mode 2) and translational repression (mode 3). Thus, additional experiments should be designed to allow full discrimination between all the mechanisms. Overall, the model comparison suggests that the total levels of Sirt1 mRNA would strongly differ in different experimental conditions for modes 1 and 4 (mRNA cleavage and pretranscriptional repression), while they would keep similar to wild-type conditions for modes 2 and 3. The model simulations also suggest that the expression levels of Sirt1 protein are more affected in the simulated experimental conditions for the translational repression model (mode 3).

To further investigate the mechanism of regulation, we verified whether any of the discussed regulatory mechanisms was able to reproduce the available dynamical data concerning the dynamics of the mRNA and protein of Sirt1 under conditions of wild type and mutated p53 expression (34). We found that silencing mechanisms encoded in mode 1 and mode 4 were not able to reproduce the experimental dynamics of Sirt1 and the simulated Sirt1 decreased too slowly compared with the experimental data after stimulation with 5-FU for wild type cells, p53+/+ (see Appendix Fig. A2). Furthermore, the time-series data concerning cell with mutated p53 expression (p53−/−) could be fitted only by mode 3, the unique model able to display unaffected levels of Sirt1 after 5-FU stimulation for at least 48 h posttreatment (see Fig. 7). Taken together, the model simulations suggest that miR-34a-mediated silencing of Sirt1 is implemented in the studied system through translational repression, the most common miRNA silencing mechanism in animals (47). We notice that this result is in agreement with the conclusion emerging from the experiments conducted by Yamakuchi and colleagues under similar experimental conditions (22).

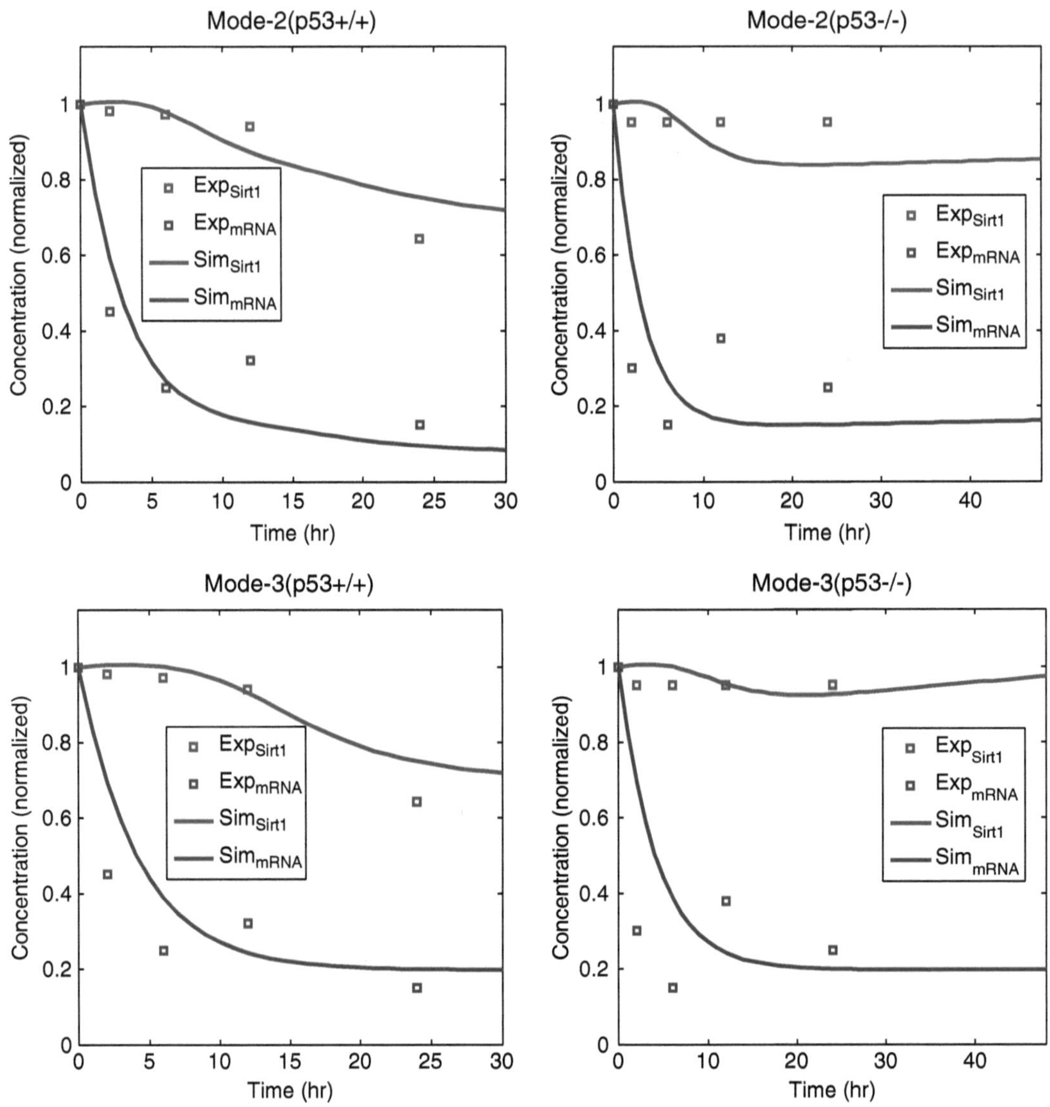

Fig. 7. The predictions of different miRNA silencing mechanisms. Here we illustrate the plots of mode 2 and mode 3, the plots of the other two modes can be found in the appendix (see Appendix Fig. 1). *Solid lines*: The model predictions for the protein (*upper line*) and mRNA (*lower line*) levels of Sirt1. *Square markers*: Experimental data for the protein (*upper square*) and mRNA (*lower square*) levels of Sirt1.

The regulatory effects of miR-34a and DBC1 on the activation of p53. We further simulated the effects that combined modulation of the two inhibitors of Sirt1 considered in the model (miR-34a and DBC1) have on the levels of active (*p53**) and non-active (*p53*) p53. Toward this end, we perturbed the parameter values accounting for the basal expression of the Sirt1 inhibitors in the

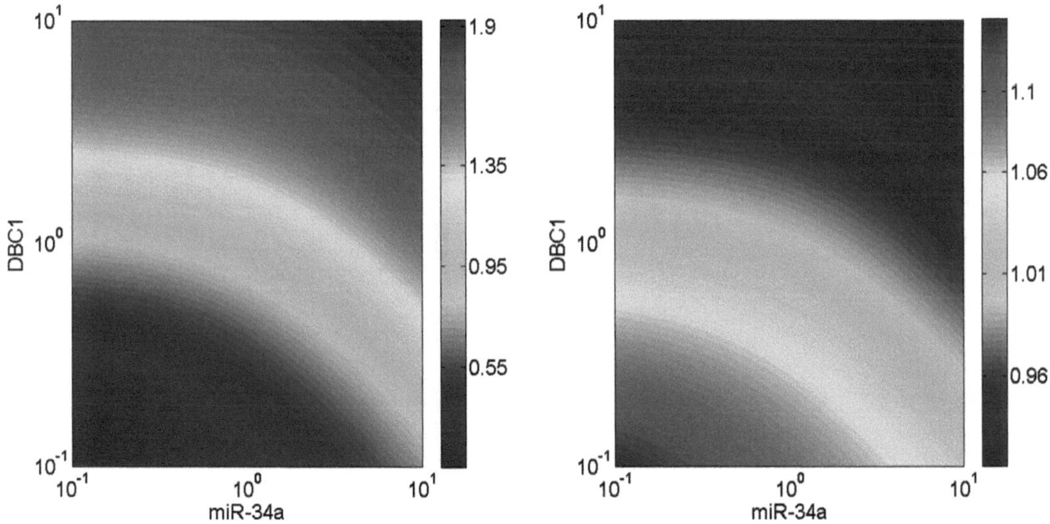

Fig. 8. The predictive simulations accounting for the steady-state levels of active p53 (*left*), non-active p53 (*right*) under up- and down-regulation of miR-34a and DBC1. Every point in the gird represents an steady-state level (the basal level is 1) of p53 for a given protein level ranging from extreme repression (10^{-1}) to extreme activation (10^{1}).

interval $[10^{-1}, 10^{1}]$ (where 1 accounts for basal expression) and compute the steady-state levels of *p53* and *p53** (see Fig. 8). Overall, the simulations indicate that repression of the Sirt1 inhibitors strongly (positive feedback loop-like) affects the levels of active p53, reducing its value to 20% of the basal level for the combined repression of the Sirt1 regulators (see Fig. 8, left). The model simulations also show that overexpression of the Sirt1 regulators may enhance nearly twofold the fraction of active p53. Interestingly enough, the simulations show that, although the expressions of DBC1 and miR-34a are perturbed in the same normalized interval, changes in the expression of DBC1 induce bigger variations in the levels of active p53 (see Fig. 8, left) than modulation of miR-34a. This suggests a stronger role for DBC1 as an enhancer of p53 activation compared to miR-34a. A similar but reversed pattern appears for inactive p53 (see Fig. 8, right).

In a new series of model simulations we investigated whether DBC1 and miR-34a are able to compensate the loss of active p53 associated with experimental settings in which Sirt1 appears overexpressed, a condition proved to be related with tumor progression in the literature (44). Our simulations show that in case of intermediate upregulation ($10^{0.5}$) of Sirt1 the loss of activity in p53 could be feedback-loop regulated by upregulating the expression of either DBC1 or miR-34a (see Fig. 9). However, only the intensive enhancement of DBC1 is able to cancel out the (feedback loop induced) loss of active p53 caused by the extreme upregulation of Sirt1 (see Fig. 9, left). Taken together, the simulation results

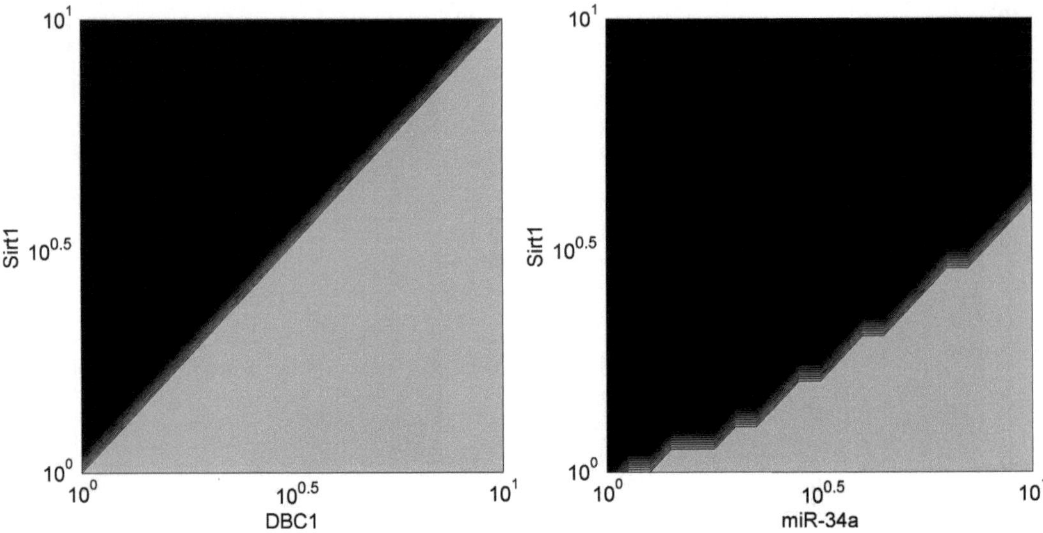

Fig. 9. The ability of DBC1 and miR-34a to compensate the loss of active p53 for the abnormal overexpression of Sirt1 in cancerous cells. *Dark color area*: The steady-state levels of active p53 are unable to reach its basal level. *Light color area*: The steady-state levels of active p53 are able to reach its basal level.

predicted different regulatory effects on the activation of p53 by DBC1 and miR-34a, and this can be explained by their different regulation routines in the signalling module. DBC1, which enhances the activation of p53 through simple regulation (DBC1-|Sirt1-|p53*), shows stronger regulatory effect than miR-34a, which enhances the activation of p53 through the positive feedback loop (p53*->miR-34a-|Sirt1-|p53*).

4. Discussion

In this book chapter we discussed and illustrated the use of data-driven mathematical modeling in the analysis of miRNA regulation in signaling pathways (see also ref. 48). The rational is that in complex regulatory systems (involving multiple positive or negative loops) the integration of experimental evidences, hypotheses, and quantitative data through mathematical modeling becomes mandatory to acquire a deep understanding of the regulatory dynamics.

To illustrate this idea, we developed a quantitative data-based mathematical model that describes the Sirt1-mediated deacetylation (deactivation) of p53, and the role of its some other regulators including the proteins DBC1 and Mdm2 and the small regulatory RNA miR-34a. Although this is not the unique regulatory loop associated with p53 signaling regulation (32, 49), this system constitutes a perfect example of positive feedback loop because active p53 induces the synthesis of miR-34a, which in turn represses Sirt1. The model was constructed using well established biomedical knowledge, which was translated into a power-law model, and characterized using quantitative experimental data in a process called parameter estimation.

After qualitative model validation to check its ability to reproduce the basic dynamics of the system, the model was used to investigate two main features of the system. Firstly, we tested whether the model allows discriminating the precise silencing mechanism exerted by miR-34a on Sirt1, for which four versions of the model accounting for different hypothese were formulated and simulated. Our analysis reveal that a quite basic experimental design allows the discrimination between some of the silencing motifs, but more sophisticated experiments would be required to reach full discrimination. Secondly, we performed a number of predictive simulations to elucidate the dynamics of the system and the role of the regulators DBC1 and miR-34a. Our computer simulations confirm that both DBC1 and miR-34a play the role as positive regulators of active p53 due to their ability to inhibit Sirt1 in two different manners. Furthermore, the results suggest that the upregulated expression of DBC1 is a much more efficient way to compensate the loss of active p53 under the condition of aberrant overexpression of Sirt1, which is found to be associated with cancer progression (44).

The results that we show here concerning the signaling module p53/miR-34a/Sirt1/DBC1 are far from giving a realistic and complete picture of the system. Our primary goal was, however, to illustrate the use of systems biology approach to investigate miRNA regulation. Further extensions of the model are required to give new insights into the dynamics of the system. Finally, is it well known that miRNA regulation is very often exerted by a set of (independent or interacting) different miRNAs rather than isolated or unique ones. Additional analysis should be carried out to investigate whether other small regulatory miRNAs can intensify (in a summative or synergistic manner) the effect of miR-34a.

Acknowledgments

J.V. and X.L. are funded by the German Federal Ministry of Education and Research (BMBF) as part of the project CALSYS-FORSYS under contract 0315264 (www.sbi.uni-rostock.de/calsys). O.W. was supported by the Helmholtz Foundation as part of the Systems Biology Network.

Appendix

Table A1
Parameter intervals and values. Top: Estimation intervals for corresponding parameters. Bottom: The parameter values used in the model. The values of kinetic orders (g_i) are fixed and other parameters are estimated. Values and intervals with stars: they are characterized by using biological protein half life. Others are estimated from the published experimental data (22, 34, 35)

Parameter	Interval	Parameter	Interval
K_{smd1}	[0, 3]	K_{smd2}	[0, 3]
K_{dmd}	[1, 1.5]*	K_{sp53i1}	[0, 3]
K_{sp53i2}	[0, 3]	K_{tp53i}	[0, 20]
K_{dp53i}	[1.5, 2.5]*	K_{tp53p}	[0, 3]
K_{dp53p}	[0, 3]	K_{smi1}	[0, 3]
K_{smi2}	[0, 3]	K_{dmi}	[0.02, 0.038]*
K_{ssi}	[0, 3]	K_{dsi}	[0.34, 0.47]*
K_{sdb1}	[0, 3]	K_{sdb2}	[0, 3]
K_{ddb}	[0.02, 0.025]*	τ_1	[0, 5]
τ_2	[0, 5]	τ_3	[0, 5]

Parameter	Value	Parameter	Value
K_{smd1}	1.317/h	K_{smd2}	0.068/h
K_{dmd}	1.386*/h	K_{sp53i1}	2.136/h
K_{sp53i2}	0.957/h	K_{tp53i}	16.242/h
K_{dp53i}	2.079*/h	K_{tp53p}	0.282/h
K_{dp53p}	0.078/h	K_{smi1}	0.010/h
K_{smi2}	0.018/h	K_{dmi}	0.028*/h
K_{ssi}	0.582/h	K_{dsi}	0.577*/h
K_{sdb1}	0.023/h	K_{sdb2}	0.003/h
K_{ddb}	0.023*/h	g_1	−1
g_2	1	g_3	1
τ_1	~1 h	τ_2	~1 h
τ_3	~1 h		

Table A2
Expanded equations of four hypothetical miR-34a silencing mechanisms

Mode 1:
$\frac{d}{dt} mRNA^*_{SIRT1} = k_1 - k_2 \cdot mRNA^*_{SIRT1} - k_3 \cdot miR34(t - \tau_3) \cdot mRNA^*_{SIRT1}$
$\frac{d}{dt} SIRT1 = K_{ssi} \cdot \frac{1}{1 + e^{-k_5(mRNA^*_{SIRT1} - k_6)}} - K_{dsi} \cdot SIRT1$
Mode 2:
$\frac{d}{dt} mRNA^*_{SIRT1} = k_1 - k_2 \cdot mRNA^*_{SIRT1} - k_3 \cdot miR34(t - \tau_3) \cdot mRNA^*_{SIRT1} + k_4 \cdot mRNA_{SIRT1}$
$\frac{d}{dt} mRNA_{SIRT1} = k_3 \cdot miR34(t - \tau_3) \cdot mRNA^*_{SIRT1} - k_4 \cdot mRNA_{SIRT1} - k_2 \cdot mRNA_{SIRT1}$
$\frac{d}{dt} SIRT1 = K_{ssi} \cdot \frac{1}{1 + e^{-k_5(mRNA_{SIRT1} - k_6)}} - K_{dsi} \cdot SIRT1$
Mode 3:
$\frac{d}{dt} mRNA^*_{SIRT1} = k_1 - k_2 \cdot mRNA^*_{SIRT1}$
$\frac{d}{dt} SIRT1 = \frac{K_{ssi}}{miR34(t - \tau_3)^{g_4}} \cdot \frac{1}{1 + e^{-k_5(mRNA^*_{SIRT1} - k_6)}} - K_{dsi} \cdot SIRT1$
Mode 4:
$\frac{\mathrm{d}}{\mathrm{d}t} mRNA^*_{SIRT1} = \frac{k_1}{miR34(t - \tau_3)^{g_4}} - k_2 \cdot mRNA^*_{SIRT1}$
$\frac{d}{dt} SIRT1 = K_{ssi} \cdot \frac{1}{1 + e^{-k_5(mRNA^*_{SIRT1} - k_6)}} - K_{dsi} \cdot SIRT1$

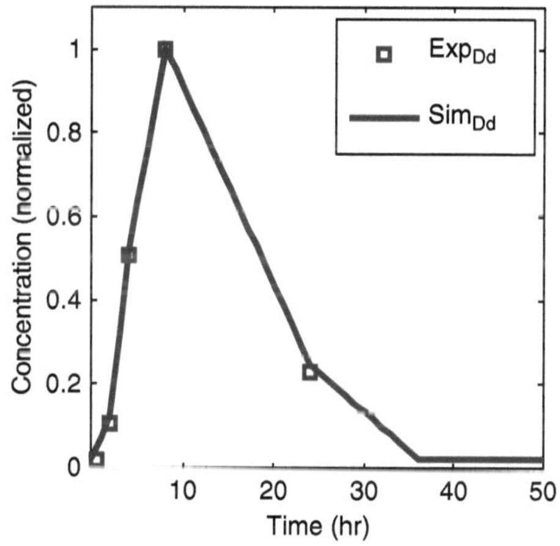

Fig. A1. Interpolation fitting plot of the experimental data regarding the dynamics of DNA damage. *Solid line*: Model simulation. *Square marker*: Experimental data.

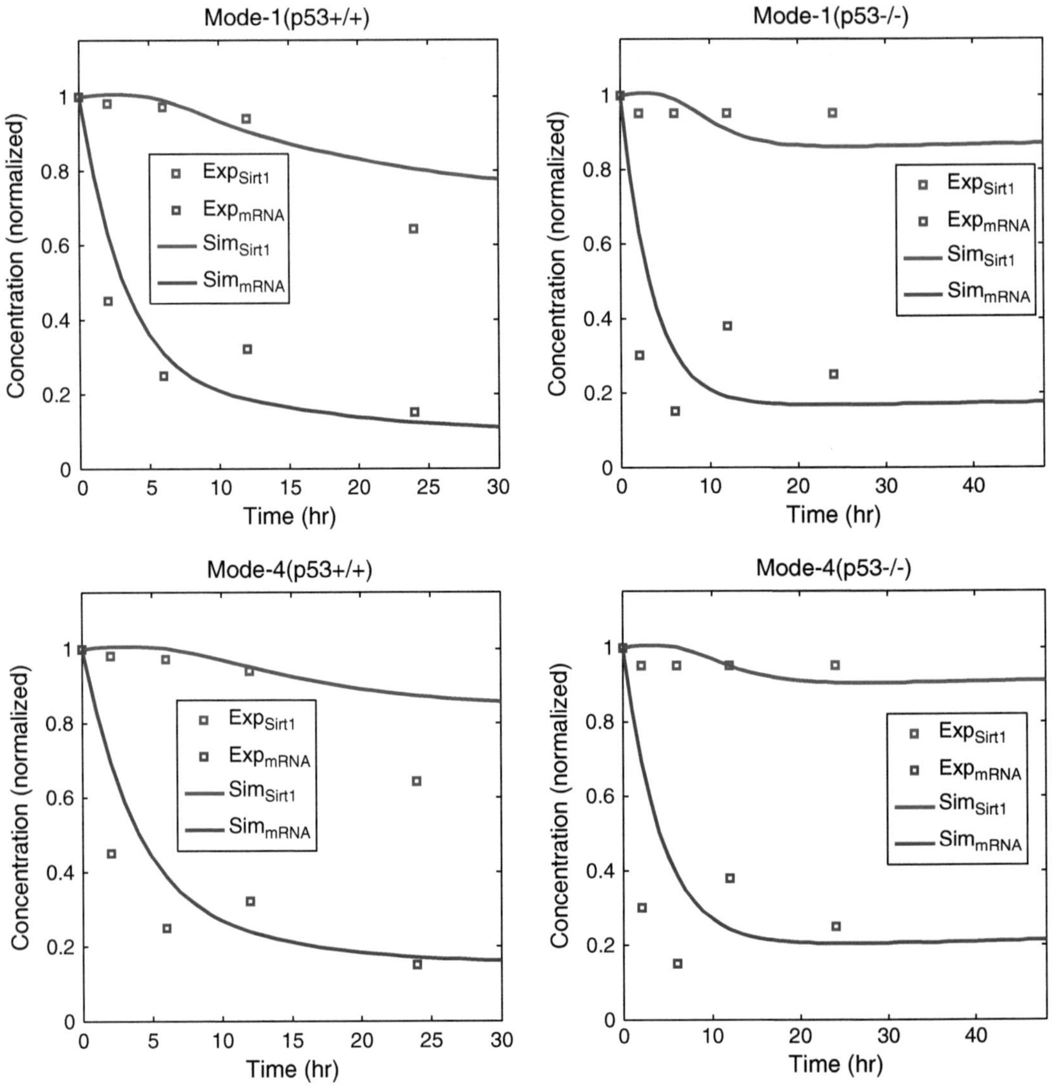

Fig. A2. The predictions of different miRNA silencing mechanism. Here we only illustrate the plots of mode 1 and mode 4. *Solid lines*: The model predictions for the protein (*upper line*) and mRNA (*lower line*) of Sirt1. *Square markers*: Experimental data for the protein (*upper square*) and mRNA (*lower square*) of Sirt1.

References

1. Nelson DE, Ihekwaba AEC, Elliott M, Johnson JR, Gibney CA, Foreman BE, Nelson G, See V, Horton CA, Spiller DG, Edwards SW, McDowell HP, Unitt JF, Sullivan E, Grimley R, Benson N, Broomhead D, Kell DB, White MRH (2004) Oscillations in NF-kappaB signaling control the dynamics of gene expression. Science 306:704–708
2. Aguda BD, Kim Y, Piper-Hunter MG, Friedman A, Marsh CB (2008) MicroRNA regulation of a cancer network: consequences of the feedback loops involving miR-17-92, E2F, and Myc. Proc Natl Acad Sci U S A 105:19678–19683
3. Chendrimada TP, Finn KJ, Ji X, Baillat D, Gregory RI, Liebhaber SA, Pasquinelli AE,

Shiekhattar R (2007) MicroRNA silencing through RISC recruitment of eIF6. Nature 447:823–828
4. Zhao W, Kruse J-P, Tang Y, Jung SY, Qin J, Gu W (2008) Negative regulation of the deacetylase SIRT1 by DBC1. Nature 451:587–590
5. Bueno MJ, Pérez de Castro I, Malumbres M (2008) Control of cell proliferation pathways by microRNAs. Cell Cycle 7:3143–3148
6. Khanin R, Vinciotti V (2008) Computational modeling of post-transcriptional gene regulation by microRNAs. J Comput Biol 15:305–316
7. Nissan T, Parker R (2008) Computational analysis of miRNA-mediated repression of translation: implications for models of translation initiation inhibition. RNA 14:1480–1491
8. Filipowicz W, Bhattacharyya SN, Sonenberg N (2008) Mechanisms of post-transcriptional regulation by microRNAs: are the answers in sight? Nat Rev Genet 9:102–114
9. Seitz H (2009) Redefining microRNA targets. Curr Biol 19:870–873
10. Chan JA, Krichevsky AM, Kosik KS (2005) MicroRNA-21 is an antiapoptotic factor in human glioblastoma cells. Cancer Res 65:6029–6033
11. Schultz J, Lorenz P, Gross G, Ibrahim S, Kunz M (2008) MicroRNA let-7b targets important cell cycle molecules in malignant melanoma cells and interferes with anchorage-independent growth. Cell Res 18:549–557
12. Akao Y, Nakagawa Y, Naoe T (2006) let-7 microRNA functions as a potential growth suppressor in human colon cancer cells. Biol Pharm Bull 29:903–906
13. Johnson SM, Grosshans H, Shingara J, Byrom M, Jarvis R, Cheng A, Labourier E, Reinert KL, Brown D, Slack FJ (2005) RAS is regulated by the let-7 microRNA family. Cell 120:635–647
14. Takamizawa J, Konishi H, Yanagisawa K, Tomida S, Osada H, Endoh H, Harano T, Yatabe Y, Nagino M, Nimura Y, Mitsudomi T, Takahashi T (2004) Reduced expression of the let-7 microRNAs in human lung cancers in association with shortened postoperative survival. Cancer Res 64:3753–3756
15. Iorio MV, Ferracin M, Liu C-G, Veronese A, Spizzo R, Sabbioni S, Magri E, Pedriali M, Fabbri M, Campiglio M, Ménard S, Palazzo JP, Rosenberg A, Musiani P, Volinia S, Nenci I, Calin GA, Querzoli P, Negrini M, Croce CM (2005) MicroRNA gene expression deregulation in human breast cancer. Cancer Res 65:7065–7070
16. Shalgi R, Lieber D, Oren M, Pilpel Y (2007) Global and local architecture of the mammalian microRNA-transcription factor regulatory network. PLoS Comput Biol 3:e131
17. Brosh R, Shalgi R, Liran A, Landan G, Korotayev K, Nguyen GH, Enerly E, Johnsen H, Buganim Y, Solomon H, Goldstein I, Madar S, Goldfinger N, Børresen-Dale A-L, Ginsberg D, Harris CC, Pilpel Y, Oren M, Rotter V (2008) p53-Repressed miRNAs are involved with E2F in a feed-forward loop promoting proliferation. Mol Syst Biol 4:229
18. Vera J, Wolkenhauer O (2008) A system biology approach to understand functional activity of cell communication systems. Methods Cell Biol 90:399–415
19. Riley T, Sontag E, Chen P, Levine A (2008) Transcriptional control of human p53-regulated genes. Nat Rev Mol Cell Biol 9:402–412
20. Rodier F, Campisi J, Bhaumik D (2007) Two faces of p53: aging and tumor suppression. Nucleic Acids Res 35:7475–7484
21. Chang T-C, Wentzel EA, Kent OA, Ramachandran K, Mullendore M, Lee KH, Feldmann G, Yamakuchi M, Ferlito M, Lowenstein CJ, Arking DE, Beer MA, Maitra A, Mendell JT (2007) Transactivation of miR-34a by p53 broadly influences gene expression and promotes apoptosis. Mol Cell 26:745–752
22. Yamakuchi M, Ferlito M, Lowenstein CJ (2008) miR-34a repression of SIRT1 regulates apoptosis. Proc Natl Acad Sci U S A 105:13421–13426
23. Kim E-J, Kho J-H, Kang M-R, Um S-J (2007) Active regulator of SIRT1 cooperates with SIRT1 and facilitates suppression of p53 activity. Mol Cell 28:277–290
24. Balsa-Canto E, Alonso AA, Banga JR (2008) Computational procedures for optimal experimental design in biological systems. IET Syst Biol 2:163–172
25. Banga JR, Balsa-Canto E (2008) Parameter estimation and optimal experimental design. Essays Biochem 45:195–209
26. Vera J, Balsa-Canto E, Wellstead P, Banga JR, Wolkenhauer O (2007) Power-law models of signal transduction pathways. Cell Signal 19:1531–1541
27. Kopelman R (1988) Fractal reaction kinetics. Science 241:1620–1626
28. Savageau MA (1998) Development of fractal kinetic theory for enzyme-catalysed reactions and implications for the design of biochemical pathways. BioSystems 47:9–36
29. Alvarez-Vasquez F, Sims KJ, Cowart LA, Okamoto Y, Voit EO, Hannun YA (2005) Simulation and validation of modelled sphingolipid metabolism in *Saccharomyces cerevisiae*. Nature 433:425–430

30. Domijan M, Kirkilionis M (2009) Bistability and oscillations in chemical reaction networks. J Math Biol 59:467–501
31. Vera J, Bachmann J, Pfeifer AC, Becker V, Hormiga JA, Darias NVT, Timmer J, Klingmüller U, Wolkenhauer O (2008) A systems biology approach to analyse amplification in the JAK2-STAT5 signalling pathway. BMC Syst Biol 2:38
32. Vera J, Schultz J, Ibrahim S, Raatz Y, Wolkenhauer O, Kunz M (2010) Dynamical effects of epigenetic silencing of 14-3-3sigma expression. Mol Biosyst 6:264–273
33. Atkinson MR, Savageau MA, Myers JT, Ninfa AJ (2003) Development of genetic circuitry exhibiting toggle switch or oscillatory behavior in *Escherichia coli*. Cell 113:597–607
34. Ford J, Ahmed S, Allison S, Jiang M, Milner J (2008) JNK2-dependent regulation of SIRT1 protein stability. Cell Cycle 7:3091–3097
35. Suzuki HI, Yamagata K, Sugimoto K, Iwamoto T, Kato S, Miyazono K (2009) Modulation of microRNA processing by p53. Nature 460:529–533
36. Schmidt H, Jirstrand M (2006) Systems Biology Toolbox for MATLAB: a computational platform for research in systems biology. Bioinformatics 22:514–515
37. Appella E, Anderson CW (2001) Post-translational modifications and activation of p53 by genotoxic stresses. Eur J Biochem 268:2764–2772
38. Vaziri H, Dessain SK, Ng Eaton E, Imai SI, Frye RA, Pandita TK, Guarente L, Weinberg RA (2001) hSIR2(SIRT1) functions as an NAD-dependent p53 deacetylase. Cell 107:149–159
39. Yamakuchi M, Lowenstein CJ (2009) MiR-34, SIRT1 and p53: the feedback loop. Cell Cycle 8:712–715
40. Fujita Y, Kojima K, Hamada N, Ohhashi R, Akao Y, Nozawa Y, Deguchi T, Ito M (2008) Effects of miR-34a on cell growth and chemoresistance in prostate cancer PC3 cells. Biochem Biophys Res Commun 377:114–119
41. Kato M, Paranjape T, Müller RU, Ullrich R, Nallur S, Gillespie E, Keane K, Esquela-Kerscher A, Weidhaas JB, Slack FJ (2009) The mir-34 microRNA is required for the DNA damage response in vivo in *C. elegans* and in vitro in human breast cancer cells. Oncogene 28:2419–2424
42. Cha EJ, Noh SJ, Kwon KS, Kim CY, Park B-H, Park HS, Lee H, Chung MJ, Kang MJ, Lee DG, Moon WS, Jang KY (2009) Expression of DBC1 and SIRT1 is associated with poor prognosis of gastric carcinoma. Clin Cancer Res 15:4453–4459
43. Kim J-E, Chen J, Lou Z (2008) DBC1 is a negative regulator of SIRT1. Nature 451:583–586
44. Kwon H-S, Ott M (2008) The ups and downs of SIRT1. Trends Biochem Sci 33:517–525
45. Li Z, Chen L, Kabra N, Wang C, Fang J, Chen J (2009) Inhibition of SUV39H1 methyltransferase activity by DBC1. J Biol Chem 284:10361–10366
46. Ramalingam S, Honkanen P, Young L, Shimura T, Austin J, Steeg PS, Nishizuka S (2007) Quantitative assessment of the p53-Mdm2 feedback loop using protein lysate microarrays. Cancer Res 67:6247–6252
47. Bartel DP (2004) MicroRNAs: genomics, biogenesis, mechanism, and function. Cell 116:281–297
48. Nikolov S, Vera J, Schmitz U, Wolkenhauer O (2011) A model-based strategy to investigate the role of microRNA regulation in cancer signalling networks. Theory Biosci 130:55–69
49. Vera J, Nikolov S, Lai X, Singh A, Wolkenhauer O (2011) A model-based investigation of the transcriptional activity of p53 and its feedback loop regulation via 14-3-3sigma. IET Syst Biol 5:293–307

Chapter 7

Design of Experiments to Investigate Dynamic Cell Signaling Models

Samuel Bandara and Tobias Meyer

Abstract

This chapter describes approaches to make use of dynamic models of cell signaling systems in order to optimize experiments in cell biology. We are particularly focusing on the question of how small molecule inhibitors or activators can best be used to get the most information out of a limited number of experiments when only a handful of molecular species can be measured. One goal addressed by this chapter is to find time course experiments to discriminate between rivaling molecular mechanisms. The other goal is to find experiments that are useful for inferring rate constants, binding affinities, concentrations, and other model parameters from time course data. Both are treated as optimal control problems in which rapid pharmacological perturbation schemes are identified *in silico* in order to close an experimental cycle from modeling back to the laboratory bench.

Key words: Parameter estimation, Model discrimination, Parameter uncertainty, Experimental design, Optimization, Dynamical systems

1. Introduction

Specifically, we address the question which small molecules or ligands should be added in what combination, time, and concentration when only a limited number of signaling molecules can be measured in a time course experiment. The goal of such experiments is to improve our understanding of biological signaling systems by clarifying the structure of a dynamic model or by revealing model parameters such as rate constants with higher accuracy. We consider systems of ordinary differential equations of the form

$$\frac{dx(t)}{dt} = f(x(t), p, u(t)), \tag{1}$$

where $x(t)$ are the differential state variables, p the parameters, and $u(t)$ the concentration–time profiles of experimental perturbations.

Xuedong Liu and Meredith D. Betterton (eds.), *Computational Modeling of Signaling Networks*,
Methods in Molecular Biology, vol. 880, DOI 10.1007/978-1-61779-833-7_7,

The first section recapitulates briefly the concept of model calibration since model predictions will be needed in the process of experiment design. The second section deals with the goal of model discrimination for which we want to find an experimental protocol $u(t)$ that allows us to falsify at least one of two alternative models. The third section describes how a model can be used to find another experiment $u(t)$ that reveals model parameters p like rate constants, binding affinities, and concentrations with increased accuracy. In both cases, one or more models that have been calibrated against preliminary measurements are desired because the predictions of these models are needed to assess the merit of a specific experimental protocol.

2. Method

2.1. Parameter Estimation

The process of model calibration to actual data is also known as parameter estimation because values for unknown model parameters are being identified that result in good agreement between model predictions and measurement data. In other words, this type of approach infers the previously unknown values of model parameters p by minimizing the discrepancy between model predictions h and N measurement data η_i that were taken with some standard deviation σ_i at time t_i:

$$\sum_{i=1}^{N} \left(\frac{\eta_i - h(x(t_i; p, u), p)}{\sigma_i} \right)^2. \tag{2}$$

Here, $x(t; p, u)$ describes the solution of the initial value problem, i.e., the state of the system for given parameters p and experimental treatment u, evaluated at point t. The observation function h predicts measurements on the basis of state variables but may also involve parameters for calibration. Expression (2) is referred to as a weighted least squares functional: agreement between model prediction and data is enforced more strongly where measurement error is believed to be small. In many cases, h can simply be one of the state variables, like the expression level of a protein as one can measure in a Western blot experiment. For a more complicated case, consider live cell measurements using a calcium dye. The concentration of calcium may be one of the state variables x, while h will model the binding interactions that change the fluorescent properties of the dye.

Figure 1a is the schematic of an estimation problem with two free parameters. Ideally, one and only one region on the parameter surface exists (shaded area) in which the model produces a satisfying fit to the data. If multiple regions exist, or if a single region is large, not much can be learned about the true value of a parameter.

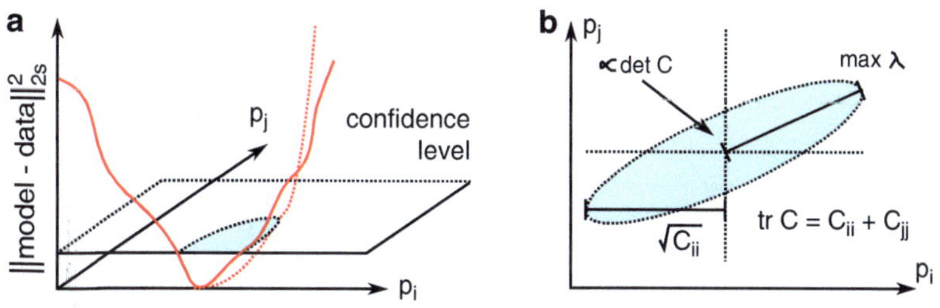

Fig. 1. (**a**) Parameter estimation illustrated for the case of two parameters, p_i and p_j. Parameter values are inferred by minimizing the discrepancy between model predictions and measurements (*red contour*). For an informative set of measurements, this discrepancy falls below the confidence level only in a small region of the parameter plane. (**b**) Confidence ellipsoid projected from (**a**). Different properties of the covariance matrix *C*, such as the trace, the largest eigenvalue maxλ, or the determinant of *C*, correspond to geometric properties of the uncertainty region (see Subheading 2.3). Figures from (6).

The third section of this chapter addresses this problem of large parameter uncertainties by describing how experiments can be designed that sensitize model predictions to parameter values. In fact, large parameter uncertainties are a commonly found feature in models of biological signaling networks. Interestingly, this means also that in some cases, model predictions can be reliable even if individual parameter values remain very uncertain after model calibration. We recommend the reading of an article by Gutenkunst et al. (1) who provide a thorough discussion of this issue. We recapitulated the concept of parameter estimation here for two reasons. First, for planning subsequent experiments by numerical optimization methods, a calibrated model is desired. Second, after each of such experiments has been performed, the model can be refined by collective fits to all of the data collected so far. Chapter 8 of this book provides a practical tutorial to parameter estimation software.

2.2. Experiment Design for Model Discrimination

Given a set of experimental observations, different alternative hypotheses, or models, may be able to explain all data previously generated. In such a case, the goal is to find the next experiment that allows one to discriminate between the competing models. A common approach to this problem is to find an experiment that maximizes contradiction between alternative model predictions. In effect, if one model remains satisfactory, the other model may no longer be able to account for all data.

For example, consider a preliminary experiment in which cells are stimulated with a ligand for 8 min, and the activation of a close downstream effector is measured every minute (Fig. 2a). The sigmoidal trajectory of effector activation lets us ask whether this process is possibly ultrasensitive, and two alternative mechanistic models are formulated. Model A postulates that activation of the effector by the receptor–ligand complex is highly cooperative.

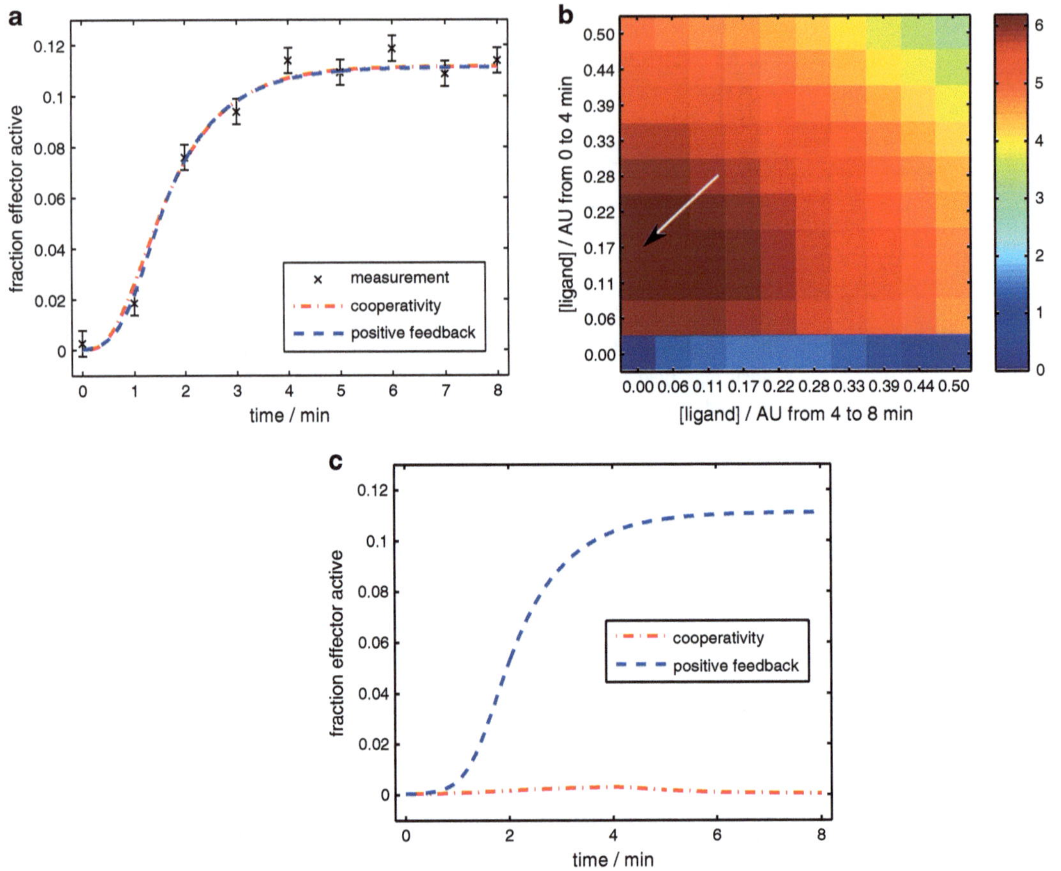

Fig. 2. Experiment design for model discrimination. (**a**) Both models, the cooperative mechanism and the feedback mechanism, can explain the measurements from a continuous stimulation experiment (data simulated with measurement noise). (**b**) To find an experiment for distinguishing between the two models, the discrepancy of model predictions can be maximized. The largest discrepancy is expected for a mild stimulation for the first 4 min, followed by a complete washout of the ligand (*arrow*). (**c**) Prediction of each model for this optimized experiment.

Model B instead speculates that the effector itself can activate the receptor in a positive feedback loop. In both models we denote the concentration of the ligand with $u(t)$. The first differential state variable x_1 reflects active receptor, and the second one x_2 reflects activated effector. Further, let the total concentration of each species be normalized. We can write out model A, the cooperative mechanism, then by

$$\frac{d\tilde{x}_1(t)}{dt} = k_1 u(t)(1 - \tilde{x}_1(t)) - k_2\tilde{x}_1(t) \tag{3}$$

$$\frac{d\tilde{x}_2(t)}{dt} = k_3\tilde{x}_1^3(t)(1 - \tilde{x}_2(t)) - k_4\tilde{x}_2(t) \tag{4}$$

with the measurement function $h_A(t) = \tilde{x}_2(t)$. The alternative model B which considers positive feedback between receptor and effector may look like this:

$$\frac{d\hat{x}_1(t)}{dt} = (k_5 u(t) + k_6 \hat{x}_2(t))(1 - \hat{x}_1(t)) - k_7 \hat{x}_1(t) \tag{5}$$

$$\frac{d\hat{x}_2(t)}{dt} = k_8 \hat{x}_1(t)(1 - \hat{x}_2(t)) - k_9 \hat{x}_2(t). \tag{6}$$

Again, the second differential state variable can be measured directly, with $h_B(t) = \hat{x}_2(t)$. Both models can be fit with similar quality to the preliminary time course experiment (Fig. 2a). This means that the data collected so far are not sufficient to distinguish between the two alternative models.

To find a more suitable experiment, we want to identify a stimulus pattern $u(t)$ that maximizes any disagreement between the two alternative models, hoping that at least one of the models will no longer be able to recapitulate the observed dynamics:

$$\sum_{i=1}^{N} (h_A(\tilde{x}(t_i; u)) - h_B(\hat{x}(t_i; u)))^2. \tag{7}$$

At the same time, the stimulus pattern $u(t)$ should be attainable by a simple experimental protocol such as by replacing culture medium once at a particular point in time. Thus, let the stimulus pattern $u(t)$ be discretized as a piecewise constant function with two intervals, one from 0 to 4 min, and the other from 4 min to the end of the experiment at 8 min. Practically, this means that the concentration of ligand is set once at the beginning of the experiment, and then adjusted in the middle of it. Figure 2b shows the objective functional, i.e., the expression to be maximized in (7), for ranges of ligand concentrations in the first and second segment of the experiment (see MATLAB code in the appendix). The result suggests that adding the ligand at a low concentration in the beginning of the experiment and then washing it out for the second segment of the experiment is the most promising approach to discriminating between the two models.

Figure 2c shows why this also makes sense intuitively. The feedback model B predicts that a low threshold can be crossed after which effector activation will be self-reliant. In contrast, the cooperativity model A predicts for the same stimulus only a negligible degree of effector activation, which furthermore decays upon withdrawal of the ligand. In essence, with a purely numerical approach we rediscover and exploit that bistability can distinguish positive feedback from a merely cooperative mechanism. Moreover, the solution for $u(t)$ also proposes an actual ligand concentration that might be sufficient to trigger the bistable switch.

While it can be highly valuable to derive experiments such as the ones discussed above, some challenges need to be mentioned. First, the outlined approach aims to eliminate alternative mechanisms by planning for results that can only be reproduced by one of the models and not the other. However, it is often difficult to definitely prove that no parameters can be found for a model to result in a satisfying fit. Here, reliable methods for parameter estimation are much needed, such as those based on multiple shooting (see e.g. (2), or (3) for a recent review). Second, although the models used for experiment design can be calibrated against preliminary data, they often remain uncertain in their predictions in novel experimental contexts due to uncertainty of parameter estimates. As a consequence, both models may still survive such an optimized experiment if some of their loosely defined parameters are subsequently adjusted. It is then not uncommon to again end up with two possible models. To address this problem, Chen and Asprey (4) describe how estimates of parameter uncertainty can be taken into consideration already when experiments are planned (4).

2.3. Experiment Design for Parameter Estimation

In a more comprehensive approach, experiments can be designed to infer rate constants, binding affinities, or abundances of signaling intermediates from time course data. This is particularly useful when one suspects that some of these parameters are altered by preceding experimental perturbations. The challenge with this approach is that a wide range of parameter values in a model often results in similarly good fits to the data, such that parameter values often remain highly uncertain (Fig. 1a). One particular reason for this is that deviations from the true value of one parameter can often be compensated for by changes in the estimates of other parameters, which adds large errors to both estimates.

Hence, good experiments are only those for which the objective functional (2) is very sensitive to changes in the parameter values p. This idea can be approximated as the sensitivity of least-squares terms in (2) to the parameter vector, and since only the model predictions h depend on the parameter estimates p, we obtain the sensitivity matrix

$$J = -\Sigma^{-1}\frac{dh}{dp}, \tag{8}$$

where the diagonal elements of Σ are the standard deviations σ_i and h is a vector with model predictions $h_i(t_i, x(t_i, p, u), p)$, for $i = 1, \ldots, N$. The covariance matrix $C = (J^T J)^{-1}$ then describes the uncertainty of the parameter estimates. In particular, the diagonal elements of C are the approximations of their variances. Note that the covariance matrix is a function of the experimental design $u(t)$. Therefore, experimental protocols $u(t)$ may exist that are much better than others at revealing parameter values from observable dynamics. This allows one to search for an optimal experimental protocol.

Figure 1b illustrates the relationship between properties of C and the confidence ellipsoid projected from Fig. 1a. These properties correspond to commonly used design criteria when one is interested in reducing the uncertainty of model parameters. For example, one objective could be to identify an experimental protocol $u(t)$ that minimizes the trace of C, the sum of all parameter variances. Bauer et al. (5) developed an efficient method to solve this kind of optimization problem in an approach that was found to be useful for many problems in chemical engineering (5). The usefulness of this approach for cell biology applications was recently first tested by our group to infer kinetic parameters of phospholipid signaling in transformed fibroblasts from confocal fluorescence microscopy experiments (6). In that study, endogenous phosphoinositide 3-kinase (PI3K) was activated by recruiting a regulator peptide (CFP-FKBP-p85) to the plasma membrane through chemically induced heterodimerization with a membrane anchor construct (Lyn-FRB), see Fig. 3a:

$$\frac{d[iRap :: p85]}{dt} = k_r[p85][iRap] - \frac{d[Lyn :: iRap :: p85]}{dt} \tag{9}$$

$$\frac{d[Lyn :: iRap :: p85]}{dt} = k_t([Lyn]_{tot} - [Lyn :: iRap :: p85]) \times [iRap :: p85] \tag{10}$$

with $[p85] = [p85]_{tot} - [iRap :: p85] - [Lyn :: iRap :: p85]$. In the absence of the inhibitory drug LY29, PI3K produces phosphatidylinositol 3,4,5-trisphosphate (PIP_3) in the plasma membrane:

$$\frac{d[PIP_3]}{dt} = -\phi[PIP_3] + \frac{[Lyn :: iRap :: p85]}{1 + \frac{[LY29]}{K_{LY}}}. \tag{11}$$

Many signaling proteins are recruited to the plasma membrane in response to elevated concentrations of PIP_3 by direct binding through pleckstrin homology (PH) domains. Thus, a fusion construct of YFP and the PH domain of Akt (YFP-PH) was used to monitor signaling downstream of PI3K:

$$\frac{d[PH]_{mem}}{dt} = -\alpha[PH]_{mem} + k_a([PH]_{tot} - [PH]_{mem})[PIP_3]. \tag{12}$$

Together, the model was comprised of four differential states and two observables: the localization of each of the fluorescent constructs, CFP-FKBP-p85 and YFP-PH. The concentrations of LY29 and iRap on the other hand served as control inputs. The time course experiments were segmented into three intervals for each of which the concentrations of the two drugs, iRap and LY29, were optimized to reveal six parameters with higher accuracy: the rate

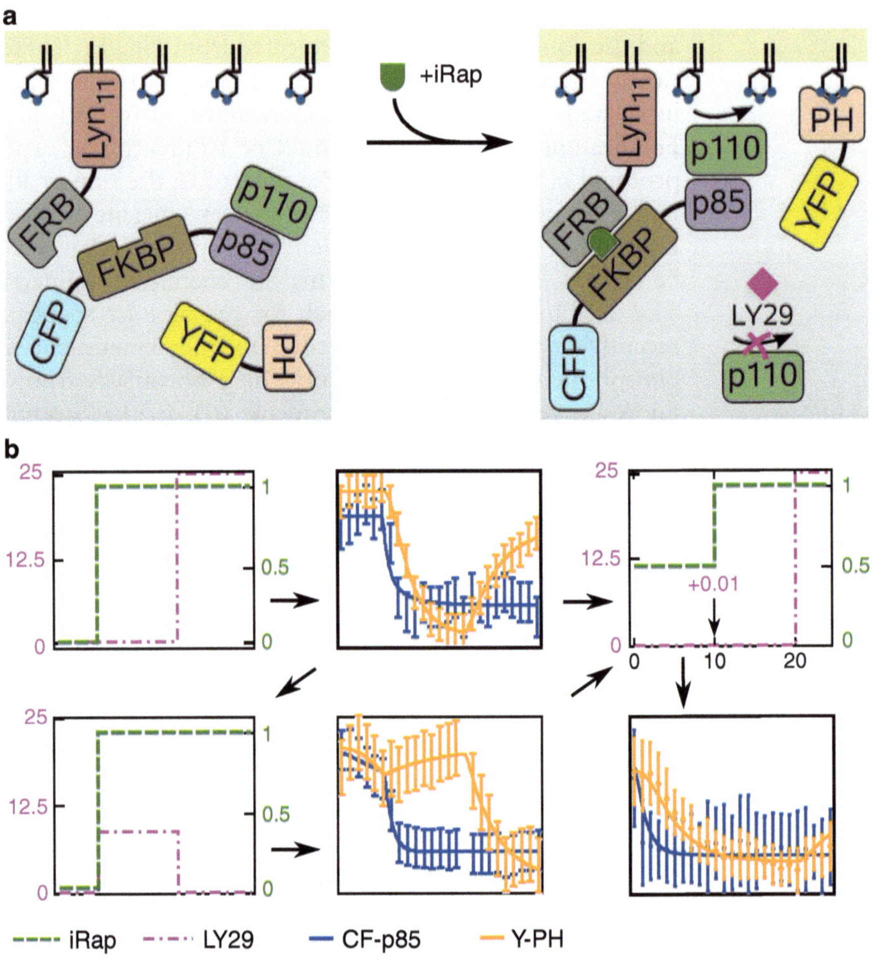

Fig. 3. Experiment design for parameter estimation. (**a**) Experimental setup for the study of PH domain translocation in response to PI3K activation. iRap-induced heterodimerization of FRB and FKBP domains recruits CFP-FKBP-p85 to the plasma membrane where it stimulates PI3K activity. A fusion construct of YFP and the PH domain of Akt can be used for observing the accumulation of PIP_3 in the plasma membrane. (**b**) Genealogy of experiments shows how experiments were derived from previous data. Step-like traces illustrate experimental protocols as concentration–time profiles of iRap (*green*) and LY29 (*pink*). Traces with error bars show the experimental results as fluorescence intensities measured in the cytosol of single cells over time. Decline of cytosolic fluorescence indicates translocation of CFP-FKBP-p85 (*blue*) or YFP-PH (*orange*) to the plasma membrane. Figures from (6).

constants k_r and k_t for the recruitment of the regulator peptide, the degradation constant ϕ of PIP_3, the rate constants k_a and α for the recruitment of the PH domain of Akt to PIP_3-enriched membranes, and the inhibitor constant K_{LY}. The other parameters, $[Lyn]_{tot}$, $[p85]_{tot}$, and $[PH]_{tot}$, were direct results of transfection efficiency, and varied from cell to cell.

The initial experiment, as a reference, followed first an intuitively chosen protocol, in which iRap was added after 5 min. iRap-induced recruitment of CFP-FKBP-p85 to the plasma membrane stimulated the production of PIP_3 as was observable by accumulation of YFP-PH. Ten minutes later, inhibition of PI3K by LY29 caused the dissociation of YFP-PH into the cytosol (Fig. 3b). The model was then calibrated against these trajectories, and used for planning a subsequent experiment. This next experimental protocol was found by minimizing the trace of the predicted parameter covariance matrix as a function of the different concentrations of drugs in the three segments of a time course. Interestingly, the optimized protocol involved the recruitment of CFP-FKBP-p85 in the presence of LY29, effectively separating the translocation kinetics of the regulator peptide from the production of PIP_3 which occurred only after LY29 was washed out in the final segment of the experiment (Fig. 3b). The parameter estimates of the model were then refined by a collective fit against the data from the intuitive and the first optimized experiment. The refined model was then used to plan a third experiment, illustrating how previous knowledge can be used in a sequential approach to narrow down the range of possible parameter values. Figure 3b shows the genealogy of how each experimental protocol was derived from previous data. Starting from the intuitively planned experiment, the sum of parameter variances was reduced 60-fold by conducting two computationally designed experiments alone.

3. Notes

The two approaches outlined here aim to close a cycle from model construction back to the laboratory bench, where the quantitative nature of the processes studied begins to require more than what intuition can provide. The numerical optimization of experimental designs based on dynamic models can result in experimental protocols that are useful to distinguish between alternative hypotheses or to reveal kinetic properties that are difficult to measure directly. The major challenge of these techniques is that only uncertain model predictions are available for the design of experiments. It is therefore important to understand these approaches as steps of an iterative process in which models are being selected, refined, or discarded with every emerging set of experimental data.

Acknowledgements

We would like to thank Johannes Schlöder, Sean Collins, Meredith Betterton, and Xuedong Liu for helpful comments on this chapter.

References

1. Gutenkunst RN, Waterfall JJ, Casey FP, Brown KS, Myers CR, Sethna JP (2007) Universally sloppy parameter sensitivities in systems biology models. PLoS Comput Biol 3(10):1871–1878
2. Bock HG (1987) Randwertproblemmethoden zur Parameteridentifizierung in Systemen nichtlinearer Differentialgleichungen. Bonner Mathematische Schriften, 183
3. Bock HG, Kostina E, Schlöder JP (2007) Numerical methods for parameter estimation in nonlinear differential algebraic equations. GAMM Mitt 30(2):352–375
4. Chen BH, Asprey SP (2003) On the design of optimally informative dynamic experiments for model discrimination in multiresponse nonlinear situations. Ind Eng Chem Res 42(7):1379–1390
5. Bauer I, Bock HG, Körkel S, Schlöder JP (2000) Numerical methods for optimum experimental design in DAE systems. J Comput Appl Math 120:1–25
6. Bandara S, Schlöder JP, Eils R, Bock HG, Meyer T (2009) Optimal experimental design for parameter estimation of a cell signaling model. PLoS Comput Biol 5(11):e1000558

Chapter 8

Mathematical Modeling of Biochemical Systems with PottersWheel

Thomas Maiwald, Oliver Eberhardt, and Julie Blumberg

Abstract

The program PottersWheel has been developed to provide an intuitive and yet powerful framework for data-based modeling of dynamical systems like biochemical reaction networks. Its key functionality is multi-experiment fitting, where several experimental data sets from different laboratory conditions are fitted simultaneously in order to improve the estimation of unknown model parameters, to check the validity of a given model, and to discriminate competing model hypotheses. New experiments can be designed interactively. Models are either created text-based or using a visual model designer. Dynamically generated and compiled C files provide fast simulation and fitting procedures. Each function can either be accessed using a graphical user interface or via command line, allowing for batch processing within custom Matlab scripts. PottersWheel is designed as a Matlab toolbox, comprises 250,000 lines of Matlab and C code, and is freely available for academic usage at www.potterswheel.de.

Key words: Computer Simulation, Systems Biology, Computational Biology, Biological models, Signal Transduction, Methods, Software, Modeling framework, Matlab toolbox, Parameter estimation, Multi-experiment fitting, ODE

1. Introduction

Modeling aims to formulate a set of mathematical equations which are able to predict computationally the dynamic behavior of a real system, e.g., of a biochemical network. Entities of the system should preferably have a representation within the model in contrast to black-box modeling. This approach deepens the understanding about the network during the process of hypothesis generation, mathematical formalization, and comparison with experimental data potentially leading to new hypotheses. At the same time, laboratory measurements too complex to be understood by visual inspection are brought into relation. Beyond that, a formalized

Xuedong Liu and Meredith D. Betterton (eds.), *Computational Modeling of Signaling Networks*, Methods in Molecular Biology, vol. 880, DOI 10.1007/978-1-61779-833-7_8,

model is helpful to unambiguously communicate a hypothesis to colleagues. Finally, a suitable model renders possible new approaches to identify therapeutic strategies and to assess the feasibility of a research agenda minimizing the cost in terms of animals, time, and money.

Modeling approaches differ in the choice of possible model structures. Regression models use algebraic equations (1), Boolean models comprise logical gates (2), Bayesian networks (3), and stochastic models (4, 5), which are based on probability distributions and mechanistic models are expressed using either ordinary differential equations (ODEs) if spatial effects are neglected (6) or partial differential equations (PDEs) in the more general case (7). Each approach implies assumptions about the system at hand, requires a certain type of measurements, and permits conclusions at different levels of detail and accuracy. Moreover, the burden of computational implementation and analysis varies strongly between the modeling concepts.

In this chapter, we focus on mechanistic modeling of biochemical networks using ODEs and the PottersWheel modeling framework (8). This approach assumes a sufficiently large number of molecules not below 1,000 and no spatial effects, requires time-resolved measurements, allows predicting transient system responses to changed environmental conditions, and often simplifies interpretation due to a one-to-one relationship to physiological entities. The numerical effort lies between regression/Boolean models and stochastic/PDE models and can usually be handled by a single PC or a small computer cluster, given that suitable algorithms are applied.

In the following, we introduce key concepts and the PottersWheel terminology based on the JAK/STAT signal transduction pathway as an example for a biochemical reaction network (9).

1.1. Reactions, Variables, and Parameters

In the simple irreversible reaction "A to B" assuming mass action kinetics, the reaction rate v_R is proportional to the time-dependent concentration of species A, denoted as [A](t), often abbreviated as A(t), neglecting the square brackets, or just A. The proportionality is expressed using a rate coefficient k, leading to the following ODE for the flux of the reaction:

$$v_R = \frac{d[A]}{dt} = [\dot{A}] = -k.\ [A] \text{ or just } \dot{A} = -k.\ A.$$

The time-dependent concentration for A is thus simply an exponential decay starting from an initial concentration at $t = 0$:

$$A(t) = A_0\ .\ e^{-k\tau}.$$

In a more complex situation, the above analytical solution is not feasible. Instead, *numerical integration* is used to determine the

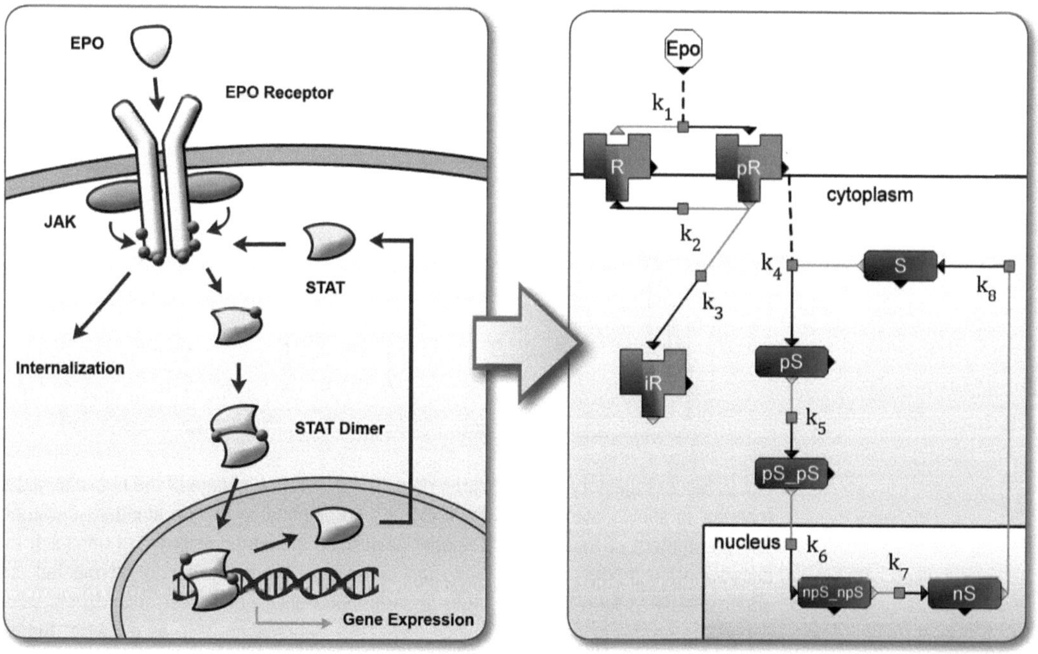

Fig. 1. Mathematical formalization of a biological hypothesis. The *left panel* shows a cartoon of the JAK/STAT signal transduction pathway. The *right panel* displays the related formalized and simplified mathematical model created and visualized using the PottersWheel model designer. In contrast to the cartoon, the receptor appears in the model three times: once for each modification state: the free receptor (R), the phosphorylated receptor (pR), and the internalized receptor iR. Every reaction is shown as a *small rectangle* with incoming fluxes in *light gray* and outgoing flux to the products in *dark grey*. *Dashed lines* indicate the effects of enzymes which are not consumed or produced within the reaction, also called modifiers. S denotes STAT. The prefixes p and np define phosphorylation in the cytoplasm and the nucleus, and the underscore "_" represents the bond of two species.

concentration–time profiles. Several kinetic laws apart from mass action can be used, e.g., Michaelis–Menten and Hill kinetics (10).

Figure 1 visualizes the formalization of a biological hypothesis into a reaction network. PottersWheel automatically creates a corresponding set of differential equations, in this case

$$\dot{R} = -k_1 \times R \times \text{EPO} + k_2 \times \text{pR},$$

$$\dot{\text{pR}} = -k_2 \times \text{pR} + k_1 \times R \times \text{EPO} - k_3 \times \text{pR},$$

$$\dot{\text{iR}} = +k_3 \times \text{pR},$$

$$\dot{S} = -k_4 \times S \times \text{pR} + k_8 \times \text{nS},$$

$$\dot{\text{pS}} = -2 \times k_5 \times \text{pS} \times \text{pS} + k_4 \times \text{S} \times \text{pR},$$

$$\text{npS}_{\text{npS}} = -k_6 \times \text{npS}_{\text{npS}} + k_5 \times \text{pS} \times \text{pS},$$

$$\text{npS}_{\text{npS}} = -k_7 \times \text{npS}_{\text{npS}} + k_6 \times \text{npS}_{\text{npS}},$$

$$\dot{\text{nS}} = -k_8 \times \text{nS} + k_7 \times 2 \times \text{npS}_{\text{npS}}.$$

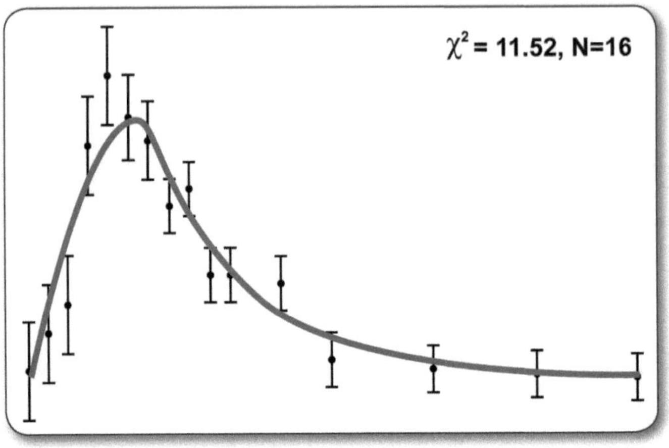

Fig. 2. Model trajectory and measured data. The experimental data of the phosphorylated receptor is shown over time with error bars corresponding to one standard-deviation. The parameters of the JAK/STAT model have been calibrated in order to minimize the distance of the model trajectory (*solid line*) to the data. The model cannot be rejected as a wrong model as long as the χ^2 (Chi-Square) value does not exceed a threshold value depending on the desired significance level (see Subheading 3.3). As a rule of thumb, the Chi-Square value should not exceed the number of fitted data points *N*. Whether the model is "true" cannot be concluded, as every mathematical formulation of processes in nature is only an idealization.

The network is equivalent to the differential equations and describes the structure of the mathematical model, also called its *topology*. Together with specific values for the dynamic parameters $k_1, \ldots, k_8$, initial values for each species $R_0 = R(t = 0), \ldots, \text{nS}_0 = \text{nS}(t = 0)$, and information about the external EPO stimulation for each time point, EPO(t), the model can be integrated, i.e., the concentration–time profiles can be determined computationally as shown in Fig. 2. Table 1 provides an overview of important model elements.

1.2. Key Modeling Questions

Often, the underlying model structure of a real biochemical network is unknown. In addition, usually only a subset of system variables can be measured. And since experiments are time-consuming and expensive, the question arises which measurements will provide the most information. This setting leads to several challenges:

1. Model discrimination: How can we distinguish competing model hypotheses?
2. Parameter calibration: Which parameter values are most probable given the experimental data? Do some parameters compensate each other's effect?
3. Experimental design: What is the most informative next experiment given the research question?

A key approach to address the above challenges is multi-experiment fitting, where a single model is fitted to several experiments at the same time.

Table 1
Overview of important model elements

Element	Name	Description	General example	Examples in JAK/ STAT model
X	Dynamic variable	All model species which occur as a reactant or product	Cytoplasmic and nuclear proteins	R, pR, iR, S, pS, pS_pS
U	Driving input variable	Externally controlled species only occurring as modifiers	Ligands, drugs	EPO
Y	Observation variable (observable)	Sum of dynamic variables that can be measured experimentally	Proteins detected using Western blotting	pS_obs = pS + 2 * pS_pS
Z	Derived variable	Dynamic variables or functions thereof of special interest	Group of proteins which has not been measured but is useful for analysis	total_nS = nS + 2 * npS_npS
C	Compartment	A distinct reaction space	Cell compartments like cytoplasm or organs like liver	Cytoplasm, Nucleus
K	Dynamic parameter	A parameter occurring in a reaction	Rate and dissociation constants	$k_1, k_2, k_3, \ldots$
X_0	Start value parameter	Value of dynamic variables at time point zero	Initial protein concentration before stimulation	$R(t=0)$, $S(t=0)$
S	Scaling parameter	Scaling between observable units in experiment and model	Relative quantification of a western blot or qPCR	scale_pS_obs

Variables are time-dependent quantities, whereas parameters have a time-independent value which can be calibrated by PottersWheel

2. Material

The PottersWheel software requires Matlab 2006a or higher on any Linux, Mac, or Windows PC. The Matlab optimization toolbox and a few external programs are recommended, depending on the intended use of PottersWheel. A detailed installation description is given at *www.potterswheel.de*, where the software can freely be downloaded for academic usage.

3. Methods

The basic use of PottersWheel covers seven phases from model creation to the prediction of new experiments as shown in Fig. 3. In the following, we summarize all necessary steps for each phase assuming a simple case scenario. The use of advanced modeling techniques is described in the PottersWheel manual and in the online help, both available at *www.potterswheel.de*.

3.1. Notation

Menu items are identified using the scheme "Window | Menu-main item | menu-subitem" where Window corresponds to either PW for the PottersWheel main user interface, EQ for the equalizer, and MD for the model designer. To create a new model, use, e.g., the menu-item "PW | Models | New model. . .."

Configuration settings have a special notation which equals their command line usage for experienced users. For example, "config.integration.useFastIntegration" corresponds to the checkbox "useFastIntegration" in the dialog "Configuration | Fields | integration." To open the configuration dialog itself, use "PW | File |

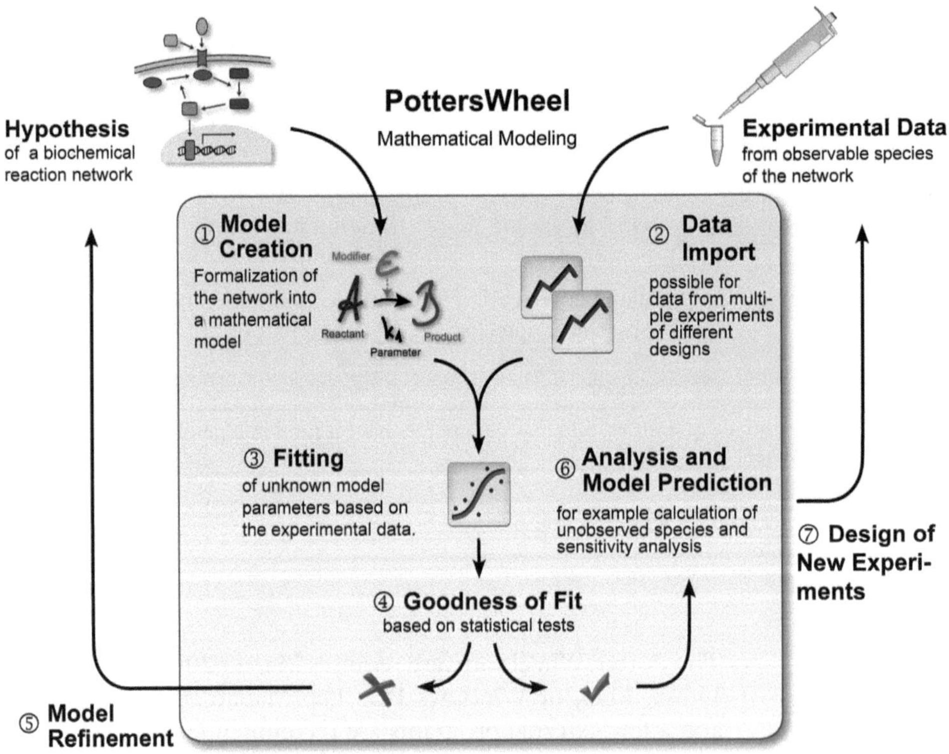

Fig. 3. The seven modeling phases from model creation to the design of new experiments.

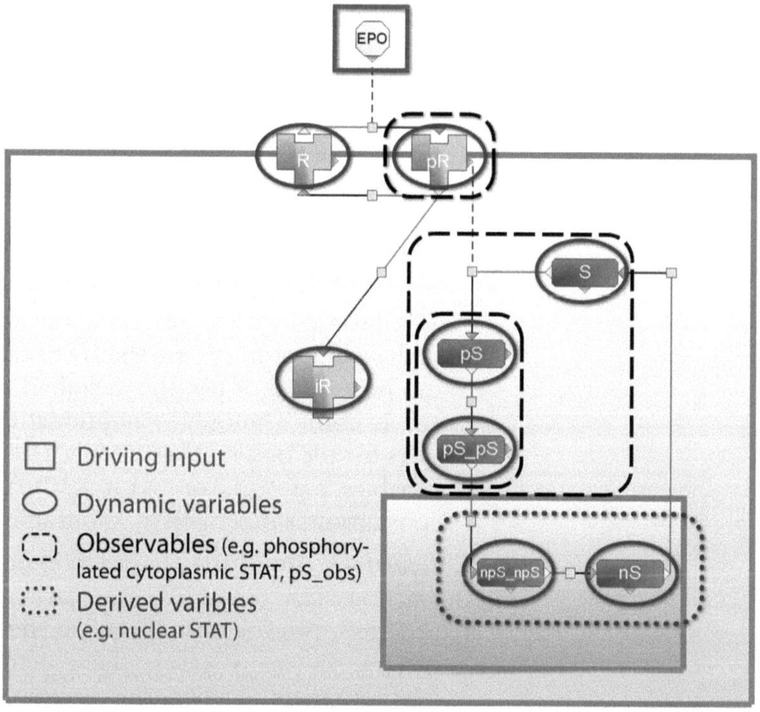

Fig. 4. Variables classes. The concentration of dynamic variables (*solid circles*) is not externally controlled like driving inputs (*solid box*). Observables (*dashed boxes*) are the sum of one or more species, potentially multiplied by the number of related bounding partners, e.g., pS_obs = pS + 2 * pS_pS. Derived variables (*dotted box*) have no special experimental meaning, but can be plotted separately within PottersWheel, e.g., to focus on a subset of interesting variables.

Configuration dialog," the corresponding button, or the command `pwConfiguration`.

A type-writer font is used to refer to PottersWheel functions, e.g., `pwAddData`. Further information is available within the online function reference at www.potterswheel.de or using "help <functionName>" in the Matlab command window, in this case "`>> help pwAddData`."

3.2. Model Creation

Based on a hypothesis regarding a biochemical reaction network, a mathematical model can be created and loaded into PottersWheel. We describe here the approach using first the model creation wizard and then the visual model designer. It is assumed that all reactions follow mass action kinetics. As shown in Fig. 4, as a preparation the different variable classes must be identified, especially the dynamical, driving input, and observation variables.

1. *Starting the model wizard.* Open the model creation wizard using "PW | Models | New model. . . ." Enter a model ID using only letters, numbers, and the underscore and select a reaction-based model type.

2. *Number of variables and parameters.* Specify the number of dynamic variables X. These are all species in the model which are not controlled externally. Note that further species can be added later. Then, specify the number of dynamic parameters k comprising, e.g., rate constants, the number of compartments C, e.g., for cytoplasm and nucleus, driving inputs U, i.e., experimentally controlled species as, e.g., a drug concentration, and the number of observables Y corresponding to measurable species. See Table 1 for further details.
3. *IDs.* Proceed with "Create and open in designer." A sequence of dialogs appears, where the ID of each variable and parameter can be entered, if not the default ID as, for example, X1, X2, etc. is used. The only nonoptional entries are the right hand sides for the observables, usually a single or the sum of a few X variables, e.g. "X1" or "X1 + X2." The recommended naming convention is described in Subheading 3.8.
4. *Reactions.* Afterward, the reaction designer opens and a set of reactions can be added. For each reaction, select reactants, products, possible modifiers like enzymes, and involved parameters. In many cases, the reaction type can be left to "A (automatic)," where the reaction rate is assumed to be the product of all reactants and modifiers multiplied by the first parameter. If two parameters are given, the reaction is assumed to be reversible and the back-reaction rate is given by the product of all products and the second parameter. Checkmark "None" if no reactant, product, or modifier takes part in the reaction.
5. *Model designer.* After closing the dialog, the model appears in the model designer. Here, the position of each species and reaction can be changed via drag and drop. The context menu of each object allows specifying or refining object properties like color, initial value, and fit setting.
6. *Model editor.* Saving (PW | Model | Save) and opening the model file in the Matlab text editor (PW | Model | Edit model file) allows to investigate the textual representation of the model. As shown in Fig. 5, changes in the editor are reflected synchronously in the model designer and vice versa using the buttons "reload" and "save" in the visual model designer.
7. *Model repository.* The model is listed in the repository list of the main user interface. Now, either experimental data can be added (see Subheading 3.2) or synthetic data is simulated.

Advanced modeling techniques comprise rule-based modeling, where placeholders are used within a reaction representing several modification states of the same species. PottersWheel

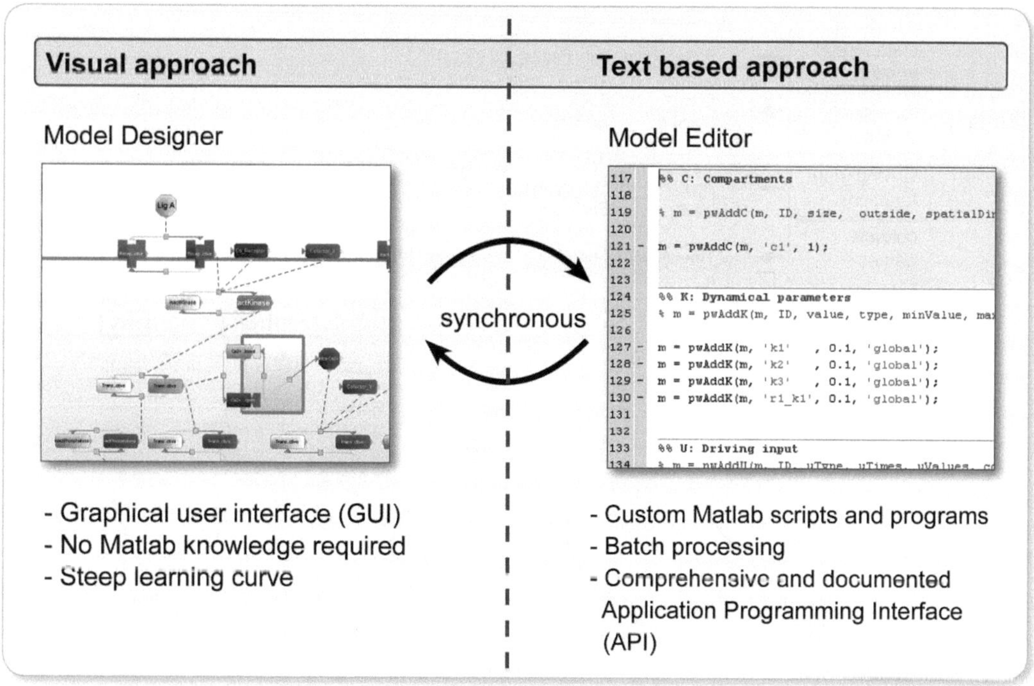

Fig. 5. Visual and text-based approach. PottersWheel functionalities can be accessed either using graphical user interfaces or by command line to support batch processing and creation of custom Matlab scripts based on the PottersWheel function library. The model designer is, for example, tightly connected to the textual representation of a model and the user can decide any time to change the model in the designer or editor.

automatically creates all reactions that match the used pattern (see Subheading 3.8).

3.3. Data Import

External data saved in text or Excel files can be added to a loaded model. In contrast to Excel files (*.xls), in text files (*.txt) all lines without numeric data have to be marked with a "#" sign as the first character.

File preparation. As shown in Fig. 6, each measurement file contains two sections: meta-information and a data matrix. The meta-information comprises the used data format version, individual comments, formatted information about the experimental setting, and the column names of the data matrix.

3.3.1. Meta Information

Excel cells in the first column in the meta-information section may only consist of text, e.g., "1. Foo bar" but not "1." The first line of the meta-information describes the used data format. Here, "PottersWheel Data Format 2.0.55" denotes that PottersWheel version 2.0.55 is required to load the data. Afterwards, individual comments can be used to describe the measurements and will be

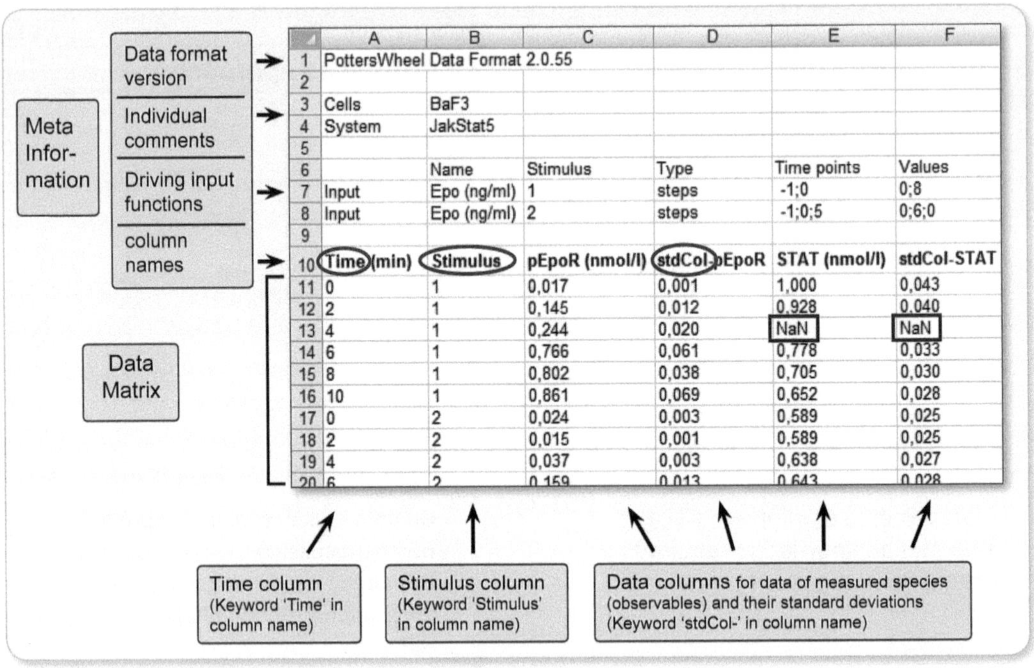

Fig. 6. Structure of a PottersWheel data file. The header of the data file contains meta-information about the experimental data consisting of comments, the experimental stimulation settings, and the data column names, whereas the data matrix comprises the values for time, stimulus number, and the measurement observables. In the example illustrated, the stimulation settings describe one input function starting with 0 (no stimulation) and switching to 1 (full stimulation) a time 0 with no subsequent changes expressing a constant stimulation. Please note that it is possible to define several input functions, which are distinguished by their stimulus number and will be assigned to the experimental data in the stimulation column. Only rows starting with the keyword "Input" will be recognized as information for the stimulus input function. In order to be recognized as a time and stimulus column, the column names have to start with the keywords "Time" and "Stimulus," respectively. Data points with no values are denoted by "NaN" (not a number).

ignored by PottersWheel unless a cell in the first column contains the keyword "Input" or "MCode" (case insensitive).

In the input paragraph, the experimental setting can be defined represented by a driving input functions for all externally controlled system species. The expected cells of an "Input" line are:

"Input" | <Name> | <Stimulus> | <Type> | <Time points> | <Values>

In order to improve readability, a header line above the input paragraph may be used as shown in the example sheets. This line will be ignored by PottersWheel. The <Name> text field is the name of the driving input function, often the same as in the model (see pwAddU). The <Stimulus> integer field defines the stimulus number of the described setting. The <Type> text field denotes the type of the input function, e.g., a "linear" or "steps" interpolation. The two fields <Time points> and <Values> comprise a semicolon separated list of numbers defining points of the interpolation: When has the driving input which value? The number of time-points and values is arbitrary but must be the same.

This approach assumes that the input is given as an analytic function. If measured values exist for the external species, they are saved as a normal data column and mapped during data loading to the corresponding input variable u.

For example, a pulse-wash experiment of 10 min length and an amplitude of 5 would be formally written with type "steps," time points "−1; 0; 10" and (step-) values "0; 5; 0." Please note that Excel requires a blank before a minus sign, so that the cell content cannot be mistaken as a mathematical formula.

Finally, the last row of the meta-information should be the column names of the data. Allowed characters for the species names are A-Za-z0-9, blank, and the underscore. The measurement unit is optional and has to be specified after the species name in round brackets, as in "pSTAT (nmol/l)."

3.3.2. Data Matrix

The first line with a number in the first column marks the beginning of the data matrix. It is not allowed to have a "NaN" (not a number) entry in the upper left corner of the matrix. This can be avoided by using "Time" as the first column.

One row of the data matrix represents the measured protein concentration for one time point and one stimulus number. One column comprises the time points and should have a column name starting with "Time," again with an optional specification of the time unit, e.g., "Time (min)". Another column named "Stimulus" refers to the stimulus number. If no stimulus column is specified, all lines are assumed to belong to stimulus 1.

The remaining columns contain the measured data. Missing values can be denoted as "NaN."

When are two Excel files used to describe two experimental settings instead of one Excel file with two stimulations? Answer: Two Excel files can always be used. One file with two stimulations can only be used if both experimental settings have been measured using the same (relative or absolute) units. Then, two data values of the Excel file for the same species can be compared. This situation occurs, e.g., when different experimental conditions are measured using the same Western blotting gel.

3.3.3. Measurement Error

Each data-point has an individual weight, i.e., importance, during parameter calibration. Data points with a larger uncertainty, i.e., standard deviation, "attract" the model trajectory less strong. Assuming Gaussian distributed errors, the weight is proportional to the inverse standard deviation. The statistical interpretation of a fit based on the Chi Square value strongly relies on realistic standard deviations. They are either given explicitly in the data matrix using the prefix "stdCol-" in the corresponding column name, e.g., "stdCol-pEpoReceptor," are determined using an error model as given in the model file, or will be estimated by PottersWheel based on the measured data (see `pwAddData`). Please note that PottersWheel

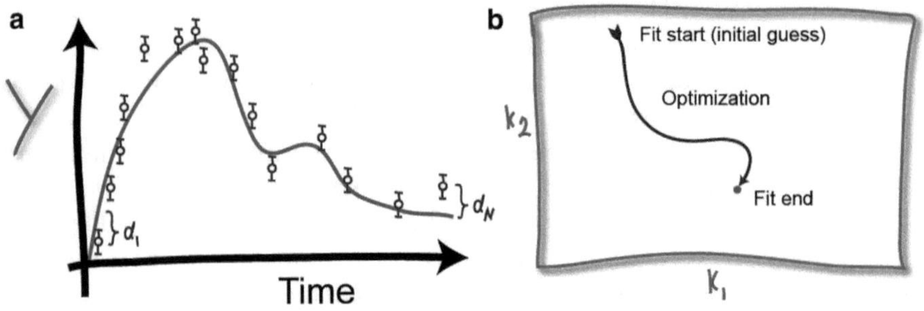

Fig. 7. Parameter calibration. Given an initial guess, unknown model parameters, e.g., k1 and k2 (panel (**b**)), are changed during optimization in order to decrease the distance of *N* measured data points $d_1, \ldots, d_N$ to the model trajectory (*solid line, panel a*), quantified using the Chi-Square value (see text).

currently assumes Gaussian distributed errors, so that data with a different error distribution must be transformed accordingly (11).

3.3.4. Creating a Model-Data Couple

After loading the model, use "PW | Data | Add data" to add the data file to the currently selected model. PottersWheel will import all given data columns to the model observables. If the columns in the external file differ from the observable IDs, a mapping dialog appears. Here, the mapping has to be specified by the user and can be saved for automatic name mapping in the future.

3.4. Fitting

All unknown model parameters comprising start values x_0, dynamic parameters k, and scaling factors s which are not marked as "fix" in the model definition file or in the "PW | Couples | Fit settings" dialog will be optimized by PottersWheel based on an initial guess until the experimental measurements match the model trajectories as best as possible, i.e., to minimize the "Chi-Square" value, given as

$$\chi^2 = \frac{d_1^2}{\sigma_1^2} + \cdots + \frac{d_N^2}{\sigma_N^2}$$

with d_i being the distance between model trajectory and data point i, and σ_i its standard deviation (see Fig. 7). With multi-experiment fitting, it is possible to fit several data sets at once, which is more challenging for the model, i.e., unsuitable models are easier to detect and to rule out. The parameter calibration procedure has many knobs to be twiddled depending on the given model and data. A good starting point is to use the RADAU5 (12) or CVODES (13) integrator in conjunction with the trust region optimizer in logarithmic parameter space.

1. *Integration settings.* Open the configuration dialog and change "config.integration.integrator" to either RADAU5 or CVODES.

2. *Assembly creation.* Select one or more model-data-couples in the upper list of the main PottersWheel user interface. Use "PW | Couples | Combine" to combine the selection into one assembly which appears in the lower list.
3. *Fit settings and optimizer.* Choose "PW | Couples | Fit settings" to define which parameters should be fitted. Then, open the equalizer and select "EQ | View | Fitting." Here choose, e.g., the trust-region optimizer in logarithmic parameter space.
4. *Applying a fit.* Arrange all figures using "EQ | Figures | Arrange primary figures" and start fitting (Button: Single fit). Depending on the model size, the fit will take a few seconds up to several minutes. The calibrated model can be saved using "PW | Couples | Save fitted model."

3.5. Goodness-of-Fit

The quality of a fit is characterized by the Chi-Square value (see Subheading 3.3). As a rule of thumb, for N fitted data points and p calibrated parameters, the Chi-Square value should have a similar value as N–p or at least N. Statistically this is expressed in a p-value above a significance threshold of, e.g., 0.05. For lower p values, the model is not able to explain the data and has to be refined (see Subheading 3.5), the standard deviation of the data points should be larger, or the used fitting strategy was not successful and the fit was trapped in a local minimum. A model-fit usually depends on the initial guess for the parameters, as depicted in Fig. 8. If the Chi-Square value of each fit is similar, the model might suffer from a so-called non-identifiability, where several parameters compensate each other's effect, e.g., an increased value for k_1 and a decreased value for k_2 may lead to the same observation time profile.

1. *Calculating the goodness-of-fit.* After each fit or parameter change, PottersWheel displays several goodness-of-fit measures on the command line. The values p-value(N) and p-value(N–p) are p-values based on Chi-Square tests with N or N–p degrees of freedom, with N being the number of fitted data-points and p the number of fitted parameters. If p-value(N) is smaller than a significance level, e.g., 0.05, the model has to be rejected. If this is only the case for p-value(N–p), the model has to be rejected if p equals the number of degrees of freedom in the fitted model. Further measures are the Akaike (14) and Bayesian information criterion (15).
2. *Improving the fit-quality.* To achieve a better fit, try one or more of the following:
   (a) Improve the fitting accuracy by reducing the tolerance for changes in the Chi-Square value "TolFun" and in the parameters "TolX" to 1e–8 and increasing the maximum

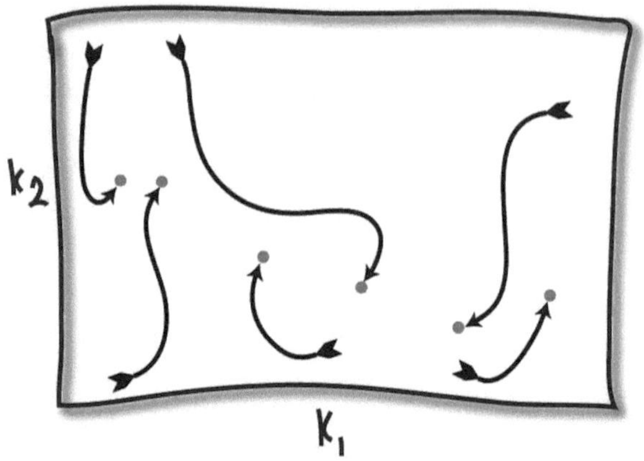

Fig. 8. Fit-sequence. Depending on the initial guess, the optimization procedure may end up in different positions in parameter space if the model comprises nonidentifiable parameters.

number of iterations "maxIter" in the configuration dialog ("optimization.trustregion").

(b) Switch between the RADAU5 and CVODES integrators.

(c) Use the Matlab symbolic math toolbox to enable PottersWheel to determine analytical derivatives for integration and optimization.

(d) Use the sliders of the equalizer to manually find a better initial guess for the parameters.

(e) Fit with simulated annealing (ASAMIN (16, 17)), scatter search (SSM (18)), or a boosted fit, selectable in the fitting view of the equalizer.

(f) Use trust-region in logarithmic parameter space within a fit-sequence (see Fig. 8) from equally sampled initial guesses throughout the parameter space ("optimization. useQRNG").

3. *Pairwise linear identifiability analysis.* Based on a valid fit, the so-called identifiability of the calibrated parameters should be investigated. In many cases, the amount of measured data is not sufficient to unambiguously determine parameter values, i.e., a forced change to one parameter may be compensated perfectly by other parameters leading to the same goodness-of-fit. "View | Fit sequence analysis," a linear identifiability analysis is applied using the fit-sequence F2 with $n > 100$ fits, preferably >300 fits.

4. *Multivariate nonlinear identifiability analysis.* The MOTA algorithm can be used to identify groups of nonlinearly related parameters after a fit sequence has been applied (19). The PLE method provides a means to distinguish between practical and

structural non-identifiabilities (20). In the former case, a better quality of measurements may resolve the identifiability problem. For the structural case, the model should be reparameterized or additional species should be measured, preferably under new experimental conditions. PLE is available under "View | Single Fit Analysis" in the equalizer.

3.6. Model Refinement

If the model structure is not able to explain the experimental measurements, create a set of physiologically reasonable alternative models. In order to avoid redundant model paragraphs and copy-and-paste errors, this could be done using a common core-model which is the same for all variants. Then, "daughter"-models are created (see `pwGetEmptyModel`). Each model variant should be fitted to the data, preferably using batch processing strategies based on PottersWheel macros (see Subheading 3.8). As a starting point to envision suitable model variants, use the PottersWheel equalizer to understand the dynamic behavior of the original system (see Subheading 3.6).

3.7. Analysis and Model Prediction

A mathematical model can be used for example to display the concentration–time profile of unobserved species, to determine sensitive parameters which could serve as a target within a clinical setting, and to calculate further model characteristics like the half-life of a species. A common analysis could be:

1. *Open the PottersWheel equalizer.* Combine one or more model-data-couples into the lower list of the main user interface. Then open the equalizer using "PW | Assembly | Open equalizer." The PottersWheel equalizer is the key user interface for interactive model analysis.
2. *Changing parameter values.* Each parameter can be changed using a pair of sliders. The right slider changes the order of magnitude and the left slider can be used for fine-tuning—it changes the mantissa. Alternatively, the parameter value can be entered directly using the text field. The original values of the parameters are available via the reset parameter button (see `pwReset`).
3. *Displaying A, U, X, Y, Z variables.* PottersWheel allows for plotting different classes of variables at the same time, which are algebraic variables A, driving inputs U, dynamic variables X, observables Y, and derived variables Z (see Subheading 3.1). Use the corresponding toggle buttons.
4. *Grouped and hidden variables.* Using "EQ | View | Plotting," several variables can be grouped and will be shown within the same subplot, offering direct comparison of trajectories. Uninteresting variables can be hidden. The manual plot settings can be saved to hard disk or into the original model file for later use.

5. *Phase space analysis.* Using "EQ | View | Phase space," the concentration profile of two variables can be plotted against each other in a phase space diagram. For related variables, a potentially complex dynamic behavior can often be seen in a simpler form, e.g. a circle or a line.
6. *Figure handling.* The figure menu provides several functions to arrange or copy figures and subplots. Especially when working with many trajectories, it can be useful to copy a single subplot into a separate figure. Use the figure saving button to store the current figures into various formats which can be specified in the configuration, e.g. "config.plotting.saveJPG." Alternatively, activate the figure toolbar and menu for individual saving (config.plotting.toolBar / menuBar).
7. *Sensitivity analysis.* Using "PW | Models | Sensitivity Analysis" in the main user interface, several sensitivity analyses are available to determine how strong a variable or a characteristic value like the time of the maximum depend on a parameter or an initial value.
8. *Residual analysis.* Using "EQ | View | Single Fit Analysis" in the equalizer, the distribution of residuals can e.g. be compared to a Gaussian distribution. A suitable model should not suffer from systematic errors.
9. *Chi-Square landscape.* Again using "EQ | View | Single Fit Analysis," the Chi-Square can be calculated systematically in two or three dimensions of the parameter space around the current position. This is a useful visualization of non-identifiability manifolds.

3.8. Design of New Experiments

An experimental setting corresponds to specific characteristics of the driving input functions and initial concentrations. In the JAK/STAT system the concentration of EPO is controlled experimentally. The driving input designer allows investigating the effect of a continuous, ramp, or pulse stimulation as shown in Fig. 9 in

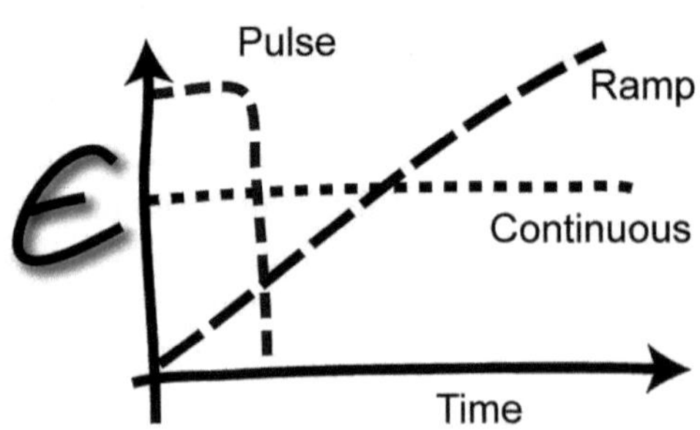

Fig. 9. Stimulations. Depending on the used cell-system continuous, pulsed and ramp ligand concentrations, here of an enzyme E, may be feasible. In order to obtain different responses by the system, it is important to apply qualitatively different driving inputs, i.e., stimulations.

combination with varying initial concentrations using the equalizer (EQ | View | Driving input designer). In order to discriminate competing model hypotheses, the next experiment should have as different observable time-profiles as possible which can be verified computationally.

3.9. Combinatorial Complexity, Rule Based Modeling, and the PottersWheel Naming Convention

Suppose an enzyme E in n different states $E_1, \ldots, E_n$ which triggers the reaction "A to B," independent of the current state of E itself, leading to n similar reactions. If protein A possesses m different states, the number of reactions explodes to $n \times m$, which is called combinatorial complexity (21, 22). The corresponding model definition file would be difficult to maintain and is prone to copy-and-paste errors. Therefore, PottersWheel supports rule based modeling, where reactions may contain placeholders that are replaced automatically by all matching species. This approach requires that variable IDs follow the recommended PottersWheel naming convention:

1. Basic, i.e., unmodified and unbound species start with a capital letter, e.g. "Erk."
2. Modifications are lowercase prefixes, e.g., "ppErk."
3. Complexes-bounds are denoted by an underscore "_."
4. The order of species in a complex is relevant, i.e., "ppMek_Erk" is not equivalent to "Erk_ppMek."

For further information, see `pwAddR` and `pwTutorial_Rule_based_modeling`.

3.10. Configuration

To allow for a detailed individual configuration of PottersWheel and the applied algorithms, hundreds of settings can be adjusted either within the configuration dialog or directly from command line. Elements in the configuration dialog share the same name as required for command line usage, which enables the user to quickly becoming an experienced command line user and thereby accelerating creation of custom Matlab script, e.g., for batch processing. Changes to entries are affecting PottersWheel instantaneously. Closing of the dialog is not necessary, but consequently canceling is not possible either. Use the file saving and opening menu to store preferred settings for use. Within own programs or in the command line, all configuration settings are accessible, e.g.:

```
>> config = pwGetConfig;
>> config.integration.useJacobian = true;
>> pwSetConfig(config);
```

This is especially useful if you are working with PottersWheel macros, since all settings, modeling, fitting, and analysis commands are documented in the same file.

3.11. Macros and Custom Matlab Programs

Since PottersWheel is created as a Matlab toolbox, the user is not restricted to graphical user interfaces and by creating own Matlab programs employing the PottersWheel function library (API—application programming interface) he or she may gain and improve his or her competence in a major and widespread numerical framework. As exemplified for the model designer and the model editor in Fig. 5, PottersWheel allows for applying mathematical modeling using a visual or text based approach, targeting the beginning as well as the advanced user. All PottersWheel functions are available as Matlab functions named as "pw...." A macro, i.e., a script based on the PottersWheel library to fit a model to a dataset could be written as follows:

```
pwAddModel('Model1.m');
pwAddData('Experiment1.xls');
pwCombine;
pwArrange;
pwFit;
```

When saving these lines into a file named "myFittingScript.m," each command will be executed sequentially after typing in the Matlab command window

```
>>myFittingScript
```

Using macros is not only useful to apply batch-processing but also ensures that each modeling step is documented and can be reproduced at any later time.

3.12. Reporting

Each analysis or visualization step can be added as a section to a report. The Report Designer user interface allows removing and reordering of sections, as well as displaying only those figures which are related to the section. Finally, a report can be created either as a Latex-based PDF, a Microsoft Word document, or as an HTML file. Several predefined reports can be created in the report menu (PW | Report), e.g., to summarize the model structure.

4. Notes

The trajectories, i.e., the time profile of differential equation based models are approximated computationally using numerical integration. The accuracy of the integration depends on the *stiffness* of the system at hand: the stiffer the ODE system, the more difficult and time-consuming the integration (12). Most biochemical networks are stiff, which is based on very fast and slow processes happening at the same time. Specialized integrators have been developed for this situation. PottersWheel supports several of them: ode15s, ode23s, RADAU, RADAU5, and CVODES. In our experience, ode15s is the slowest but also most robust integrator. It is used automatically if the current integrator fails.

A model may be integrable for the current set of parameter values and also for the optimal set of parameters, but during optimization the model may become extremely stiff, because a certain region of parameter values may be entered. Consequently the integration fails. As a remedy, the limits of parameter optimization could be narrowed to avoid the stiff region. In addition, a deterministic optimizer like trust-region can be used instead of a more "jumping" stochastic optimizer like ASAMIN (16, 17). Finally, reformulation of the mathematical model may be useful in order to avoid too many extremely fast reactions.

In order to learn more about a function, e.g. pwAddData, use either the online function reference or type

```
>> help pwAddData
```

In order to quickly search for a function given a keyword, use

```
>> pwSearch keyword
```

References

1. Seber GAF, Wild CJ (1989) Nonlinear regression. Wiley, New York
2. Saez-Rodriguez J, Alexopoulos LG, Epperlein J, Samaga R, Lauffenburger DA, Klamt S et al (2009) Discrete logic modelling as a means to link protein signalling networks with functional analysis of mammalian signal transduction. Mol Syst Biol 5:331
3. Jensen FV, Jensen FVV (1996) An introduction to Bayesian networks. UCL, London
4. Turner TE, Schnell S, Burrage K (2004 Jul) Stochastic approaches for modelling in vivo reactions. Comput Biol Chem 28(3):165–178
5. Gillespie DT (2007) Stochastic simulation of chemical kinetics. Annu Rev Phys Chem 58:35–55
6. Aldridge BB, Burke JM, Lauffenburger DA, Sorger PK (2006 Nov) Physicochemical modelling of cell signalling pathways. Nat Cell Biol 8(11):1195–1203
7. Murray JD (2003) Mathematical biology: spatial models and biomedical applications. Springer, Berlin
8. Maiwald T, Timmer J (2008 Sep 15) Dynamical modeling and multi-experiment fitting with PottersWheel. Bioinformatics 24(18):2037–2043
9. Swameye I, Muller TG, Timmer J, Sandra O, Klingmuller U (2003) Identification of nucleocytoplasmic cycling as a remote sensor in cellular signaling by databased modeling. Proc Natl Acad Sci U S A 100(3):1028–1033
10. Bowden AC (1995) Fundamentals of enzyme kinetics, revised edn. Portland, London
11. Kreutz C, Bartolome Rodriguez MM, Maiwald T, Seidl M, Blum HE, Mohr L et al (2007) An error model for protein quantification. Bioinformatics 23(20):2747–2753
12. Hairer E, Wanner G (2010) Stiff and differential-algebraic problems. Solving ordinary differential equations II. Springer, Berlin
13. Hindmarsh AC, Brown PN, Grant KE, Lee SL, Serban R, Shumaker DE et al (2005) SUNDIALS: Suite of nonlinear and differential/algebraic equation solvers. ACM Trans Math Softw (TOMS) 31(3):396
14. Akaike H (1974) A new look at the statistical model identification. IEEE Trans Automat Contr 19(6):716–723
15. Schwarz G (1978) Estimating the dimension of a model. Ann Stat 6:461–464
16. Ingber L (1993) Simulated annealing: Practice versus theory. Mathematical Computer Modelling 18(11):29–57 http://www.ingber.com/asa93_sapvt.pdf
17. Sakata S (2009) ASAMIN: a Matlab gateway routine to adaptive simulated annealing (ASA)

18. Egea JA, Rodríguez-Fernández M, Banga JR, Martí R (2007) Scatter search for chemical and bio-process optimization. J Global Optim 37 (3):481–503
19. Hengl S, Kreutz C, Timmer J, Maiwald T (2007 Oct 1) Data-based identifiability analysis of non-linear dynamical models. Bioinformatics 23(19):2612–2618
20. Raue A, Kreutz C, Maiwald T, Bachmann J, Schilling M, Klingmüller U et al (2009 Aug 1) Structural and practical identifiability analysis of partially observed dynamical models by exploiting the profile likelihood. Bioinformatics 25(15):1923–1929
21. Blinov ML, Faeder JR, Goldstein B, Hlavacek WS (2004) BioNetGen: software for rule-based modeling of signal transduction based on the interactions of molecular domains. Bioinformatics 20(17):3289
22. Borisov NM, Chistopolsky AS, Faeder JR, Kholodenko BN (2008 Sep) Domain-oriented reduction of rule-based network models. IET Syst Biol 2(5):342–351

Chapter 9

Rule-Based Modeling of Signal Transduction: A Primer

John A.P. Sekar and James R. Faeder

Abstract

Biological cells accomplish their physiological functions using interconnected networks of genes, proteins, and other biomolecules. Most interactions in biological signaling networks, such as bimolecular association or covalent modification, can be modeled in a physically realistic manner using elementary reaction kinetics. However, the size and combinatorial complexity of such reaction networks have hindered such a mechanistic approach, leading many to conclude that it is premature and to adopt alternative statistical or phenomenological approaches. The recent development of rule-based modeling languages, such as BioNetGen (BNG) and Kappa, enables the precise and succinct encoding of large reaction networks. Coupled with complementary advances in simulation methods, these languages circumvent the combinatorial barrier and allow mechanistic modeling on a much larger scale than previously possible. These languages are also intuitive to the biologist and accessible to the novice modeler. In this chapter, we provide a self-contained tutorial on modeling signal transduction networks using the BNG Language and related software tools. We review the basic syntax of the language and show how biochemical knowledge can be articulated using reaction rules, which can be used to capture a broad range of biochemical and biophysical phenomena in a concise and modular way. A model of ligand-activated receptor dimerization is examined, with a detailed treatment of each step of the modeling process. Sections discussing modeling theory, implicit and explicit model assumptions, and model parameterization are included, with special focus on retaining biophysical realism and avoiding common pitfalls. We also discuss the more advanced case of compartmental modeling using the compartmental extension to BioNetGen. In addition, we provide a comprehensive set of example reaction rules that cover the various aspects of signal transduction, from signaling at the membrane to gene regulation. The reader can modify these reaction rules to model their own systems of interest.

Key words: Computational systems biology, Mathematical modeling, Combinatorial complexity, Signal transduction, Formal languages, BioNetGen, Stochastic simulation, Ordinary differential equations, Network-free simulation, Compartmental modeling, Compartmental modeling, Protein–protein interactions, Signal transduction, Kinase cascades, Signaling diagrams

1. Introduction

Biological cells have evolved complex molecular interaction networks that involve genes, proteins, and small molecules and function as information processing units. These networks are

Xuedong Liu and Meredith D. Betterton (eds.), *Computational Modeling of Signaling Networks*, Methods in Molecular Biology, vol. 880, DOI 10.1007/978-1-61779-833-7_9,

dynamic and self-modifying, responding to cues integrated from the internal and external environments. The traditional approach to understanding signal processing inside cells has been reductionist, i.e., to examine every component in isolation and characterize its microscopic behavior based on its structure, interaction partners, and physiological effect that can be observed from its overproduction or negation. Based on the relationships between components, molecules were organized into roughly linear signaling cascades that were useful in comprehending the overall role of each set of components.

However, the preponderance of experimental and structural evidence obtained in the last few decades tells a different story (1–3). Signaling cascades overlap with each other by sharing components and are not strictly delineated, resulting in significant cross talk between them (e.g., ref. 4). Given two different contexts (a broad term encompassing chemical, spatial, and temporal aspects (1–3)), the same signal can elicit different responses. Thus, in order to understand complex and nonobvious emergent phenomena, a system-wide mechanistic approach to modeling is necessary. Such an approach is complementary to the reductionist approach and builds on the available data.

When studying large networks or even large subnetworks, gross approximations are often employed that form the basis of statistical and bioinformatics approaches to network modeling (5). Explicit reaction kinetic modeling of signaling pathways may be more realistic, but has been limited by several issues (6): the high level of uncertainty in current knowledge about the key molecular players and interactions, the combinatorial complexity that arises from multiple states of each molecule and multiple molecules in a complex, the computational effort required to simulate large reaction networks using generally available methods, and the difficulty in maintaining and reusing large-scale models.

One solution to the above issues still followed by many practitioners in the field is to ignore the many combinations of complexes and many possible molecular states in order to achieve a model size that can be understood, simulated, and analyzed. Such approximations on model structure may be subjective and are usually difficult to test.

Recently, several groups have developed new modeling languages and approaches designed to enable the development of large-scale biochemical models without the need to make ad hoc assumptions. In these so-called rule-based approaches (reviewed in refs. 7, 8), molecules are treated as objects with a defined substructure and reactions are described using reaction rules. Unlike the standard reaction description where the entire reactant species is named distinctly, in a reaction rule, it suffices to represent only those parts of a reactant that actually participate in the reaction or affect the ability of the reaction to occur or its rate.

Since these substructures can be shared between many reactants, a reaction rule effectively represents multiple reactions with identical kinetic descriptions and parameters. This also results in a reduction in the number of essential parameters required by the model. By taking advantage of the modularity exhibited by protein interactions, one can effectively represent a combinatorially complex (potentially infinite) reaction network by a much smaller and finite set of reaction rules. By providing a formal language to intuitively represent detailed biophysical and biochemical information and providing algorithms to automatically mathematize and simulate such models, rule-based modeling languages (such as BioNetGen (BNG) (9, 10) and Kappa (11, 12)) form a much-needed bridge between descriptive reductionist biology and exact mathematical modeling and simulation.

BNG (9, 10) is a rule-based modeling framework and language that has previously been described in detail by Faeder et al. (9). While Faeder et al. focused on presenting a comprehensive description of the modeling language and software capabilities, we present here a self-contained tutorial on developing models following the rule-based approach. Although some aspects of the process are specific to the BNG modeling language (BNGL) and tools, much of the information presented is broadly applicable to developing detailed reaction kinetic models of signaling.

We show that the rule-based syntax and approach provide an excellent way not only to develop models and simulations, but also to encode biochemical knowledge in a concise and modular way. We also do not assume any specialized knowledge on the part of the reader. We provide suggestions about common difficulties in the creation and representation of large reaction models in general and rule-based models in particular. Most importantly, we provide a comprehensive set of scenarios that can be easily studied and modified by a novice modeler and applied to any biological system of interest.

The remainder of this chapter is structured as follows. Subheading 2 indicates where to obtain BNG and related software. Subheading 3 introduces the elements of a rule-based model and the basic syntax of BNGL. Subheading 4 illustrates the basic techniques and nuances behind writing reaction rules. Subheading 5 discusses theory and methods necessary for large-scale reaction models. Subheading 6 shows how to construct, simulate, and analyze a complete BNGL model using biological knowledge and hypotheses. Subheading 7 discusses the more advanced concept of rule-based models in a structured hierarchy of compartments. Subheading 8 talks about challenges involved in parameterizing signal transduction models. Subheading 9 builds on these technical aspects and provides examples of common biological mechanisms of signal transduction, including complex assembly at the membrane, secondary messenger activation, kinase cascades, and gene regulation. Finally, a summary of key points is provided in Subheading 10.

2. Software

In order to follow the examples and gain hands-on experience with the software, it is highly recommended that the reader obtain access to BNG before proceeding. Two convenient interfaces are available. First, a stand-alone graphical user interface called RuleBender can be obtained free of charge from http://rulebender.org. RuleBender includes a model editor, network browser, simulation interface, and simulation results viewer. Binary distributions of RuleBender are provided free of charge for Windows, Mac, and Linux, and these also include both binaries and source code for BNG. RuleBender can be used to construct, simulate, and plot the output of all examples presented in this chapter. The latest developmental version of the BNG source code as well as additional documentation and model examples can be found at http://bionetgen.org.

Second, a Java-applet-based interface is available free of charge through the Virtual Cell modeling platform (http://vcell.org). The interface to BNG contains an editor, simulation interface, and plotting tools. Not all advanced features of BNG are available, but this is a good way to try modeling with BNGL without having to download or install any software.

3. Model Representation

3.1. Molecules and Components

In the context of signal transduction, a reaction kinetic network typically involves multiple functional complexes of biological entities, such as proteins and small molecules. In a traditional reaction kinetic model, each configuration of the complex is given a unique name and treated as a single unit that can participate in reactions. The modeler would need to identify every unique configuration and assign it a name or a label. From this list of names, the modeler would have to pick out every possible combination of reactants and products and write out the reactions by hand. Due to the combinatorial nature of association between proteins, this approach can be cumbersome and limits not only the modeling process, but also the computational memory required to store, manipulate, and simulate a model.

Rule-based modeling takes a more structured approach. Indivisible entities in the model that associate with other indivisible entities are called *molecules*. Multiple functional attributes for each *molecule type* are described using a number of *components*. This mirrors the hierarchical nature of real biological entities, for example, proteins have multiple functional substructures such as domains, motifs, and binding sites. These components can be unchanging (called *stateless*) or exist in one of many different internal *states*. For example, certain binding motifs may have different behaviors depending on whether they are unphosphorylated or phosphorylated.

Consider a receptor with the following functional substructures: a ligand-binding site and a channel that is either open or closed. The BNG *molecule type* blueprint for such a molecule would be `R(lig,ch~open~closed)`. `R` is the name of the receptor molecule, `lig` is the ligand-binding site, and `ch` is the channel. `lig` is stateless, whereas `ch` can take states `open` or `closed`, indicated using the tilde symbol (~).

Multiple components within a molecule type can have the same name and will be treated identically. For example, the *molecule type* `A(x,x,y)` has two identical but independent `x` components and one `y` component. When defining the molecule type, one must ensure that identically named components have identical sets of allowed states. For example, `A(x~a~b,x~a~b,y)` is a valid molecule type, but `A(x~a~b,x~b~c,y)` is invalid since the two `x` components have been assigned different sets of allowed states.

The *molecule* itself (created from the *molecule type* blueprint) can only take one state for each component at any given time. Therefore, given the molecule type `R(lig,ch~open~closed)`, the following molecules can be created: `R(lig,ch~open)` and `R(lig,ch~closed)`. In the traditional modeling framework, we might refer to them by names, such as R_o and R_c. The order in which the components are written within the molecule is immaterial and `R(lig,ch~open)` means the same as `R(ch~open,lig)`.

3.2. Bonds and Species

A BNG molecule can associate with another molecule through a *bond* between their components. Bonds can be formed between both stateless components and components with states. The two "bonded" components are indicated by an exclamation mark followed by a unique label that marks the two ends of the bond. A component cannot participate in more than one bond at any given time.

For example, the ligand molecule `L(rec)` and receptor molecule `R(lig,ch~open)` can associate using a bond between the `rec` and `lig` components, represented as `L(rec!0).R(lig!0,ch~open)`. The bond label "`0`" after the exclamation marks indicates that `rec` and `lig` are bound. The dot between `L()` and `R()` indicates that they are in the same complex. A good convention to follow is to label bonds with numbers. In the absence of identifying information about the structural domains or motifs involved, label components based on the molecule that they are supposed to bind.

The same bond between the same pair of components can be represented using different labels in different places as long as the right pair of components is matched, for example, `L(rec!0).R(lig!0,ch~open)` is the same as `L(rec!2).R(lig!2,ch~open)`.

In the traditional modeling framework, this complex may be named R_oL. The corresponding complex with the channel closed, i.e., `L(rec!0).R(lig!0,ch~closed)`, may be referred to as R_cL. By providing a systematic way of linking molecules together, BNGL enables the systematic representation of any arbitrary complex.

For example, the following is a valid complex:

```
A(x,y!0).B(p!0,q~abc!1).C(r~def!1,s!2).A(x!2,y)
```

This complex has two `A` molecules, one each of `B` and `C` molecules. One `A` binds `B` using a `y-p` bond labeled `0`. The other `A` binds `C` using an `x-s` bond labeled `2`. `B` binds `C` using a `q-r` bond labeled `1`. `q` and `r` are in `abc` and `def` states, respectively. Dots separate the molecule names and all bonds have unique labels. The order in which the molecules are written is immaterial as long as the same pairs of components are bonded and the same internal states are assigned. The same complex can also be written as:

```
B(p!0,q~abc!1).C(r~def!1,s!2).A(x!2,y).A(x,y!0)
```

Such structured representation provides an important advantage over the traditional modeling approach. The modeler does not need to assign a unique name to every complex configuration. The name space that the modeler has to comprehend is limited to the number of molecules and components and not the possibly infinite number of complex configurations. Where finite, the construction of the full set of configurations can be easily automated.

BNG uses the term *species* to refer to any unique configuration of one or more molecules in a complex. The critical elements of a unique species definition are:

- Identifying every molecule in the species (including repeated ones)
- Identifying internal states taken by every component (including repeated ones)
- Identifying every component pair linked by a bond

3.3. Transformations and Reactions

In the BNGL, a reaction is simply one or more transformations applied simultaneously to one or more species. The following transformations are allowed:

- Forming a bond, e.g. `A(b) + B(a) -> A(b!0).B(a!0)`
- Breaking a bond, e.g. `A(b!0).B(a!0) -> A(b) + B(a)`
- Changing of component state, e.g. `X(y~0) -> X(y~p)`
- Creating a molecule, e.g. `A(b) -> A(b) + C(d)`

- Destroying a molecule, e.g. `A(b) + B(a) -> A(b)`

A single reaction may involve any number of transformations. For example,

```
A(b) + B(a!0).C(d!0) -> A(b!0).B(a!0) + C(d)
```

Here, the `a-d` bond is broken and the `b-a` bond is formed simultaneously. Another example is:

```
A(b) + B(a!0).C(d!0) -> A(b!0).B(a!0)
```

Here, the `a-d` bond is broken, the `b-a` bond is formed, and the `C(d)` molecule is destroyed simultaneously. Usually, it is advisable to restrict the number of transformations in a reaction to one or two.

4. Writing Reaction Rules

4.1. Combining Multiple Reactions

Consider the ligand and receptor defined in the Subheading 3.1. Let us assume that ligand binding and channel opening/closing are independent of each other. This means that both open and closed forms of the receptor bind the ligand with the same rate constant k.

$$R_o + L \rightarrow R_oL \quad k$$
$$R_c + L \rightarrow R_cL \quad k.$$

The ordinary differential equation (ODE) description for this set of reactions is:

$$-\frac{d[R_o]}{dt} = +\frac{d[R_oL]}{dt} = k[R_o][L].$$
$$-\frac{d[R_c]}{dt} = +\frac{d[R_cL]}{dt} = k[R_c][L].$$
$$-\frac{d[L]}{dt} = k[R_o][L] + k[R_c][L] = k([R_o] + [R_c])[L].$$

Since the rate constant k is common in the first two ODEs, we can add them up.

$$-\frac{d([R_o] + [R_c])}{dt} = +\frac{d([R_oL] + [R_cL])}{dt} = k([R_o] + [R_c])[L]$$
$$-\frac{d[L]}{dt} = k([R_o] + [R_c])[L].$$

Let us create two new terms:

$$[R_{o/c}] = [R_o] + [R_c]$$
$$[R_{o/c}L] = [R_oL] + [R_cL].$$

Using these two new terms, we can compress the original set of ODEs as:

$$-\frac{d[R_{o/c}]}{dt} = -\frac{d[L]}{dt} = +\frac{d[R_{o/c}L]}{dt} = k[R_{o/c}][L].$$

Note that we have simply summed the binding rates for open and closed receptors. The fact that the two reactions have an identical rate constant k enables us to use the summed terms in a compressed manner. Representing this compressed description using traditional labels, we might write it as:

$$R_{o/c} + L \rightarrow R_{o/c}L \quad k.$$

BNGL provides an intuitive way to formulate such compressed representations using reaction rules. In rule-based modeling, creating a new label such as $R_{o/c}$ or $R_{o/c}L$ is not necessary. A *reaction rule* can be created from many reactions (that have identical kinetics) simply by omitting the components that do not influence the reactions. For example, consider the same reactions represented in BNGL:

```
R(lig,ch~open)+L(rec)->\
 R(lig!0,ch~open).L(rec!0) k
R(lig,ch~closed)+L(rec)->\
 R(lig!0,ch~closed).L(rec!0) k
```

Here, the state of the `ch` component does not affect the rate constant for the `lig-rec` binding. Therefore, we can omit the `ch` component to create a *reaction rule* that represents *both* of the above reactions:

```
R(lig)+L(rec)->R(lig!0).L(rec!0) k
```

Thus, a reaction rule can be thought of as a generator of reactions, where each generated reaction must have some substructure that corresponds exactly to the reaction rule. The generated reactions have identical kinetic descriptions at the microscopic level.

Suppose the open and closed channels had two different rate constants of association with ligand, i.e.,

$$R_o + L \rightarrow R_oL \quad k_1$$
$$R_c + L \rightarrow R_cL \quad k_2.$$

Since the kinetics of the two reactions are fundamentally different, we *cannot* create a single reaction rule from these two reactions. We can only create two different reaction rules (which happen to be identical to the reactions):

```
R(lig,ch~open)+L(rec)->\
 R(lig!0,ch~open).L(rec!0) k1
R(lig,ch~closed)+L(rec)->\
 R(lig!0,ch~closed).L(rec!0) k2
```

Thus, identifying independent interactions is the key to writing a rule-based model. If within a molecule every component influences every other component and if the same is true for all molecules, then the number of reaction rules is identical to the number of reactions. As more and more independent interactions are identified, the number of reaction rules required to represent a reaction network decreases.

Bidirectional reaction rules can be combined into a single line using bidirectional arrows and two reaction rate constants. However, BNG still interprets them as two different reaction rules. For example, the bidirectional reaction rule

```
R(ch~closed) <-> R(ch~open) k_open,k_close
```

will be interpreted as two reaction rules:

```
R(ch~closed) -> R(ch~open) k_open
R(ch~open) -> R(ch~closed) k_close
```

4.2. Patterns and Species

Consider the reaction rule

```
R(lig) + L(rec) -> R(lig!0).L(rec!0) k
```

As we already saw in the previous subsection, if the molecule types were `R(lig,ch~open~closed)` and `L(rec)`, then this reaction rule would generate the following reactions:

```
R(lig,ch~open) + L(rec) ->\
  R(lig!0,ch~open).L(rec!0) k
R(lig,ch~closed) + L(rec) ->\
  R(lig!0,ch~closed).L(rec!0) k
```

Note how the complexes are fully specified in the reactions, i.e., all the components of all the molecules present and all the corresponding binding and internal states are mentioned. As mentioned in the previous section, a fully specified complex definition is called a BNG *species*. In the reaction rule, however, the complexes are not fully specified. A partially specified complex is called a BNG *pattern*.

The reactants of a reaction rule are called the *reactant patterns* and they are said to match the *reactant species* of the reactions. The reactant pattern `R(lig)` matches the reactant species `R(lig,ch~open)` and `R(lig,ch~closed)`. Similarly, the reactant pattern `L(rec)` matches the reactant species `L(rec)`. Pattern matching to generate reactions is summarized in Fig. 1.

Patterns lie at the heart of rule-based modeling. Patterns can be said to select species that possess a matching substructure. *A reaction rule is effectively a set of reactant patterns and a set of transformations applied to those patterns.* A reaction is generated by selecting species that match the reactant patterns and applying the transformation to them to get the product species.

4.3. Observables

Many biological experiments involve following the time courses of experimental outputs, often referred to as *trajectories*. Each output is usually specific to a single collection of species. For example, in a FRET experiment on receptor aggregation, one might design the

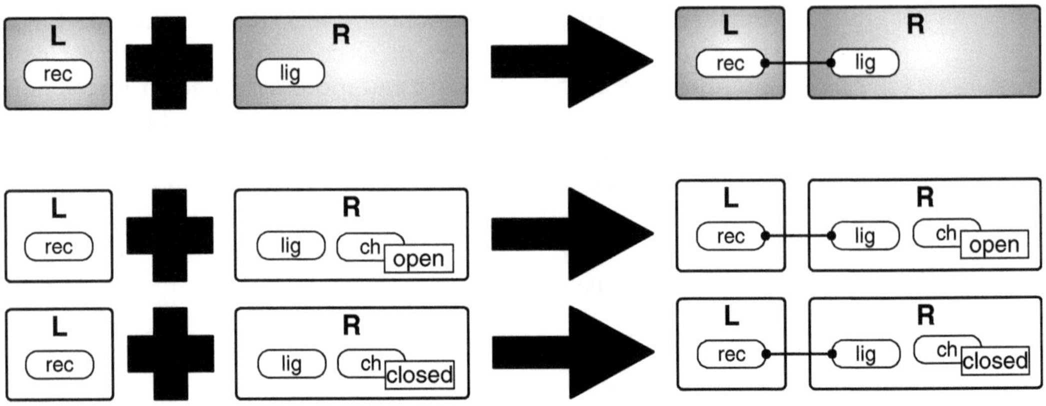

Fig. 1. Pattern matching to generate reactions. Patterns are incompletely defined species, such as those in *gray*. The pattern is said to match to a species if the species possesses a substructure that is equivalent to the pattern. In this case, the pattern R(lig) matches the species R(lig,ch~open) and R(lig,ch~closed). Reactions on patterns are called reaction rules. When the patterns match to full species, reaction rules match to reactions. The reactions all are identical with respect to their microscopic kinetics, i.e., they have identical per-site asymmetric rate constants. The symmetry and multiplicity factors are automatically detected and assigned by BioNetGen. In this case, the reaction is the formation of a bond between the lig component of R and the rec component of L.

experiment such that only dimers can produce fluorescence, and the fluorescent output is a function of the total concentration of all dimer species. Similarly, in a Western blot experiment using an antibody specific to a particular protein, the measurement at different time points is proportional to the total concentration of the protein available for binding. In a BNG model, one can similarly define specific sums of concentrations of species that are of interest and need to be monitored or tracked through the simulation of the model. Such sums are called *observables* and are of two types: molecules-observables and species-observables.

Consider the molecule types `A(b,b,c)`, `B(a)`, and `C(a)`. A can bind two B and one C. This results in the following species:

B	`B(a)`
C	`C(a)`
A	`A(b,b,c)`
AB	`A(b!0,b,c).B(a!0)`
BAB	`A(b!0,b!1,c).B(a!0).B(a!1)`
AC	`A(b,b,c!2).C(a!2)`
ABC	`A(b!0,b,c!2).B(a!0).C(a!2)`
BABC	`A(b!0,b!1,c!2).B(a!0).B(a!1).C(a!2)`

A *species-observable* is an unweighted sum of the concentrations of species matched to a pattern (or a list of patterns). For example, let us define a species-observable to be the pattern `A(b)`. This pattern matches any species that contains an `A` molecule with an unbound `b` component. The species A and AC have two unbound b components each, and the species AB and ABC have one unbound b component each. Thus, the observable would match all four of these species (but not the species BABC). The value of this observable would be given by the sum [A] + [AB] + [AC] + [ABC], which would vary over the course of a simulation.

A *molecules-observable* weights the concentrations of the species by the number of matches to a pattern (or a list of patterns). For example, let us define a molecules-observable with the same pattern `A(b)`. This would match species AB and ABC once each and the species A and AC twice (since they have two unbound `b` sites). The value of this observable would be given by the sum $2 \times [A] + 2 \times [AC] + 1 \times [AB] + 1 \times [ABC]$.

4.4. Reaction Center and Context

Consider the reaction rule:

```
R(lig)+L(rec)->R(lig!0).L(rec!0) k
```

The transformation applied in this rule is the creation of the `lig-rec` bond. The components that participate in the transformation(s) carried out by a reaction are collectively called the *reaction center*. Here, the reaction center is the set of components `lig` and `rec`.

Now, consider another reaction rule:

```
R(lig,ch~open)+L(rec)->\
  R(lig!0,ch~open).L(rec!0) k1
```

Here, the same transformation is being applied, i.e., the `lig-rec` bond, and therefore it has the same reaction center (underlined for emphasis). However, the additional `ch` component in the `open` state is required to indicate that if the channel is open then the reaction rate should be k_1. The components that do not participate in the reaction, but are nevertheless required and influence the reaction rate constant, are collectively called the *reaction context*.

Finally, consider the reaction rule:

```
R(lig,ch~closed)+L(rec)->\
  R(lig!0,ch~closed).L(rec!0) k2
```

This rule also has a reaction center identical to the previous rules (underlined for emphasis). However, the reaction context is now `ch~closed` and the rate constant is k_2.

When a reaction rule is used to generate reactions, the reaction context provides minimum compulsory conditions that need to be obeyed by the generated reactions. In the absence of any reaction context, the reaction rule is general and matches all possible reactions having the same transformation. Adding more context "specializes" the reaction rule to match fewer reactions that obey the conditions imposed by the context. The reaction itself can be considered as a highly specialized reaction rule with every component in the interacting species added as context.

Note that in bidirectional reaction rules, the context is preserved in both directions. For example, consider the bidirectional rule (with reaction center underlined for emphasis):

```
R(lig,ch~closed)+L(rec)<->\
  R(lig!0,ch~closed).L(rec!0) k2,kr
```

This is equivalent to the two reaction rules, both with `ch~closed` as context:

```
R(lig,ch~closed)+L(rec)->\
  R(lig!0,ch~closed).L(rec!0) k2
R(lig!0,ch~closed).L(rec!0)->\
  R(lig,ch~closed)+L(rec) kr
```

4.5. Synthesis and Degradation

When writing a reaction rule in BNGL, one must make sure that there is at least one reactant pattern and one product pattern. However, there are a number of situations, where the number of molecules on either the reactant side or the product side of the reaction is zero. To indicate zero stoichiometry, one can use the symbol 0 (zero).

When the reactant side of a reaction rule has zero stoichiometry, the reaction is treated as a zero-order synthesis reaction, for example:

```
0 -> L(rec) k_syn
```

This reaction rule should produce the ligand molecule at a constant zero-order rate, which is equal to `k_syn` (typical units M/s). Zero-order synthesis is useful to model the case, where a reacting agent is continuously flowing into the system from the outside, for example the continuous movement of growth factors from the blood to any body tissue or the controlled flow of nutrients in a bioreactor.

Where the synthesis depends on the concentrations of specific molecules, it is better to use first-order and second-order elementary reactions. For example, the production of mRNA from a gene depends on the number of actively transcribing copies of the gene in a cell:

```
Gene() -> Gene() + mRNA() k_transcription
```

When the product side of a reaction rule has zero stoichiometry, it has no bearing on the rate of the reaction since the rate is determined only by the reactant concentrations. Such reactions are typically used to model spontaneous or background degradation, which occurs in some form or the other with most biomolecules. For example, to model spontaneous degradation of the receptor:

```
R() -> 0 k_degr
```

The internal and binding states of components can be mentioned to restrict the degradation to only certain types of molecules. For example, if the only receptors degraded were those in the closed state:

```
R(ch~closed) -> 0 k_degr
```

It is important to note that in a degradation reaction rule, the *complete matched reactant species* is degraded and not just the molecules mentioned in the rule. The above reaction rule will generate the following reactions:

```
R(lig,ch~closed) -> 0 k_degr
R(lig!0,ch~closed).L(rec) -> 0 k_degr
```

To delete only the molecules concerned (and not the whole species), use the `DeleteMolecules` keyword:

```
R(ch~closed) -> 0 k_degr DeleteMolecules
```

In this case, the connected molecules are not deleted and this would generate the following reactions:

```
R(lig,ch~closed) -> 0 k_degr
R(lig!0,ch~closed).L(rec) -> L(rec) k_degr
```

4.6. Symmetry, Multiplicity, and Rate Addition

Consider the following situation: The A molecule can dimerize. It exists in two different isoforms A_1 and A_2 and both hetero- and homo-dimerization occur. Expressing this in the traditional framework, this implies the existence of three different reactions:

$$A_1 + A_2 \rightarrow D_{12}$$
$$A_1 + A_1 \rightarrow D_{11}$$
$$A_2 + A_2 \rightarrow D_{22}.$$

Even if both hetero- and homo-dimerization have identical interaction strengths (affinities), the homo-dimerization would proceed at half the rate of the hetero-dimerization due to the symmetry in picking out a homo-dimerizing molecule pair. Therefore, the rate constants are usually expressed as:

$$A_1 + A_2 \rightarrow D_{12} \quad k$$
$$A_1 + A_1 \rightarrow D_{11} \quad k/2$$
$$A_2 + A_2 \rightarrow D_{22} \quad k/2.$$

Here, one can think of “k” as the asymmetric reaction rate constant and the (1/2) multiplying factor as being due to the *symmetry effect*. Expressing these reactions using BNG species and representing the two isoforms as two different states:

```
A(iso~a1) + A(iso~a2) ->\
 A(iso~a1!0).A(iso~a2!0) k
A(iso~a1) + A(iso~a1) ->\
 A(iso~a1!0).A(iso~a1!0) 0.5*k
A(iso~a2) + A(iso~a2) ->\
 A(iso~a2!0).A(iso~a2!0) 0.5*k
```

BNG automatically detects reaction symmetries and applies the symmetry factor even if the reaction rules themselves are asymmetric. When writing a reaction rule, the modeler must always provide the *asymmetric reaction rate constant*. All three reactions above can be generated with the correct reaction rate constants using only the single rule:

```
A(iso!0) + A(iso!0) -> A(iso!0).A(iso!0) k
```

The states of the `iso` component are left unmentioned in the reaction rule since they do not affect the asymmetric reaction rate constant. BNG detects that some of the generated reactions are symmetric and some are not. It then automatically assigns the (1/2) multiplying factor to the symmetric reactions.

The *multiplicity effect* is seen when interactions are multivalent. Suppose the molecule A has two identical sites for binding the molecule B. Then, in the traditional framework, one would write the following reactions:

$$A + A \rightarrow AB$$
$$AB + B \rightarrow BAB.$$

Even though the two binding sites have identical and independent chemical interactions, the first reaction should proceed at twice the rate of the second because A has twice the number of binding sites as AB. This is usually expressed as

$$A + A \rightarrow AB \quad 2 \times k$$
$$AB + B \rightarrow BAB \quad k.$$

Since BNGL allows multiple identical components in a molecule, it automatically accounts for the multiplicity factor when generating the reactions. The modeler must always provide the *per-site reaction rate constant*. If the molecule types are `A(b,b)` and `B(a)`, then it is sufficient to write the following reaction rule:

```
A(b) + B(a) -> A(b!0).B(a!0) k
```

This would automatically generate the following reactions:

```
A(b,b) + B(a) -> A(b,b!0).B(a!0) 2*k
A(b,b!0).B(a!0) + B(a) -> A(b!1,b!0).B(a!0).B(b!1) k
```

Lastly, if multiple reaction rules generate two versions of the same reaction, both are kept in the reaction network because they may have different rate constants or rate laws. During simulation, the net effect of the two identical reactions is additive, i.e., the rate of the transformations proceeds using the sum of the rates of the two copies. This is important when considering events such as dephosphorylation that might occur through general and specific mechanisms. For example, consider the following set of rules:

```
A(y~p) -> A(y~0) k_bkgrnd
A(y~p).E() -> A(y~0).E() k_enzyme
```

The second rule requires complexation with E for dephosphorylation to occur, whereas the first does not. These rules will generate two copies of the same reaction with different rate constants:

```
A(y~p,e!1).E(a!1) -> A(y~0,e!0).E(a!1) k_bkgrnd
A(y~p,e!1).E(a!1) -> A(y~0,e!0).E(a!1) k_enzyme
```

During simulation, the effect of these two reactions on the species `A(y~p,e!1).E(a!1)` is a unimolecular transformation with a net rate constant equal to the sum `k_bkgrnd+k_enzyme`.

4.7. Wildcards and Context Modification

BNGL provides syntax to tailor the context of a reaction rule to only match specific sets of reactions. The absence of an exclamation mark indicates that a component *must* be unbound. For example, the pattern `A(b,c)` will *not* match the species `A(b,c!1).C(a!0)`.

If the presence of a bond is contextually important but the identity of the binding partner is not relevant, then the + wildcard is used with the bond symbol. For example, the pattern `A(b!+,c)` will match both `A(b!0,c).B(a!0)` and `A(b!0,c).P(q!0)` and any other species with a bound `b` component and an unbound `c` component.

To match both unbound and bound components, one can use the `?` wildcard with the bond symbol. For example, the pattern `A(b!?,c)` matches both species `A(b,c)` and `A(b!0,c).B(a!0)`.

If the internal state of a component is not relevant to the context, the `~state` definition can be omitted. The presence or absence of a `!` symbol is still used for context. For example, the pattern `R(s)` matches both species `R(s~x1)` and `R(s~x2)` but not the species `R(s~x1!0).S(r!0)`.

The dot symbol can be used to indicate the presence of another molecule in a complex even if the explicit bond links are not shown. For example, the pattern `C().A(b)` will match both species `C(x!0).A(b,c!0)` and `C(x!0).A(b,c!1).X(a!1,c!0)`, even though `C` is connected to `A` directly in the first species and indirectly through `X` in the second species. This is useful when the same molecule can be linked to a complex in different ways but performs the same function, for example a kinase.

The bond and state wildcards, as well as the dot operator, can be used to tailor patterns in observables as well as reaction rules.

In reaction rules, more flexibility in defining context is provided by the `include_[reactants/products]` and `exclude_[reactants/products]` keywords. For example, consider the reaction rule

```
A(b) + B(a) -> A(b!0).B(a!0) k_ab\
  exclude_reactants(1,E,F(g))
```

The number 1 in the `exclude_reactants` directive indicates that it applies to the *first* reactant pattern. The remaining elements within the parentheses indicate the additional context that will cause exclusion. Here, it means that if the rule matches reactants with `E` or `F(g)` in the reactant 1 position, then those reactants would be discarded. Therefore, the following reactants would be *excluded*:

```
A(b,x!1).E(a!1)
A(b,x!1).F(a!1,g)
```

However, since only an unbound `F(g)` component is excluded, a reactant with a bound `F(g)` component (underlined for emphasis) would not be excluded, for example:

```
A(b,x!1).F(a!1,g!2).G(f!2)
```

The `include_reactants()` works similarly to the `exclude_reactants()`, but here the condition imposed is that *all* the additional context must be included.

For example, consider the following reaction rule:

```
A(b)+B(a)->A(b!0).B(a!0) k_ab\
  include_reactants(2,E,F,G)
```

This rule is equivalent to the reaction rule

```
A(b)+B(a).E().F().G()->A(b!0).B(a!0).E().F().G() k_ab
```

Both rules are different ways of saying that `E()`, `F()`, and `G()` molecules should be present in the second reactant species. Note that for bidirectional reaction rules `[include/exclude]_reactants()` is treated as `[include/exclude]_products` for the reaction rule in the reverse direction.

5. Modeling Background

5.1. Law of Mass Action

By default, BNG assigns rates to reactions based on the Law of Mass Action for Elementary reactions.

This means that for a unimolecular reaction

$$\mathrm{A} \rightarrow \mathrm{Products} \quad k_{\mathrm{uni}},$$

the rate is determined by the concentration of the reactant as follows:

$$\text{Rate } r = -\frac{\mathrm{d}[\mathrm{A}]}{\mathrm{d}t} = k_{\mathrm{uni}}[\mathrm{A}] \ \mathrm{M/s}.$$

Similarly, for a bimolecular reaction,

$$\mathrm{A} + \mathrm{B} \rightarrow \mathrm{Products} \qquad k_{\mathrm{bi}}$$

$$\text{Rate } r = -\frac{\mathrm{d}[\mathrm{A}]}{\mathrm{d}t} = -\frac{\mathrm{d}[\mathrm{B}]}{\mathrm{d}t} = k_{\mathrm{bi}}[\mathrm{A}][\mathrm{B}] \ \mathrm{M/s}.$$

The reaction rate here is expressed in volume-independent *macroscopic* terms, i.e., molar per second. But to aid modeling flexibility and simulation, it is preferable to specify rate constants such that the reaction rate is given in reaction-events/second. This *microscopic* rate is related to the macroscopic rate by a factor of the volume in which the reaction occurs times Avogadro's number.

$$r = \frac{\mathrm{d}[\mathrm{A}]}{\mathrm{d}t} = \frac{\mathrm{d}(N_{\mathrm{A}}/(VN_{\mathrm{Avo}}))}{\mathrm{d}t} = \frac{1}{VN_{\mathrm{Avo}}} \times \frac{\mathrm{d}N_{\mathrm{A}}}{\mathrm{d}t} = \left(\frac{1}{VN_{\mathrm{Avo}}}\right) r'$$

$$\Rightarrow r' = r \times VN_{\mathrm{Avo}},$$

where r = macroscopic rate constant (per M/s) and r' = microscopic rate constant (per second).

k_{uni} and k_{bi} are unimolecular and bimolecular reaction rate constants, respectively, with units per second and per molar per second. In general, any nth order volume-independent *macroscopic* reaction rate constant should have the units molar^{1-n} per second. However, since BNG treats the rate as microscopic, the modeler must convert the macroscopic rate constants to microscopic ones, as follows.

For unimolecular reactions, if N_A denotes the population number of the reactant molecules, then the microscopic rate is

$$r'\,(s^{-1}) = r \times VN_{Avo} = k_{uni}[A] \times VN_{Avo} = k_{uni}\left(\frac{N_A}{VN_{Avo}}\right) \times VN_{Avo}$$

$$r' = k_{uni}\, N_A.$$

For bimolecular reactions, if N_A and N_B denote the population numbers of the two participating reactants, then the microscopic rate is

$$\begin{aligned} r'\,(s^{-1}) &= r \times VN_{Avo} = k_{bi}[A][B] \times VN_{Avo} \\ &= k_{uni}\left(\frac{N_A}{VN_{Avo}}\right)\left(\frac{N_B}{VN_{Avo}}\right) \times VN_{Avo}. \\ r' &= \left(\frac{k_{bi}}{VN_{Avo}}\right) N_A\, N_B. \end{aligned}$$

If we define the microscopic reaction rate constants k'_{uni} and k'_{bi} as

$$\text{For unimolecular reactions,} \quad r' = k'_{uni}\, N_A$$
$$\text{For bimolecular reactions,} \quad r' = k'_{bi}\, N_A\, N_B,$$

then the relationship between macroscopic and microscopic rate constants is given by

$$k'_{uni} = k_{uni}.$$

$$k'_{bi} = \frac{k_{bi}}{VN_{Avo}}.$$

In general,

$$k'_{nthorder} = \frac{k_n}{(VN_{Avo})^{n-1}}, \quad n \geq 1.$$

Thus, when writing a unimolecular reaction rule in BNGL, one does not need to worry about converting the unimolecular reaction rate constant as long as it is in per second:

```
A(x~on) -> A(x~off) k_uni
```

where `k_uni` is defined as a parameter with units per second.

When writing a bimolecular reaction rule in BNGL, however, one should convert the bimolecular reaction rate constant to per second:

```
A(b)+B(a)->A(b!0).B(a!0) k
```

where the parameter `k` can be defined by an expression `k_bi/(V*N_Avo)` involving other parameters: `k_bi` with units molar per second, `V` with units liters, and `N_Avo` is defined as 6.022×10^{23}/mol. If bimolecular and higher order reactions occur in different volumes, then the reaction rate constant needs to be scaled differently for the reaction occurring in each volume. The modeler has to explicitly include the volume in the rate constant or use compartmental BNG (cBNG), as described in Subheading 7. Unimolecular reactions do not need to be scaled by volume.

The following assumptions are made under this paradigm: The reaction occurs in a fixed volume reactor at constant temperature and pressure and the molecules are distributed uniformly through the volume, i.e., there are no spatial gradients. Although these seem to be strong assumptions, they are often reasonable for biological models of events on the plasma membrane and in the cytosol (13). Diffusion in these compartments is often sufficiently fast that spatial effects remain minimal.

5.2. Reaction Order and Transformation Reversibility

Reaction orders higher than two are rare at the molecular level since the probability of three or more freely diffusing molecules finding each other simultaneously and in the right orientation is small. Although BNG supports reactions of any nonnegative integer order, it is prudent to restrict the model to a set of unimolecular and bimolecular reactions (i.e., first- and second-order reactions) in order to maintain physical realism and keep model assumptions to a minimum.

It is advisable to model every associative complex-forming interaction as a bimolecular association reaction between a pair of reactants. Also, in the context of protein–protein and protein–ligand associations, most association reactions are due to the cumulative effect of many weak non-covalent interactions that can be broken and re-formed. Therefore, unless an interaction is specifically known to be irreversible, it is recommended that every bimolecular association reaction be paired with the complementary unimolecular dissociation reaction. Such complementary modeling is advisable for all transformation pairs, such as synthesis/degradation and phosphorylation/dephosphorylation.

Such pairing is important to prevent system blowup or depletion. This happens when a unidirectional transformation is initiated and all the relevant molecules in the system get transformed in bulk, regardless of the magnitude of the initial stimulus. This is usually not intended or a reasonable biological behavior. For example, synthesis alone will lead to artificially high numbers and degradation alone will lead to complete depletion of a molecule type or a set of molecule types. Similarly, modeling phosphorylation alone will lead to runaway phosphorylation of the entire system (instead of a graded response) unless paired with dephosphorylation mechanisms.

5.3. Catalysis and Sigmoidal Kinetics

5.3.1. Michaelis–Menten Kinetics

The kinetic formulation of a catalytic process dates back to Michaelis and Menten (14) and Briggs and Haldane (15). The enzyme/catalyst binds to the substrate in a reversible fashion. A fraction of the bound complex causes the substrate to be converted to the product and dissociate simultaneously in a unimolecular fashion. The canonical formulation is

$$\mathrm{E} + \mathrm{S} \underset{k_f}{\overset{k_r}{\rightleftharpoons}} \mathrm{ES} \xrightarrow{K_{cat}} \mathrm{E} + \mathrm{P}.$$

Applying reaction kinetic theory, reaction flux through ES is given by

$$\frac{\mathrm{d[ES]}}{\mathrm{d}t} = k_f[\mathrm{ES}] - (k_r + k_{cat}[\mathrm{E}][\mathrm{P}]).$$

Applying a quasi-steady-state approximation on [ES] and using mass conservation laws,

$$\frac{\mathrm{d[ES]}}{\mathrm{d}t} \approx 0$$

$$\Rightarrow [\mathrm{ES}] \approx \frac{[\mathrm{E}]_\mathrm{T}[\mathrm{S}]}{\left(\frac{k_r + k_{cat}}{k_f}\right) + [\mathrm{S}]},$$

where $[\mathrm{E}]_\mathrm{T}$ is total enzyme concentration (i.e., [E] + [ES]) which can be assumed to be constant.

The rate of conversion of substrate (and formation of product) is then given by

$$r = -\frac{\mathrm{d[S]}}{\mathrm{d}t} \approx +\frac{\mathrm{d[P]}}{\mathrm{d}t} \approx k_{cat}[\mathrm{ES}] = \frac{k_{cat}[\mathrm{E}]_\mathrm{T}[\mathrm{S}]}{\left(\frac{k_r + k_{cat}}{k_f}\right) + [\mathrm{S}]} = \frac{V_{max}[\mathrm{S}]}{K_\mathrm{M} + [\mathrm{S}]},$$

where the Michaelis constant $K_\mathrm{M} = (k_r + k_{cat})/k_f$ and maximum rate $V_{max} = k_{cat}[\mathrm{E}]_\mathrm{T}$.

According to this formulation, the reaction rate should saturate when there is excess substrate, i.e.,

$$[\mathrm{S}] \gg K_\mathrm{M} \Rightarrow r \approx V_{max} = k_{cat}[\mathrm{E}]_\mathrm{T}.$$

Note that the rate is still first order with respect to enzyme concentration. The Michaelis–Menten rate law can be thought of as an extension to the elementary rate law, where the microscopic rate "constant" is not really a constant but a function that depends on the concentration of the substrate, i.e.,

$$r = \frac{k_{\text{cat}}[\text{E}]_{\text{T}}[\text{S}]}{K_{\text{M}} + [\text{S}]} = \text{f(S)} \times [\text{E}]_{\text{T}}, \quad \text{where f(S)} = \frac{k_{\text{cat}}[\text{S}]}{K_{\text{M}} + [\text{S}]}.$$

Note that the [S] term in the reaction indicates free substrate concentration which is equal to the total substrate concentration minus the substrate sequestered with the enzyme, i.e., $[\text{S}]_{\text{T}} - [\text{ES}]$. Often, the [ES] term is quite small relative to $[\text{S}]_{\text{T}}$ and can be neglected, i.e.,

$$[\text{ES}] \ll [\text{S}]_{\text{T}} \Rightarrow [\text{S}] \approx [\text{S}]_{\text{T}} \Rightarrow \text{f(S)} \approx \frac{k_{\text{cat}}[\text{S}]_{\text{T}}}{K_{\text{M}} + [\text{S}]_{\text{T}}}.$$

As a shortcut, for a single enzyme–substrate pair, the Michaelis–Menten rate law can be implemented in BNGL using the `MM` or `Sat` keywords. The substrate is given as the first reactant and the enzyme is given as the second reactant:

```
S(x~0)+E()->S(x~p)+E() MM(kcat,KM)
```

or

```
S(x~0)+E()->S(x~p)+E() Sat(kcat,KM)
```

The `MM` rate law calculates the free substrate concentration [S] and is more accurate than the `Sat` rate law which approximates it as the total substrate concentration $[\text{S}]_{\text{T}}$.

5.3.2. General Sigmoidal Kinetics

In general, if there are multiple reactants and it is required to saturate the rate with respect to a *single* reactant, one can use the `Sat` keyword. The saturating reactant is provided as the first reactant. The rate is elementary with respect to all the other reactants that follow. For example, if the reaction rule is

```
S(x~0)+A()+B()+C()->S(x~p)+A()+B()+C()
  Sat(k,K)
```

then the rate of the reaction is calculated as

$$r = \text{f(S)} \times [\text{A}][\text{B}][\text{C}], \quad \text{where f(S)} = \frac{k[\text{S}]}{K + [\text{S}]}.$$

The `Sat` is a special case of the Hill rate law, another common approximation used to model cooperative phenomena (72). The Hill rate law can be employed in BNGL using the `Hill` keyword.

```
S(x~0)+A()+B()+C()->S(x~p)+A()+B()+C()
  Hill(k,K,n)
```

The rate is calculated as follows:

$$r = \text{f(S)} \times [\text{A}][\text{B}][\text{C}], \text{ where f(S)} = \frac{k[\text{S}]^n}{K^n + [\text{S}]^n}.$$

f(S) is a sigmoidal function which exhibits a switch-like behavior, moving from zero to maximum over a range of [S]. Figure 2 shows

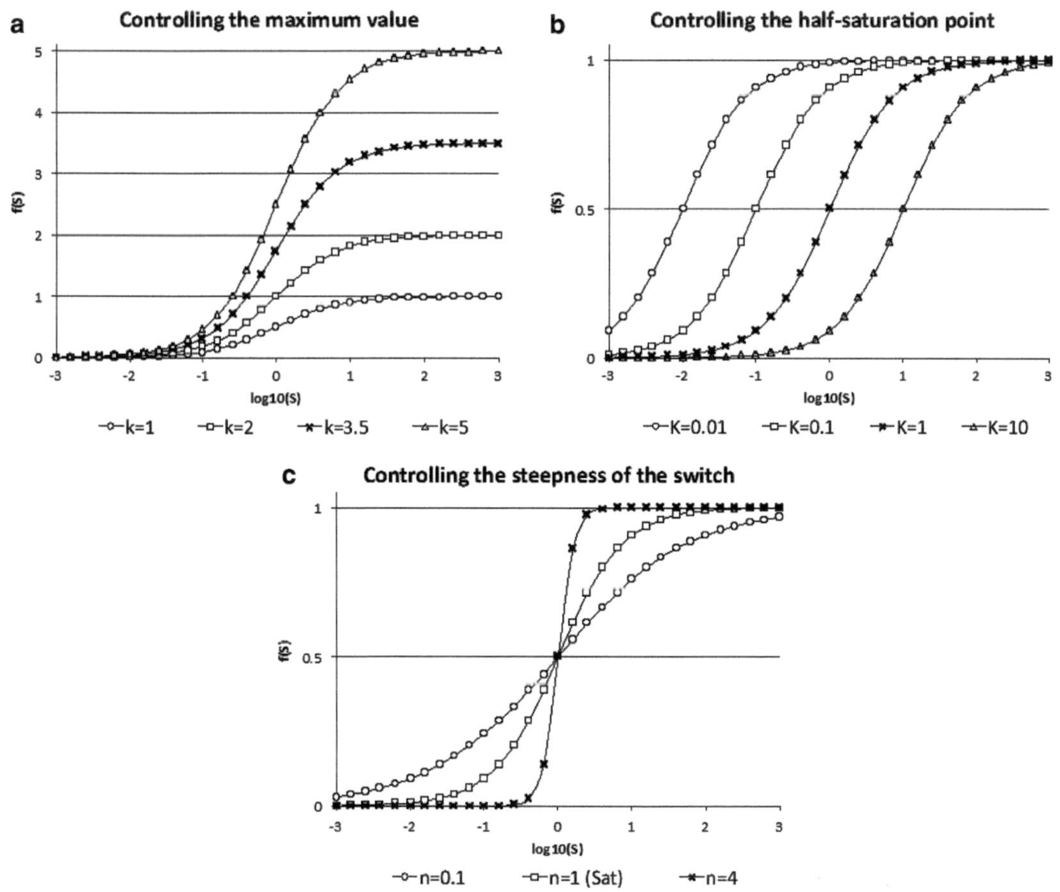

Fig. 2. Use of Hill/Sat rate laws. The Hill rate law function *f*(*S*) is plotted for different parameter sets. (**a**) $K = 1$, $n = 1$. *k* is varied from 1 to 5 and it controls the maximum value attained at high concentrations. (**b**) $k = 1$, $n = 1$. *K* is varied from 0.01 to 10 and it controls the point at which half the maximum value is attained. (**c**) $K = 1$, $k = 1$. *n* is varied from 0.1 to 4 and it controls the steepness of the switch from zero to maximum. $n = 1$ is the special case of the Sat rate law. *n* is also called the Hill coefficient.

how the function f(S) is controlled by the three parameters. The *k* parameter controls the maximum value that f(S) tends to at high [S] values. The K parameter controls the half-saturation point, i.e., the [S] value for which f(S) is half of its maximum value. The n parameter is also called the Hill coefficient and controls the steepness of the switching behavior. The Hill coefficient should be a positive number, but does not have to be an integer.

Note that both `Sat` and `Hill` rate laws are first order with respect to the other reactants that follow the first reactant. `Sat` is a special case of the `Hill` rate law with the Hill coefficient set at $n = 1$.

5.3.3. Rate Law Approximations Versus Explicit Mechanisms

The advantages of using the `Sat`, `MM`, and `Hill` keywords are that they do not introduce new species into the network for intermediate complexes. A sequence of kinetic steps is approximated into a single-step rate law using these approximations. In the traditional framework, such approximations are typically employed to reduce the size of the network. For enzymatic interactions that are fairly independent, a significant reduction is achieved and the fast time-scale of the enzyme–substrate equilibrium is removed, which can speed up model simulation. However, using a rate law approximation such as Michaelis–Menten carries the risk of modeling error, especially in densely connected catalytic networks such as signal transduction. For example, if there were two substrates that competed for binding to a single enzyme, but with identical reaction parameters, i.e.,

$$E + S_1 \underset{k_r}{\overset{k_f}{\rightleftharpoons}} ES_1 \xrightarrow{k_{cat}} E + P_1,$$

$$E + S_2 \underset{k_r}{\overset{k_f}{\rightleftharpoons}} ES_2 \xrightarrow{k_{cat}} E + P_2,$$

then the MM-like formulation for conversion for each substrate can be derived to be

$$-\frac{d[S_1]}{dt} = \frac{k_{cat}[E]_T[S_1]}{K_M + [S_1] + [S_2]}.$$

$$-\frac{d[S_2]}{dt} = \frac{k_{cat}[E]_T[S_2]}{K_M + [S_1] + [S_2]}.$$

Note how the denominator term for each conversion has both $[S_1]$ and $[S_2]$ terms. This cannot be expressed using a conventional `MM` keyword. We can get around this by manually deriving the rate function for such interactions (such as the relations above) and then assigning to the rate "constant" a custom global function (see Subheading 6.1). The concentration terms in the function need to be provided as molecules-observables. For example, one of the relations above can be rephrased as

$$-\frac{d[S_1]}{dt} = \left(\frac{k_{cat}}{K_M + [S_1] + [S_2]}\right) \times [E]_T \times [S_1] = k_{func}[E]_T[S_1].$$

k_{func} can be assigned a global function involving the parameters k_{cat}, K_M, and a molecules-observable which returns the sum $[S_1] + [S_2]$.

Such custom rate laws may be useful for simplifying reaction mechanisms and speeding up simulations, but they come at a significant price in terms of model clarity and modeler effort to derive or find the appropriate reduced model. Many reduced mechanisms are mathematically quite complex and may make the resultant model inaccessible to the average biological researcher. Common situations in signal transductions that can be modeled using such reduced mechanisms include (16):

- Multiple competitive substrates with different enzymatic parameters
- Inhibitors with one or more mechanisms (competitive, non-competitive, etc.)

- Combinations of inhibitors and allosteric enhancers of enzymatic activity
- Multiple enzymes acting on a common substrate
- Allosteric inhibition/enhancement caused by substrate or product feedback

More importantly, the approximations used to derive these reduced mechanisms may not hold in the context of a large reaction network. For example, the quasi-steady-state approximation may not be appropriate when changes in [ES] are of the same order of magnitude as changes in [S], i.e., when $d[ES]/dt \ll d[S]/dt$ is false (16). This may happen in signal transduction networks because the enzyme and substrate are often of similar concentrations (17). A second example where the MM approximation would fail is when the product is already present in significant amount in the system making the ES → E+P catalytic step a reversible process (16, 18). For some enzymes, the catalytic flux may even shift to flow in the opposite direction, converting product to substrate. Recent studies have shown that the repeated invocation of the Michaelis–Menten mechanism can cause a significant loss of model accuracy for dense biological networks (17, 19).

An alternative approach is simply to write out the explicit reaction mechanisms and use no higher order approximations, except the most basic Law of Mass Action (see Subheading 5.1). Although more accurate than the reduced mechanisms, the exact treatment can potentially increase the size of the reaction network by introducing a large number of intermediary complexes. In the traditional modeling framework, this approach was unfeasible because every complex had to be identified and labeled manually. *BNGL and other rule-based languages alleviate this problem by providing a structured way to create large numbers of complexes using relatively few reaction rules.*

Although rule-based languages simplify the task of coding large networks, they do not, by themselves, avoid the computational costs associated with simulating large networks using standard methods for chemical kinetics. However, theoretical advances in simulation methods for rule-based models (see Subheading 5.4) have produced network-free methods, which has computational costs that are independent of network size (20, 25, 30). Recent attempts at modeling large signal transduction systems have embraced the elementary reaction perspective (21, 22) for its accuracy. For a sufficiently large (or even infinite) network, instead of trying to compress the reactions using potentially inaccurate rate law approximations, it may be better to write down the full reaction mechanisms and use a network-free simulator (see Subheading 5.4) to simulate the system.

To illustrate, consider the example of two enzymes E1 and E2, both of which can act on two different substrates S1 and S2.

Assume that every enzyme–substrate pair has unique enzymatic parameters. Writing down the reaction rules to describe this situation takes a matter of minutes, whereas deriving the *correct* rate law approximation would take much longer, and the computational cost would likely not be reduced, especially for ODE-based simulations (see Subheading 5.4). The system can be explicitly written out as

```
E1(s) + S1(e,x~0) <-> E1(s!0).S1(e!0,x~0) kf11,kr11
E1(s) + S2(e,x~0) <-> E1(s!0).S2(e!0,x~0) kf12,kr12
E2(s) + S1(e,x~0) <-> E2(s!0).S1(e!0,x~0) kf21,kr21
E2(s) + S2(e,x~0) <-> E2(s!0).S2(e!0,x~0) kf22,kr22
E1(s!0).S1(e!0,x~0) -> E1(s) + S1(e,x~p)  kcat11
E1(s!0).S2(e!0,x~0) -> E1(s) + S2(e,x~p)  kcat12
E2(s!0).S1(e!0,x~0) -> E2(s) + S1(e,x~p)  kcat21
E2(s!0).S2(e!0,x~0) -> E2(s) + S2(e,x~p)  kcat22
```

5.4. Simulation Methods

Since chemical kinetic theory provides a formal way of defining network dynamics in terms of ODEs (such as those used in Subheading 4.1), the simplest way to simulate a chemical kinetic model is to initiate the model with certain concentrations of starting species and then use ***numerical integration of ODEs*** to propagate the model through time. The chemical system is assumed to exist in an isothermal reactor of constant volume with the molecules homogenously distributed throughout the system and freely diffusing. Also, the notion of the individual molecule does not exist and the concentration changes in the trajectories are "smooth," continuous, and deterministic.

The ODE integration method is fast and works well when the system is well behaved and concentrations are sufficiently large such that variations are smooth. However, when concentrations are small—on the order of tens to hundreds of molecules—stochastic noise can play a significant role. For example, turning a gene off or on can have a drastic effect on the synthesis of the corresponding protein and the timing of gene activation may be important in a gene regulation model. The continuum approximation of the ODE method is not useful in this case, since the behavior is not smooth and the notion of individual molecules becomes more important. Stochastic noise can introduce dynamic behavior that is not observable at the continuous limit, but is critical for the function at hand (e.g., see ref. 23). Another example to consider is cell differentiation, where stochastic noise could determine one of many cell fates from the same starting point. The deterministic ODE method cannot be employed in these cases.

To simulate noisy models or models where stochasticity is important, the simulation should be performed with Gillespie's stochastic simulation algorithm (SSA) (24), which provides an exact way of simulating a set of chemical ODEs in terms of such discrete random events. The SSA updates population numbers discretely, i.e., based on the firing of a single reaction event at a time. Because of this, the SSA is also typically much slower than ODE integration. When molecule numbers are high and there are no drastic events in the model, the discrete noise becomes less noticeable and the SSA trajectories are closer to the "smooth" limit of ODE trajectories. Subheading 6.3.2 and Fig. 5 provide an example of such a case.

Both SSA and ODE integration methods suffer from a significant computational limitation. They both require the entire reaction network to be stored in memory so that the respective populations and concentrations can be updated. This limits the size of the network that can be simulated (usually about 10^4 reactions and species). But biological reaction networks are typically dense and some networks can also be infinitely sized due to oligomerization (25).

For a certain subset of reaction networks (such as scaffolded interactions and linear cascades), significant reduction in network size can be achieved by exact model reduction techniques (12, 26, 27), which can systematically synthesize a compressed set of ODEs from a rule-based model specification (similar to the approach in Subheading 4.1 but on a larger scale). This enables exact simulation of the model for relevant outputs with a reduced number of differential equations. However, some types of interactions that are common in rule-based models, e.g., cooperativity between binding sites (such as in Subheading 6.2.4), preclude exact model reduction. Current model reduction methods are not powerful enough to guarantee reduction of large-scale rule-based models, and other simulation techniques are required to overcome the combinatorial bottleneck.

Another possible solution is to truncate the size of the network arbitrarily, for example, by not allowing complexes larger than ten molecules to form. However, the artificially constrained network is not guaranteed to have the same behavior as the full network. A more accurate approach is to generate only the portion of the network that is required at a particular time to advance the simulation. This forms the basis of "on-the-fly" methods (28, 29), one of which has been implemented in BNG (29). However, its performance degrades significantly for networks larger than about a thousand species (20).

The most efficient solution developed so far is to discard generating the network altogether. The molecules can be treated as particles instead of populations and particle-based stochastic simulation methods can be applied to them instead of the network-based SSA. This forms the basis of the ***network-free simulation methods***, with computational and memory requirements that scale with the number of particles, rather than the network size (25, 30). There are a number of network-free reaction kinetics simulators that have been developed for rule-based modeling. STOCHSIM (31) was the first agent-based chemical kinetics simulator, but its rule-based language is less expressive than BNGL and Kappa and it uses a sampling method that becomes inefficient as the range of rate constant values in the model becomes large (an order of magnitude or more), which is typical for large models. DYNSTOC (32) uses an extended version of the STOCHSIM algorithm and takes BNGL models as input, but still performs poorly in comparison to SSA-based simulators. RuleMonkey (33) works on BNGL models and uses a more efficient simulation algorithm based on the Gillespie's SSA. Both DYNSTOC and RuleMonkey have been validated for a broad range of rule-based models (32, 33). Network-free simulators are also available for the Kappa language (30) (see http://kappalanguage.org).

The most efficient and flexible network-free simulator compatible with BNGL that has been developed so far is *NFsim* (20) (see http://nfsim.org). NFsim and BNGL compatibly support global functional rate laws (such as those discussed in Subheading 5.3), local functions (which can be used to make rules even more concise and powerful), and logical Boolean functions. This increased functionality combined with the efficient particle-based stochastic simulator greatly expands the range of models that can be developed and simulated. The discussion of local functions is outside the scope of this tutorial and the reader is referred to Sneddon et al. (20). Defining global functions in BNGL is discussed in Subheading 6.1.

For models with less than a few hundred species, the simulation with SSA is faster than with NFsim. As the model size increases, however, the SSA rapidly becomes less efficient, whereas NFsim displays little or no decrease in performance as model size increases (20). Thus, NFsim allows the modeler to write any number of reaction rules without worrying if the underlying set of reactions is too large or infinite. One has to pay attention to the concentrations used, however, since the simulation speed is limited by the number of particles in the system. NFsim (and other network-free simulators such as RuleMonkey and the Kappa simulator) exactly samples the full chemical master equation for the system. Therefore, the trajectories produced from these simulators are indistinguishable from and equivalent to those produced by the SSA.

6. Receptor Ligand Interactions: A BioNetGen Tutorial

In this section, we walk through the process of constructing, simulating, and analyzing a BNGL model. We consider the example of the human epidermal growth factor (EGF) ligand binding specifically to the epidermal growth factor receptor (EGFR, also known as HER1 and ERBB1). This interaction is important for mammalian cells in most tissue types and functions as a signal that initiates growth and proliferation in both healthy and cancerous cells.

The model presented here can be used to investigate how hypothesized interactions affect aggregate distributions of receptors.

We formulate the model by drawing descriptions of biochemical interactions from the existing knowledge base in the biomedical literature and representing these as BNG molecules and rules. These descriptions should not be considered established facts, but rather should be considered hypotheses about the system structure, which can be tested by comparing model predictions with experimental observations and data.

The model is written in a normal text file with the extension ".bngl". The model comprises a series of text blocks, namely, parameters, molecule types, observables, seed species, reaction rules, and actions. The blocks' functions and compartments may also be provided in more complex models. All blocks are constructed as

```
begin block-name
 [blocktext]
 [blocktext]
 [blocktext]
end block-name
```

Any text on a line following a # (hash) symbol is considered a comment and is ignored. If a line becomes too long, the \ (backslash) symbol can be used to extend a statement to the next line. Subsequent tabs and spaces are treated as a single white space.

6.1. Seeding the Model

Knowledge Base:

> The ligand EGF monovalently binds the receptor EGFR. The extracellular domain of EGFR has a binding site for EGF and a domain that mediates dimerization with other receptor molecules (34). EGFR autophosphorylates at multiple amino acid positions, including Y1068 (tyrosine at position 1068 in the amino acid sequence of human EGFR) and Y1173 (reviewed in ref. 35).

Translating this information from the literature, we can construct the molecule types as follows:

`L(rec)`—ligand with a receptor-binding site.

`R(lig,dim,y1068~0~p,y1173~0~p)`—receptor with a ligand-binding site, a dimerization domain and two tyrosines that can be in unphosphorylated or phosphorylated states.

The molecule types are defined in a molecule types block, i.e.,

```
begin molecule types
  L(rec)
  R(lig,dim,y1068~0~p,y1173~0~p)
end molecule types
```

We also need to define the starting species and their concentrations for the model. This is provided using the seed species block. The complete species specifications should be provided. In this case, the starting species are simply free ligand and monomeric unphosphorylated receptor. It is a good convention to provide population numbers for the seed species and use microscopic rate constants in the reaction rules.

```
begin seed species
  R(lig,dim,y1068~0,y1173~0) R0
  L(rec) L0
end seed species
```

`R0` and `L0` are parameters that are to be defined separately in the parameters block. In the parameters block, a parameter name is assigned a real number or a standard mathematical expression involving other parameters. For example, `R0` can be defined directly as a number and `L0` can be defined in terms of concentration converted to population number. The units of the parameters can be given in comments for clarity:

```
begin parameters
  V_ext   1.6e-9     #liters
  N_Avo   6.022e23   #molecule number per mole
  R0    1e5          #molecule number per cell
  L0    L_conc*V_ext*N_Avo   #molecule number
end parameters
```

Parameters used elsewhere in this tutorial are assumed to have been defined in the parameters block either with numerical values or as expressions of other parameters.

Finally, we need to define the observables for the model. This model treats receptor aggregation and phosphorylation. Not only are we interested in the number of dimer species, but also in the number of receptors in dimers. These can be represented using species-observables and molecules-observables, respectively, in the observables block (note the use of wildcards in observables).

```
begin observables
   Molecules BoundLigand L(rec!+)
   Molecules BoundReceptor R(lig!+)
   Species Dimer R(dim!0).R(dim!0)
   Species UnligatedSpecies R(lig,dim),\
     R(lig,dim!0).R(lig,dim!0)
   Species PhosphSpecies R(y1068~p),R(y1173~p)
end observables
```

The `Molecules/Species` keyword indicates the type of observable. This is followed by the name assigned to the observable and then the list of patterns to match the species summed up in the observable.

An optional functions block can be used to define global functions of the observables. These functions can be used to track key variables in the simulation or for modified rate laws in reaction rules (see Subheading 5.3). The observables referenced by a global function must be defined in the observables block. The function itself is defined in the functions block. The observables block must *precede* the functions block for the observables to be used in the functions.

An example function involving the species-observable SubstrateSum would be defined in the functions block as:

```
begin functions
  k_func kcat*E_tot/(KM + SubstrateSum)
end functions
```

Although it is not used in this particular model, k_func defined here can be used instead of the Sat rate law in a reaction rule.

6.2. Building the Model

In this section, we create several hypotheses about the receptor aggregation mechanism and then implement reaction rules based on each hypothesis.

The rules are enclosed in a reaction rules block as

```
begin reaction rules
 [reaction_rule]
 [reaction_rule]
end reaction rules
```

6.2.1. Dimer-Dependent Phosphorylation

Knowledge Base:

> The kinase on EGFR is inactive in monomers and is activated by dimerization (36).

If both phosphorylation sites were represented as identical (e.g., if the molecule type were R(lig,dim,y~0~p,y~0~p) instead), then we would only need a single rule to represent it. The dimerization is given as context and the transformation is phosphorylation.

```
R(dim!0).R(dim!0,y~0) -> R(dim!0).R(dim!0,y~p) k_ph
```

However, we have chosen to model the two phosphorylation sites distinctly. Therefore, we would need two different rules. The kinetics can still be made identical with identical rate constants or nonidentical with different rate constants.

```
R(dim!0).R(dim!0,y1068~0)   ->\
 R(dim!0).R(dim!0,y1068~p)   k_ph1
R(dim!0).R(dim!0,y1173~0)   ->\
 R(dim!0).R(dim!0,y1173~p)   k_ph2
```

Since the only available binding partner on dim component is another dim component (i.e., we are not writing any reaction rules, where dim binds to anything else), we can shorten the context representation using the + wildcard on the bond.

```
R(dim!+,y1068~0) -> R(dim!+,y1068~p) k_ph1
R(dim!+,y1173~0) -> R(dim!+,y1173~p) k_ph2
```

Not dephosphorylating the tyrosine sites would result in runaway phosphorylation, which is undesirable. We can use a uniform first-order background dephosphorylation:

```
R(y1068~p) -> R(y1068~0) k_deph
R(y1173~p) -> R(y1173~0) k_deph
```

This is a valid assumption to make when the phosphatases are highly active and nonspecific.

6.2.2. Ligand-Dependent Dimerization

Hypothesis:

Ligand binding initiates receptor dimerization.

Expressed as a traditional equilibrium model:

$$R + L \overset{K_{\text{bind}}}{\Longleftrightarrow} RL.$$

$$RL + RL \overset{K_{\text{dim}}}{\Longleftrightarrow} LRRL$$

Here, the reactions are represented as bidirectional equilibrium reactions defined by ***equilibrium association constants***, which are ratios of the forward and reverse rate constants. BNGL requires that the model be specified in terms of individual rate constants, i.e.,

$$R + L \underset{k_r}{\overset{k_f}{\rightleftharpoons}} RL, \qquad \text{where } K_{\text{bind}} = \frac{k_f}{k_r}.$$

$$RL + RL \underset{k_{rd}}{\overset{k_{fd}/2}{\rightleftharpoons}} LRRL, \quad \text{where } K_{\text{dim}} = \frac{k_{fd}/2}{k_{rd}}.$$

The two bidirectional reactions can be directly translated into bidirectional reaction rules.

```
R(lig,dim) + L(rec) <->\
  R(lig!0,dim).L(rec!0) k_f,k_r
R(lig!+,dim) + R(lig!+,dim) <->\
  R(lig!+,dim!0).R(lig!+,dim!0)k_f_d,k_r_d
```

Note that the bimolecular association in the second rule is symmetric and in the traditional expression would require a (1/2) multiplication factor. However, as mentioned earlier, BNG requires the modeler to use the *asymmetric* reaction rate constant.

On processing the rule, BNG discovers the symmetry and automatically assigns the multiplication factor.

Knowledge Base:

> Both singly ligated and unligated dimers have been discovered to exist (37).

The model as it stands represents the typical manner in which ligand-induced dimerization and phosphorylation are modeled, e.g., in refs. 21, 38, 39. However, this model is deficient because it does not capture the full nature of the interaction between ligand-binding and dimerization processes. In the current model hypothesis, the dimer state is inextricably linked to the ligand-bound state, which is not true in the light of the above evidence. The experimental evidence of existence of singly ligated and unligated dimers enables *rejecting* the current model hypothesis.

6.2.3. Ligand-Independent Dimerization

To explain the existence of singly ligated and unligated dimers, we introduce an alternate hypothesis.

Hypothesis:

> Ligand binding and receptor dimerization are independent of each other.

Expressed as a traditional equilibrium model (with equilibrium association constants):

$$\begin{array}{ccccc} \mathrm{R,R,L,L} & \overset{K_{\text{bind}}}{\Longleftrightarrow} & \mathrm{R,RL,L} & \overset{K_{\text{bind}}}{\Longleftrightarrow} & \mathrm{RL,RL} \\ \Updownarrow K_{\text{dim}} & & \Updownarrow K_{\text{dim}} & & \Updownarrow K_{\text{dim}} \\ \mathrm{RR,L,L} & \overset{K_{\text{bind}}}{\Longleftrightarrow} & \mathrm{RRL,L} & \overset{K_{\text{bind}}}{\Longleftrightarrow} & \mathrm{LRRL} \end{array}$$

In terms of reactions:

$$\begin{array}{ccccc} \mathrm{R,R,L,L} & \underset{k_r}{\overset{k_f}{\rightleftharpoons}} & \mathrm{R,RL,L} & \underset{k_r}{\overset{k_f}{\rightleftharpoons}} & \mathrm{RL,RL} \\ k_{fd}\downarrow\uparrow k_{rd} & & k_{fd}\downarrow\uparrow k_{rd} & & k_{fd}\downarrow\uparrow k_{rd} \\ \mathrm{RR,L,L} & \underset{k_r}{\overset{k_f}{\rightleftharpoons}} & \mathrm{RRL,L} & \underset{k_r}{\overset{k_f}{\rightleftharpoons}} & \mathrm{LRRL} \end{array}$$

where $K_{\text{bind}} = \dfrac{k_f}{k_r}$ and $K_{\text{dim}} = \dfrac{k_{fd}}{k_{rd}}$.

Since the dimerization and ligand binding reactions are completely independent of each other in this model, we can model this system using two context-free rules:

```
R(lig)+L(rec)<->R(lig!0).L(rec!0) k_f,k_r
R(dim)+R(dim)<->R(dim!2).R(dim!2) k_f_d,k_r_d
```

Note that by removing context we have enabled the reaction rule to generate additional reactions. The dimerization rule in the earlier hypothesis generated only one reaction: RL with RL. Here, it generates three dimerization reactions: R with R, R with RL, and RL with RL.

Knowledge Base:

> Phosphorylation of receptor increases on ligand addition (reviewed in ref. 35).

In this model, we have delinked the dimerization from ligand binding, i.e., ligand binding does not influence the equilibrium concentrations of the dimer (and vice versa). Since phosphorylation is dimer dependent, this implies that ligand binding cannot directly influence phosphorylation. The model behavior is at odds with experimental evidence. Hence, this model hypothesis is also *rejected*.

6.2.4. Cooperative Ligand Binding and Dimerization

Two events are said to exhibit *cooperativity* if the sequence of occurrence of those two events affects the intrinsic rates at which they occur. One plausible model for dimer formation posits that the monomer, unligated dimer, and singly ligated dimer all have different affinities for the ligand (40).

Hypothesis:

> Ligand binding and receptor dimerization are mutually cooperative.

Expressing this as a traditional equilibrium model (with equilibrium association constants):

$$\begin{array}{ccccc} \mathrm{R,R,L,L} & \overset{K_{\mathrm{bind},1}}{\Leftrightarrow} & \mathrm{R,RL,L} & \overset{K_{\mathrm{bind},1}}{\Longleftrightarrow} & \mathrm{RL,RL} \\ \Updownarrow K_{\mathrm{dim},1} & & \Updownarrow K_{\mathrm{dim},2} & & \Updownarrow K_{\mathrm{dim},3} \\ \mathrm{RR,L,L} & \overset{K_{\mathrm{bind},2}}{\Leftrightarrow} & \mathrm{RRL,L} & \overset{K_{\mathrm{bind},3}}{\Longleftrightarrow} & \mathrm{LRRL} \end{array}.$$

Cooperativity imposes thermodynamic constraints on the model. If there is no external energy source or sink, a system of reversible reactions should obey mass and energy conservation.

If there are multiple paths from the same set of reactants to the same set of products, then the product of the equilibrium constants along all paths should be identical. In other words, the product of equilibrium constants over a closed loop of reversible reactions should be unity. This effect is called *detailed balance* and places constraints on the parameters and rules used to model the system (e.g., in ref. 38). In this case, there are at least two closed loops, resulting in the following constraints:

$$K_{\text{bind},1}K_{\text{dim},2} = K_{\text{bind},2}K_{\text{dim},1}.$$
$$K_{\text{bind},3}K_{\text{dim},2} = K_{\text{bind},1}K_{\text{dim},3}.$$

To simplify parameters, we describe the three ligand-binding equilibrium constants using multiplicative factors α and β on the monomer ligand-binding equilibrium constant K_{bind}, i.e.,

$$K_{\text{bind},1} = K_{\text{bind}}.$$
$$K_{\text{bind},2} = \alpha K_{\text{bind},1}.$$
$$K_{\text{bind},3} = \beta K_{\text{bind},1}.$$

Using these relations and the constraints obtained previously, we can describe the dimerization equilibrium constants as multiplicative factors on the unligated dimerization equilibrium constant K_{dim}:

$$K_{\text{dim},1} = K_{\text{dim}}.$$
$$K_{\text{dim},2} = \alpha K_{\text{dim}}.$$
$$K_{\text{dim},3} = \alpha\beta K_{\text{dim}}.$$

The network can be rewritten as

$$\begin{array}{ccccc} \text{R,R,L,L} & \overset{K_{\text{bind}}}{\Longleftrightarrow} & \text{R, RL, L} & \overset{K_{\text{bind}}}{\Longleftrightarrow} & \text{RL, RL} \\ \Updownarrow K_{\text{dim}} & & \Updownarrow \alpha K_{\text{dim}} & & \Updownarrow \alpha\beta K_{\text{dim}}. \\ \text{RR,L,L} & \overset{\alpha K_{\text{bind}}}{\Longleftrightarrow} & \text{RRL, L} & \overset{\beta K_{\text{bind}}}{\Longleftrightarrow} & \text{LRRL} \end{array}$$

Note that BNG requires reaction rate constants, whereas here we have only managed to obtain multiplicative factors for the equilibrium constants. We do not know how the factor distributes between the forward reaction and the reverse reaction and must make assumptions about it.

A general way of resolving this issue would be to distribute the multiplicative factors in the following manner (41):

$$\text{If K'} = \frac{k'_{\text{f}}}{k'_{\text{r}}} = \alpha K = \alpha\left(\frac{k_{\text{f}}}{k_{\text{r}}}\right).$$
$$k'_{\text{f}} = \alpha^{\phi}k_{\text{f}}.$$
$$k'_{\text{r}} = \alpha^{\phi-1}k_{\text{r}}.$$
$$0 \leq \phi \leq 1.$$

Now, by changing ϕ, we can control the distribution of the multiplicative factor over the forward and reverse reaction rate constants in the most general way. There is no single best value for ϕ. In this tutorial, we use $\phi = 0$, causing the multiplicative factor to affect only the reverse rate and not the forward rate, i.e.,

$$k'_{\mathrm{f}} = k_{\mathrm{f}}.$$

$$k'_{\mathrm{r}} = \alpha^{-1} k_{\mathrm{r}} = \frac{k_{\mathrm{r}}}{\alpha}.$$

Rewriting the equilibrium model in terms of reactions and implementing the multiplicative factors (for the equilibrium constants) as divisive factors on the reverse rate:

$$\begin{array}{ccccc} \mathrm{R,R,L,L} & \underset{k_{\mathrm{r}}}{\overset{k_{\mathrm{f}}}{\rightleftharpoons}} & \mathrm{R,RL,L} & \underset{k_{\mathrm{r}}}{\overset{k_{\mathrm{f}}}{\rightleftharpoons}} & \mathrm{RL,RL} \\ k_{\mathrm{fd}} \downarrow\uparrow k_{\mathrm{rd}} & & k_{\mathrm{fd}} \downarrow\uparrow k_{\mathrm{rd}}/\alpha & & k_{\mathrm{fd}} \downarrow\uparrow k_{\mathrm{rd}}/\alpha\beta\,, \\ \mathrm{RR,L,L} & \underset{k_{\mathrm{r}}/\alpha}{\overset{k_{\mathrm{f}}}{\rightleftharpoons}} & \mathrm{RRL,L} & \underset{k_{\mathrm{r}}/\beta}{\overset{k_{\mathrm{f}}}{\rightleftharpoons}} & \mathrm{LRRL} \end{array}$$

where $K_{\mathrm{bind}} = \frac{k_{\mathrm{f}}}{k_{\mathrm{r}}}$ and $K_{\mathrm{dim}} = \frac{k_{\mathrm{fd}}}{k_{\mathrm{rd}}}$.

These reactions are now consistent with detailed balance. Depending on the values chosen for α and β, one can cause the system to exhibit multiple cooperative behaviors.

Knowledge Base:

Dimerization is enhanced by ligand binding. However, the two ligand-binding sites on the dimer are negatively cooperative (42).

If the unligated dimer has a higher affinity for ligand than the monomer, the presence of ligand shifts the dimer–monomer equilibrium towards the dimer, which also increases receptor phosphorylation. In this model, ligand binding cooperates positively with dimer formation leading to an increase in dimerization when ligand is added. Another feature of this system, however, appears to be that once one receptor in a dimer binds ligand, the second receptor exhibits a considerably reduced affinity for ligand (40). In other words, ligand binding is *negatively cooperative* on the dimer. Structural evidence strongly supporting negative cooperativity has emerged recently (43).

Based on the equilibrium constants determined from experiment ((42), Fig. 3), we use the values $\alpha = 120$ and $\beta = 0.07$. These values are consistent with our previous discussion that ligand binding

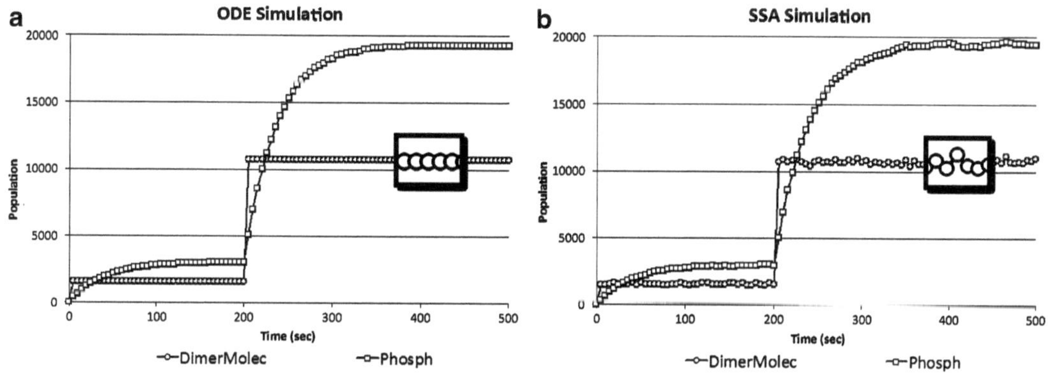

Fig. 3. Simulation of receptor dimerization model. (**a**) A model of ligand-induced dimerization and subsequent kinase activation is simulated using deterministic ODE integration. From $t = 0$ to 200 s, the system is equilibrated, and at $t = 200$ s, the system is perturbed by adding the ligand. The monomer–dimer equilibrium shifts rapidly with an increase in the number of molecules in dimers. Since kinase activity depends on dimerization, the number of phosphorylated sites also increases concurrently. Note that the trajectories are smooth (see *inset*). (**b**) The same sequence of simulations is performed using Gillespie's SSA. Note that the trajectories are noisy (see *inset*), but the noise is relatively small because of the larger population sizes and the observed behavior is similar to that of *Panel* (**a**).

enhances dimerization ($K_{\mathrm{dim},2} = 120\ K_{\mathrm{dim},1}$ and $K_{\mathrm{dim},3} = 8.4\ K_{\mathrm{dim},1}$) whereas the ligand binding to the dimer is negatively cooperative ($K_{\mathrm{bind},3}$=0.07 $K_{\mathrm{bind},2}$).

Modeling this scenario requires more reaction rules than the previous model. This is a characteristic feature of rule-based modeling. Independent processes, lacking reaction context, are the easiest to express with the fewest number of rules. Unidirectional influences, in which one process affects the rate of another but not vice versa, require additional context. Bidirectional influences, in which two processes mutually affect their rates giving rise to cooperativity, represent the most complex linkage between two components. Care must be taken in modeling such interactions to ensure that detailed balance is obeyed.

For the current hypothesis, we can write the rules as follows:

```
# Ligand association
R(lig)+L(rec) -> R(lig!1).L(rec!1) k_f
# Ligand dissociation for monomer, singly-ligated dimer and
  doubly-ligated dimer
R(dim,lig!1).L(rec!1) -> R(dim,lig)+L(rec) k_r
R(dim!2,lig).R(dim!2,lig!1).L(rec!1) ->\
R(dim!2,lig).R(dim!2,lig) + L(rec)
  k_r/alpha
R(dim!2,lig!+).R(dim!2,lig!1).L(rec!1) ->\
R(dim!2,lig!+).R(dim!2,lig) + L(rec) k_r/beta
```

```
# Dimer association
R(dim) + R(dim) -> R(dim!1).R(dim!1) k_f_d
#Dimer dissociation for unligated/singly-ligated/doubly-
  ligated dimers
R(lig,dim!1).R(lig,dim!1) -> R(lig,dim) + R(lig,dim)
  k_r_d
R(lig!+,dim!1).R(lig,dim!1) -> R(lig!+,dim) +
  R(lig,dim) k_r_d/alpha
R(lig!+,dim!1).R(lig!+,dim!1) -> R(lig!+,dim) +
  R(lig!+,dim) k_r_d/(alpha*beta)
```

This model of receptor aggregation corresponds closely with the experimental and structural evidence. Hence, we accept these rules and proceed to simulate the model. The choice of rate parameters is discussed in general in Subheading 8.

6.3. Simulating the Model

The molecule types, parameters, seed species, and reaction rules constitute the model. Various actions can then be performed on the model using the actions block. Actions are preexisting BNGL routines that can be called at will by the modeler. Typically, actions require action parameters and flags and should be terminated with the semicolon symbol.

```
begin actions
 [action];
 [action];
end actions
```

6.3.1. Network Generation

The network generation action starts with the seed species and iteratively applies the reaction rules to the species set to generate new reactions and species. This can proceed until the network size stops increasing, in which case the whole network has been generated. Or, the size of the network can be arbitrarily limited by limiting the number of iterations (using `max_iter` flag) or the maximum number of molecules (using `max_stoich`) in a complex. The network is printed out in a text file with ".net" extension. The name of the *.bngl* file is used as the base name for the *.net* file.

The different species are assigned numbers by default to compactly represent the reactions, but the full configuration can also be written out instead using a `TextReaction` flag. The `overwrite` flag is used to indicate whether to overwrite a preexisting file of the same name. A typical `generate_network` command would look like this:

```
generate_network({overwrite=>1,max_stoich=>{R,2},
  max_iter=>20,TextReaction=>1});
```

In this statement, the `generate_network` routine is called which generates the reaction network subject to the condition that the maximum number of receptor molecules is 2 in any complex and that the rules are iteratively applied at most 20 times. When the network is being written in the *.net* file, the routine is allowed to overwrite the existing files of the same name and is told to write down the full species specifications in the reactions.

For the purposes of this model, we could simply use:

```
generate_network({overwrite=>1,TextReaction=>1});
```

For large networks, it is advisable to generate the network only once using the `generate_network` command and then for subsequent simulation reuse the network using the `readFile` action or by giving the *.net* file name as a parameter in other actions (see online documentation at http://bionetgen.org for more details).

6.3.2. Equilibration and Perturbation

In any experiment, a ground state is perturbed in some defined fashion and then the ground state and perturbed states are compared using some measurable quantity. In biological systems, the ground state is typically a preexisting equilibrated control state and the perturbation is typically the addition of some chemical species or activation of some reaction. Then, after a defined time or at many time points, the two states are compared using some experimental probe.

When simulating a biological model, it is important to follow the same principle. To establish the ground state of the experiment, the model is seeded with everything, except the perturbing agent and simulated until equilibrium or steady state is achieved. This step is called the ***equilibration*** step. A "good" model, i.e., one which closely corresponds to reality, is one where all transformations are complementary and, in the absence of a perturbing agent, the system equilibrates. Following equilibration, the perturbing agent is added to the system and subsequent simulation is called ***perturbation***, or plainly ***simulation*** (which can be misleading).

For the receptor aggregation model, the ground state involves receptor association and dissociation in the absence of ligand and the perturbed state involves addition of ligand and monitoring the aggregation. The monomeric receptors that seed the model would necessarily be far away from equilibrium. Using actions, we can set the ligand concentration to zero and then use numerical ODE integration to equilibrate the model.

```
simulate_ode({suffix=>ode,t_start=>0,t_end=>200,
  n_steps=>100});
```

The `suffix` option is used to append the "_" and then "ode" to the base name, which is taken from the name of the *.bngl* file. `t_start` and `t_end` indicate the total time over which the model is equilibrated and `n_steps` indicates the number of time points at which concentrations are recorded.

Following equilibration, we wish to add ligand and perturb the model. To do this, first we need to save the concentrations from the equilibration step and then add new ligand. This is accomplished using

```
saveConcentrations();
setConcentration("L(rec)","L0_temp");
```

The `continue` flag can be used to turn concatenate a new simulation with the earlier one. One must make sure that the `t_start` of the new simulation is the same as the `t_end` of the earlier simulation and that the `suffix` value is not changed. Any number of simulations can be so concatenated.

```
simulate_ode({suffix=>ode,continue=>1,t_start=>200,
  t_end=>500,n_steps=>100});
```

Invocation of the SSA simulator follows the same syntax, but the `simulate_ssa` command is used instead. For the SSA and ODE simulations, the entire network is generated (and saved in the *.net* file) and the trajectories (concentration vs. time) of all species are recorded in the *.cdat* text file. The trajectories of the observables (which are weighted sums of specific species) are recorded in the. *gdat* text file. The NFsim method does not generate the network and, therefore, writes only a *.gdat* file. Since multiple simulations can be concatenated in a single actions block, use of appropriate suffixes is encouraged to distinguish between data files.

If the `continue` flag was set to 1 and the suffix was not changed, then the trajectory of the simulation is concatenated with the previous simulation. For a single sequence of concatenated simulations, a single *.cdat* file and a single *.gdat* file are generated. If the `continue` flag was set to 0 (which is the default value) or if the suffix used for the simulation step is changed, then each simulation output is written to a separate set of data files.

To simulate the same perturbation using SSA, we would need to reset the species concentrations to the equilibrated state. This is performed by the `resetConcentrations()` method which changes all the species concentrations to what they were during the last `saveConcentrations()` call. Following the reset, ligand is introduced using the `setConcentration()` method and SSA is used to simulate the model.

NFsim can be invoked using the `simulate_nf` command, provided the binary for NFsim exists in the bin subfolder of the BNG installation. The actions that can be used in the actions block and their respective syntax are posted with the documentation on the BNG Web site (http://bionetgen.org).

6.3.3. Execution

The actions block is used to list the sequence of commands that the modeler wishes to implement when BNG executes the model. The modeler should save the complete model as a *.bngl* file (say model. bngl), open a terminal or command prompt, and enter the directory in which the BNGL file resides. Then, the following command should be issued:

```
perl /.../BioNetGen/Perl2/BNG2.pl model.bngl
```

where /. . ./ is a placeholder for the path to the parent folder where BNG is installed. The slash convention used (backslash, forward slash) might differ depending on the terminal used. For example, on a Windows command prompt, the command would look like:

```
perl C:\BioNetGen\Perl2\BNG2.pl model.bngl
```

On a Unix terminal such as bash (found in OS X and Linux), the command would look like:

```
perl /Users/username/BioNetGen/Perl2/BNG2.pl model.
 bngl
```

Perl is required to run BNG and might need to be separately installed on certain platforms. The command can also be embedded in terminal scripts and other scripting environments, such as R, Octave, MatLab, or Mathematica.

On execution, BNG reads the BNGL file and sequentially implements the actions in the actions block. Files written by the actions (say, network generation or simulation) are saved in the same folder.

6.3.4. Analysis

The simulation data are in table format in the *.cdat* and *.gdat* text files. They can be easily imported into different software for statistical analysis and visualization, including (but not limited to) spreadsheet software (Microsoft Excel, OpenOffice Calc), plotting programs (Gnuplot, Grace), or interactive computing environments (Matlab, Octave, R, Mathematica). The models themselves can be exported to the SBML language using `writeSBML` action, as a Matlab function file using `writeMfile` action or as a Matlab Mex file using the `writeMexfile` action. The M-file uses Matlab's inbuilt ODE15 simulation engine, whereas the Mex file is a compiled ODE model that utilizes the faster CVODE libraries used in BNG. The Mex file can be used to run computationally intensive parameter scans and analysis in Matlab.

RuleBender, the integrated development environment for BNG (see Subheading 2), also provides facilities for plotting and analyzing simulation output. In addition, it provides several different global views on models, including contact maps and influence diagrams (see http://rulebender.org).

The trajectories for both ODE and SSA simulations of the same model are shown in Fig. 3. The model file used for the simulation is shown in Fig. 4. Note how the ODE simulations are smooth, but the SSA simulations exhibit some noise. Because the model involves large populations, the SSA trajectories closely resemble the ODE trajectories.

```
begin parameters
# Universal constants
 N_Avo  6.023e23 # per mol

# Biophysical considerations
# Cell is a CHO cell, dimensions & quantities from ref 1
# Volume 1.2 pL, Cytoplasm 1 pL, Radius 6.6 microns, Sphere
# Surface area 5.5e-10 m^2 or 5.5e-8 dm^2, Membrane thickness 10 nm (or) 1e-7 dm
# External volume 1.6 nL (assuming 6.15e5 cells/mL)
 V_ext  1.6e-9 # L
 V_cell 1.2e-12 # L
 V_cyt          1.0e-12 # L
 V_nuc          V_cell - V_cyt # L
 h_mem          1e-7 # dm
 S_mem          5.5e-8 # dm^2
 V_mem          h_mem*S_mem # L, comes to 5.5 fL

# Assumed association reaction rate constant, per M per s
 k_ext  1e9 # EGF, small freely diffusing ligand, so high
 k_mem  1e6 # EGFR, large bulky protein, so low. Calculated from refs 5-7
 k_cyt  1e9 # Only small molecules such as Grb2, Sos

# Equilibrium Constants (Association Constants). Calculated from ref 2
 Kdim_2D        5.3e9 # per (mol/dm^2)
 Kdim   Kdim_2D*h_mem # per (mol/dm^3) or per M
 Kbind  4.6e9 # per M
 alpha  120
 beta   0.07

# Kinetic Rate Constants
 k_f            k_ext/(N_Avo*V_ext)
 k_f_d  k_mem/(N_Avo*V_mem)

 k_r1   k_ext/Kbind
 k_r2   k_r1/alpha
 k_r3   k_r1/beta

 k_r_d1 k_mem/Kdim
 k_r_d2 k_r_d1/alpha
 k_r_d3 k_r_d1/(alpha*beta)

# Kinase & Phosphatase
 kcat           0.025 # per s

# Initial Concentrations
 R_conc 1e5 # Receptors per cell
 L_conc 1e-3    # M
 P_conc 200e-9  # M
 R0             R_conc
 L0_new L_conc*V_ext*N_Avo
 P0             P_conc*V_cyt*N_Avo
 L0             0
end parameters

begin molecule types
 R(lig,dim,tyr~u~p,tyr~u~p)
 L(rec)
 PTase(a)
end molecule types

begin seed species
 R(lig,dim,tyr~u,tyr~u) R0
 L(rec)                         L0
 PTase(a)                       P0
end seed species

begin reaction rules
# Ligand binding
R(lig) + L(rec) -> R(lig!0).L(rec!0) k_f
R(dim,lig!0).L(rec!0) -> R(dim,lig) + L(rec) k_r1
R(dim!1,lig).R(dim!1,lig!0).L(rec!0) -> R(dim!1,lig).R(dim!1,lig) + L(rec) k_r2
R(dim!1,lig!+).R(dim!1,lig!0).L(rec!0) -> R(dim!1,lig!+).R(dim!1,lig) + L(rec) k_r3

# Dimerization
R(dim) + R(dim) -> R(dim!0).R(dim!0)    k_f_d
R(dim!0,lig).R(dim!0,lig) -> R(dim,lig) + R(dim,lig) k_r_d1
R(dim!0,lig!+).R(dim!0,lig) -> R(dim,lig!+) + R(dim,lig) k_r_d2
R(dim!0,lig!+).R(dim!0,lig!+) -> R(dim,lig!+) + R(dim,lig!+) k_r_d3

# Phosphorylation
R(dim!+,tyr~u) -> R(dim!+,tyr~p) kcat
R(tyr~p) -> R(tyr~u)    kcat
#R(tyr~p) -> R(tyr~u)   Sat(vmax,KM)
end reaction rules

begin observables
 Molecules      BoundLig        L(rec!+)
 Molecules      Monomers        R(dim)
 Molecules      DimerMolec      R(dim!+)
 Molecules      Phosph  R(tyr~p)
end observables

begin actions
generate_network({overwrite=>1});
simulate_ssa({suffix=>ssa,t_start=>0,t_end=>200,n_steps=>200});
saveConcentrations();
setConcentration("L(rec)","L0_new");
simulate_ssa({suffix=>ssa,continue=>1,t_start=>200,t_end=>500,n_steps=>300});
end actions

#references
#1: Seewoster and Lehmann. Biotechnology and Bioengineering (1997) vol. 55 (5) pp. 793-797
#2: Macdonald and Pike. Proc Natl Acad Sci USA (2008) vol. 105 (1) pp. 112-7
#3: Gabdoulline and Wade.Current Opinion in Struct Biol (2002) vol. 12 (2) pp. 204-13
#4: Kholodenko. European Journal of Biochemistry (2000) vol. (267) pp. 1583-1588
#5: Mayawala et al. Biophys Chem (2006) vol. 121 (3) pp. 194-208
#6: Lauffenburger and Linderman, Oxford University Press, New York, 1993.
#7: Kusumi et al. Biophysical Journal (1993) vol. 65 (5) pp. 2021-40
```

Fig. 4. Model file. This is the model file used for simulation in Subheading 6, the results of which are plotted in Fig. 3.

Since BNGL models are merely text files, scripts written in scripting languages (such as Python or Perl) or scripting environments (such as Matlab) may be used to read, modify, and simulate BNGL models from the outside. This flexibility enables BNGL models to be processed for advanced model analysis, including fitting models to experimental data, scanning numerical ranges for parameters, checking local and global parameter sensitivity, using BNGL models as modules in a hierarchical framework, etc.

7. Compartmental Modeling

The default assumption in BNG is a reactor of unit volume. For modeling reactions in multiple volumes, the user can manually model the location as a separate component state and provide it as context for each reaction rule. The user would also need to include the correct volume-scaling factor for each reaction rule, however, and this quickly gets tedious as the number of compartments increases, especially for reaction rules that perform the same transformation in different compartments. For detailed modeling with multiple compartments, a more sophisticated approach is required.

The compartmental extension to BNG seeks to alleviate these issues by accommodating within its language a compartmental hierarchy, a compartment attribute to molecules and species, an automatic detection of possible inter-compartmental transport reactions, and an automatic scaling of reaction rules based on the locations of the reactants. cBNG was introduced and discussed in detail by Harris et al. (44). In this section, we provide the technical concepts underlying compartmental modeling in cBNG and explain them using appropriate examples.

7.1. Compartment Topology

There are two types of compartments in cBNG: 3D *volumes* and 2D *surfaces*. cBNGL supports a hierarchical topology that mimics that of the cell. Every volume can enclose one or more surfaces, whereas every surface has to enclose exactly one volume. The enclosing compartment is referred to as the parent and the enclosed compartments are referred to as children. A compartment is considered ***adjacent*** to its parent and children. Volumes cannot enclose volumes and surfaces cannot enclose surfaces.

The parent volume that encloses a surface and the child volume enclosed by the same surface are referred to as *bridged-volumes* and the intermediary surface is referred to as the *bridging surface*. Similarly, the parent surface that encloses a volume and any child surface enclosed by the same volume are referred to as *bridged-surfaces* and the intermediary volume is referred to as the *bridging volume*.

The compartments block is used to encode the compartment hierarchy and compartment volumes. It is an optional block that triggers the use of cBNGL framework and syntax. Consider the hierarchy described in Fig. 5. This can be encoded in a compartments block as follows:

```
begin compartments
  Ext 3   V_ext          # External milieu
  Plm 2   V_plm   Ext    # Plasma Membrane, enclosed
by Ext
  Cyt 3   V_cyt   Plm    # Cytosol, enclosed by Plm
  Enm 2   V_enm   Cyt    # Endosomal Membrane,
enclosed by Cyt
  Num 2   V_num   Cyt    # Nuclear Membrane,
enclosed by Cyt
  End 3   V_end   Enm    # Endosome, enclosed by Enm
  Nuc 3   V_nuc   Num    # Nucleus, enclosed by Num
end compartments
```

Each line indicates the name of the compartment, the dimension (2D or 3D), the volume of the compartment, and the name of the parent compartment. A short description can be given as a comment for clarity.

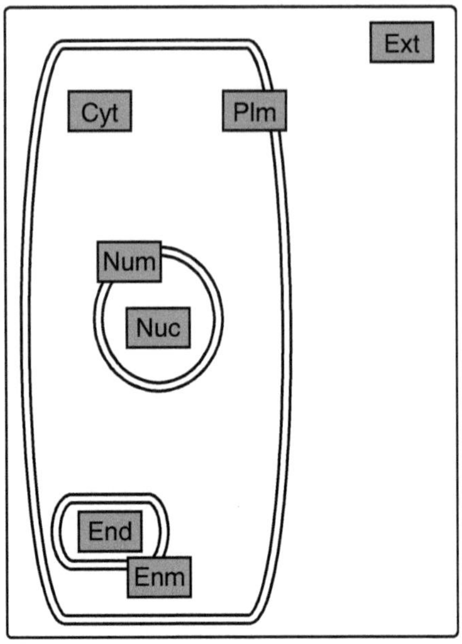

Fig. 5. Compartment hierarchy. To use cBNG, the modeler must define a compartments block with an appropriate nesting of compartments. 3D-compartments, called volumes, can only enclose 2D-compartments, called surfaces, and vice versa. A volume can enclose more than one surface, but a surface cannot enclose more than one volume. In this hierarchy, Ext, Cyt, Nuc, and End are volumes and Plm, Num, and Enm are surfaces.

7.2. Molecule Location

Molecule location is given by a postfix with the @ symbol in both patterns and species. For example, to represent a monomeric receptor present in the plasma membrane compartment (`Plm`), we use:

```
R(dim)@Plm
```

In patterns and species with multiple molecules, every molecule's location can be indicated by an @ postfix. For example, to represent a dimer of receptors present in the plasma membrane, we use:

```
R(dim!0)@Plm.R(dim!0)@Plm
```

Molecules can form bonds with molecules in the same compartments or in adjacent compartments. Thus, a single species can span multiple compartments. For example, to represent a receptor on the plasma membrane that binds a ligand outside the cell, we use:

```
R(lig!0)@Plm.L(rec!0)@Ext
```

A *topologically consistent species* does not span more than one surface and does not have bonds that need to pass through compartments. Examples are shown in Fig. 6.

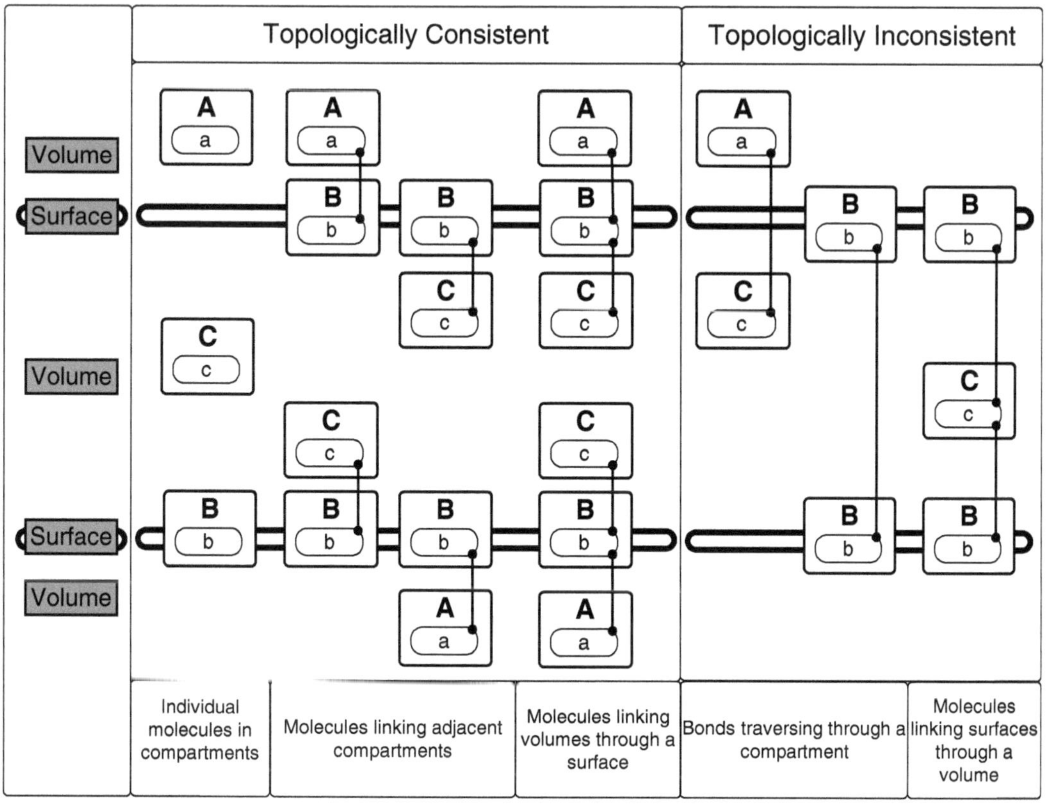

Fig. 6. Topologically consistent species. cBNG maintains topological consistency whenever a transport rule is implemented. Topologically consistent species do not have bonds traversing compartments. The species itself cannot span more than one surface. Reactions that create topologically inconsistent species are discarded.

If cBNG is invoked using the compartments block, and then the seed-species block should contain topologically consistent species with their full compartmental specification, for example,

```
begin seed species
 R(lig,dim)@Plm   R0
 L(rec)@Ext       L0
end seed species
```

7.3. Location in Patterns

cBNG provides a lot of flexibility in tailoring patterns to match species based on the location of their molecules.

The *aggregate location* of a species can be inferred from the location of its molecules.

- If all the molecules in a species are in a volume compartment, then their aggregate location is that volume compartment.
- If all the molecules in a species are in a surface compartment, then their aggregate location is that surface compartment.
- If the species spans one surface and one or two volumes, then the aggregate location is that surface compartment.

Species can be referred to by their aggregate location using an @ prefix on a pattern. For example, consider the dimer pattern (on the plasma membrane), `R(dim!0)@Plm.R(dim!0)@Plm`.

The same pattern can be referred to using the aggregate location as `@Plm:R(dim!0).R(dim!0)`.

For species that bridge compartments, the aggregate location (the surface) can still be provided as @ prefix, whereas the connected molecules in the adjacent compartments can be tagged with the @ postfix as exceptions to the rule.

For example, consider the receptor dimer on the plasma membrane connected to a single ligand molecule in the external compartment: `R(dim!0,lig!1)@Plm.L(rec!1)@Ext.R(dim!0)@Plm`.

The aggregate location is the `Plm` surface. To match this species using a pattern, we can use the `@Plm` prefix and provide the `Ext` locations as postfix (underlined for emphasis), i.e., `@Plm:R(dim!0,lig!1).R(dim!0).L(rec!1)@Ext`

Volume molecules bound to a surface can be matched by prefix location tags of the aggregate location.

For example, consider the species: `R(dim,lig!1)@Plm.L(rec!0)@Ext`.

The aggregate location of the species is `Plm`.

It can be matched by both patterns: `L(rec!+)@Ext` and `@Plm:L(rec!+)`.

The matched region is underlined for emphasis.

7.4. Reaction Rules in the Compartmental Framework

7.4.1. Automatic Volume Scaling

If cBNG is not invoked, i.e., if the compartments block is not used, the modeler must manually provide the volume factor and Avogadro number factor for the microscopic reaction rate constant, for example:

```
A(b) + B(a) -> A(b!0).B(a) k/(N_Avo*V_cyt)
```

If cBNG is invoked, i.e., the compartments and volumes are specified in the compartments block, then the volume scaling is part of the cBNG processing and the modeler needs to only provide the Avogadro number factor. For example, the same reaction rule would be modeled in the cBNG framework as:

```
A(b) + B(a) -> A(b!0).B(a) k/N_Avo
```

cBNG would automatically scale it to `k/(N_Avo*V_cyt)` if it detects that the bimolecular reaction can occur in the cytoplasm and `V_cyt` is specified as the volume of the cytoplasmic compartment in the compartments block.

In accordance with the explanation given in Subheading 5.1, unimolecular reactions are never scaled by volume, either in surfaces or volumes. Bimolecular reactions in "volumes" and "surfaces" are automatically scaled by the respective compartmental volume. Molecules in surfaces are assumed to be restricted to a small volume enveloping the surface (i.e., the surface volume is equal to the surface area multiplied by a surface thickness) provided by the modeler. It is *not* recommended to model reaction orders higher than bimolecular in the compartmental framework.

7.4.2. Universal Reaction Rules

Biological reactions can happen between reactants as long as they are in the same or adjacent compartments. Often, the identity of the compartments could be irrelevant to the rate constant (except for the volume-scaling factor). For example, multiple cell types can produce different numbers of the same receptor on their plasma membranes, but these receptors bind with the same intrinsic affinity to an external ligand. Such universal phenomena can be written as universal reaction rules, i.e., without any compartmental context.

On processing a universal reaction rule, cBNG can identify the compartments in which the reactants can occur. Then, cBNG automatically generates reactions matching the reaction rule, but applicable to the specific compartments (or pairs of adjacent compartments) where the reactants are in proximity. For bimolecular and higher reactions, cBNG automatically scales the reaction rate by the appropriate volume.

For example, receptor dimerization can occur on both plasma membranes (`Plm`) and endosomal membranes (`Enm`). This can be represented by a universal rule:

```
R(dim) + R(dim) -> R(dim!0).R(dim!0) k
```

This rule would generate dimerization reactions in both `Plm` and `Enm` compartments.

```
R(dim)@Plm + R(dim)@Plm -> R(dim!0)@Plm.R(dim!0)@Plm
  k/V_Plm
R(dim)@Enm + R(dim)@Enm -> R(dim!0)@Enm.R(dim!0)@Enm
  k/V_Enm
```

If a specific reacting pair occurs only in one location (or one pair of adjacent locations), then it is sufficient to use a universal rule to model the reaction.

7.4.3. Scope-Restricted Reaction Rules

Scope-restricted reactions are reactions whose behavior the modeler wishes to restrict depending on where the location of the reaction is. For example, the same reactions can occur in three different compartments, but the modeler might wish to assign different microscopic rates for the reaction in two compartments and disallow the reaction in the third.

Universal rules cannot be employed for these purposes. In cBNGL, by using the @ symbol, modelers have the power to specify exactly where a particular reaction occurs. For example, suppose receptors were degraded at two different rates depending on the location, i.e., rapidly in endosomes, but only very slowly in the plasma membrane, then we can use scope-restricted rules to show this:

```
R()@End -> 0 k_degr_fast DeleteMolecules
R()@Plm -> 0 k_degr_slow DeleteMolecules
```

Appropriate volume scaling is done for bimolecular and higher order reactions.

7.4.4. Transport Rules

The cBNGL specification allows molecules and species to be moved between compartments in a number of ways as long as the resulting species is *topologically consistent.* If a transport reaction generated by a transport rule creates a topologically inconsistent species, BNG will detect it and discard the reaction.

Molecules can be moved individually from their location to any adjacent compartment (volume to surface or surface to volume) simply by changing the @-postfix. For example, moving a hypothetical A molecule from the plasma membrane to the child compartment cytoplasm:

```
A(x)@Plm -> A(x)@Cyt
```

This is *adjacent-compartment molecular transport.*

Moving a molecule between adjacent compartments does not affect the postfix location tag of other molecules in the same species. For example, if a B molecule were connected to A, then it would stay in `Plm`. The above rule would generate the reaction (with transported molecule underlined for emphasis):

```
A(x,b!0)@Plm.B(a!0)@Plm -> A(x,b!0)@Cyt.B(a!0)@Plm k
```

An example where such a transport rule would be useful is to model membrane insertion of proteins, i.e., when a protein freely diffusing in the cytoplasm is transported to a membrane compartment. Adjacent-compartment molecular transport *cannot be extended to species transport* by using the @-prefix instead of the postfix.

Individual molecules can be moved from one volume to another volume through a bridging surface. For example, they can be moved between the cytoplasm and the nucleus, which are bridged by the nuclear membrane. This is *bridged-volume molecular transport.*

```
A(x)@Cyt -> A(x)@Nuc k
```

An example where such a transport mechanism would be useful is transport through a channel in the surface, such as the flow of chloride ions from the external volume to the cytoplasm through a chloride channel in the membrane.

Bridged-volume molecular transport has the potential to create topologically inconsistent complexes. Such reactions, if they could be generated from the reaction rules, will be automatically discarded. Unlike adjacent-compartment transport, bridged-volume molecular transport *can be extended* to *bridged-volume species transport* by using the @-prefix instead of the postfix. This would simply move the entire species matched by the pattern as long as it is fully contained in the bridged-volume.

For example, consider the rule:

```
@Cyt:A(x) -> @Nuc:A(x) k
```

When this rule is applied to the `@Cyt:A(x,b!0).B(a!0)` species, the following reaction is generated:

```
A(x,b!0)@Cyt.B(a!0)@Cyt -> A(x,b!0)@Nuc.B(a!0)@Nuc k
```

As one can see, the location tags of all molecules in the species have been converted from `Cyt` to `Nuc`.

The reaction mechanism does not have to be unimolecular, for example, a valid bridged-volume transport rule would be:

```
@Cyt:A(x)+B()@Num -> @Nuc:A(x) + B()@Num k
```

Here, a molecule `B()` on the bridging surface `Num` enables the transport of some species matching `A(x)` between the bridged volumes `Cyt` and `Nuc`. The rate constant k is bimolecular rate constant scaled only by the Avogadro number.

An example where a bridged-volume transport mechanism could be useful is when target proteins bind transport proteins and the resulting complex is compact enough to penetrate the pores on the nuclear membrane. Bimolecular mechanisms are useful to model a situation, where the rate of transport depends on the number of pores or channels present.

A slightly more complicated form of bridged-volume transport is when a "fixed" molecule is on a surface and is connected to a "transporting" molecule in a volume. Now, when the transporting molecule is moved across the surface, it carries over any molecules (or "cargo") that are connected to it in the volume. To describe this transport mechanism, we use the `MoveConnected` keyword.

For example, if the fixed molecule were R on the nuclear membrane, the transporting molecule were T, and cargo molecules were A and B, one can imagine a situation where T is connected to R and is able to move between cytoplasm and nucleus. This can be described with the reaction rule:

```
T(r!0)@Cyt.R(t!0)@Num -> T(r!+)@Nuc.R(t!0)@Num k
  MoveConnected
```

The `MoveConnected` keyword would ensure that any molecules connected to T (and any molecules connected to those molecules and so on) would be moved too, i.e., it would generate the following reactions:

```
A(t!1)@Cyt.T(car!1,r!0)@Cyt.R(t!0)@Num ->\
  A(t!1)@Nuc.T(car!1,r!0)@Nuc.R(t!0)@Num k
B(t!1)@Cyt.T(car!1,r!0)@Cyt.R(t!0)@Num ->\
  B(t!1)@Nuc.T(car!1,r!0)@Nuc.R(t!0)@Num k
```

This is *bridged-volume connected transport*. In the absence of `MoveConnected` option, the A and B molecules would retain their

`Cyt` tags creating a topologically inconsistent complex. BNG would then discard the reaction.

This mechanism is useful to model situations, where the transported molecule forms relatively stable associations with the pore that is necessary for the transport process. For example, glucose transporters need to form stable complexes with the glucose molecules before they can be moved across a membrane.

The final form of transport is *bridged-surface connected transport*. In BNG, this is implemented in a manner consistent with endocytosis. Compartments cannot be dynamically created in BNG; hence, endocytosis is modeled as movement between preexisting surface compartments plasma membrane and endosomal membrane bridged by the cytoplasm. When implementing a transport between surfaces bridged by a volume compartment, cBNG does the following:

- The molecules in the rule on the starting surface are moved to the destination surface.
- The connected molecules directly on the starting surface are moved to the destination surface.
- The connected molecules in the bridging volume stay in the bridging volume.
- The connected molecules in the outer volume are moved to the inner volume and vice versa.

Consider the bridged-surface transport rule from plasma membrane to endosomal membrane:

```
R()@Plm -> R@Enm k
```

Using the hierarchy of compartments in Fig. 5, cBNG interprets the rule to mean:

`Plm`— starting surface

`Enm`—destination surface

`Ext`—outer volume (parent of `Plm`)

`Cyt`—bridging volume (child of `Plm` and parent of `Enm`)

`End`—inner volume (child of `Enm`)

Let us see how the rule affects the following species that spans the plasma membrane, i.e., it has molecules in external compartment, plasma membrane, and cytoplasm:

```
A(r!0)@Ext.R(a!0,b!1)@Plm.B(r!0)@Cyt
```

Molecules in `Plm` (starting surface) are moved to `Enm` (destination surface)
Molecules in `Ext` (outer volume) are moved to `End` (inner volume).
Molecules in `Cyt` (bridging volume) remain in `Cyt`.

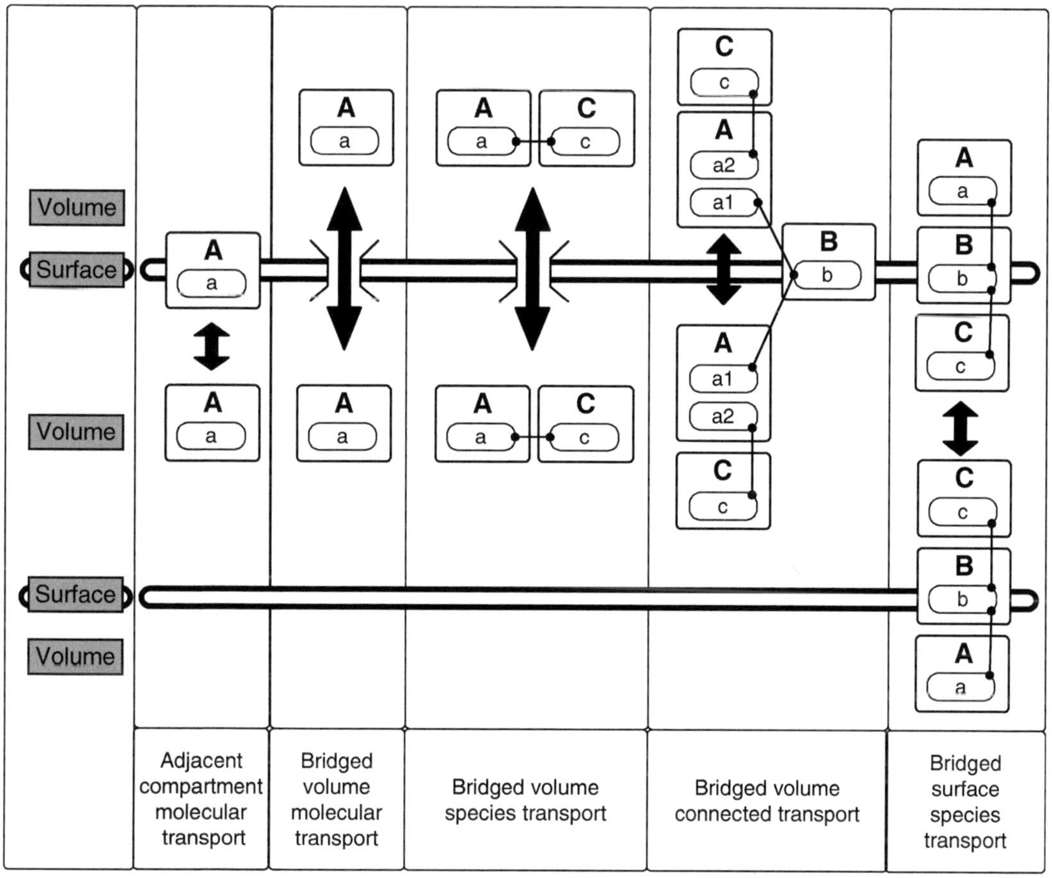

Fig. 7. Transport. The different types of transport mechanisms available in cBNG are illustrated. The details can be found in Subheading 7.4.4.

In other words, the above species would be converted to:

```
A(r!0)@End.R(a!0,b!1)@Enm.B(r!0)@Cyt
```

The different forms of allowed transport mechanisms are described in Fig. 7.

8. Biological Parameterization

One of the problems with modeling detailed biochemistry is the combinatorially large requirement of rate parameters. Rule-based modeling alleviates this requirement to a large extent by assigning identical parameters to reactions modeled by a single rule. This makes the number of required kinetic parameters proportional to the number of rules in the model. Where possible, these parameters are obtained either directly or estimated from experiments detailed in the biochemical literature.

Should the literature prove insufficient, certain limited assumptions can be made about parameters using general biophysical and biochemical knowledge. These assumptions can also be tested for their impact on a model using parameter sensitivity analysis methods (45). In this section, we demonstrate approaches to estimate, calculate, and represent important parameters of signal transduction models, such as cell geometry, concentrations, reaction rate constants, and equilibrium constants. This section can be skipped by the experienced modeler familiar with biological parameters and their relevance to reaction kinetic modeling.

8.1. Volumes

As mentioned in the Subheading 5.1, the volume of the compartment in which a reaction occurs directly affects the microscopic reaction rate constant. Therefore, before modeling a system, one must specify the volumes of the compartments in which the reactions occur. Consider an example parameterization used for the model in the Subheading 6.

We are studying the effects of an external growth factor on a eukaryotic cell line. For modeling purposes, we assume a single cell of volume V_{cell} suspended in a liquid milieu of volume V_{ext}. For Chinese Hamster Ovary cells, an experiment determining the cell size distribution (46) indicates that the mean cell volume (V_{cell}) is 1.2 picoliters (pl) and that the cell suspension has a cell density of 6.15×10^5 cells/ml. Inverting this value, we can estimate V_{ext} as 1.6 nl. The cell is assumed to be divided into the cytosol and nucleus in a 5:1 ratio (from ref. 47 similar to assumptions in ref. 48). This enables us to calculate the compartment volumes V_{cyt} as 1 pl and V_{nuc} as 0.2 pl.

Secondly, we wish to model reactions on the cell membrane, which can be assumed to be a flat sheet containing freely diffusing molecules. These surface interactions are scaled by the 2D surface area rather than the 3D volume. To arrive at these values, we can assume that the cell is a sphere.

$$\text{Volume } V_{\text{cell}} = 1.2\,\text{pl} = 1.2 \times 10^{-12}\text{L} = 1.2 \times 10^{-15}\text{m}^3$$

$$\Rightarrow \text{Radius } r = \left(\frac{V_{\text{cell}}}{4\pi/3}\right)^{(1/3)} = 6.6 \times 10^{-6}\text{m} = 6.6\,\mu\text{m}$$

$$\Rightarrow \text{Surface Area } S_{\text{mem}} = 4\pi r^2 = 5.5 \times 10^{-10}\,\text{m}^2 = 5.5 \times 10^{-8}\,\text{dm}^2.$$

To calculate a surface "volume," one can approximate membrane thickness h, say 10 nm (or 10^{-7} dm), encompassing the membrane and all the proteins attached to the membrane on either side.

$$h = 10\,\text{nm} = 10 \times 10^{-9}\,\text{m} = 10^{-7}\,\text{dm}$$

$$\Rightarrow \text{Membrane Volume } V_{\text{mem}} = S_{\text{mem}}\,h = 5.5 \times 10^{-15}\,\text{dm}^3$$

$$= 5.5 \times 10^{-15}\,\text{L} = 5.5\,\text{fl}.$$

Although they are 2D interactions, surface reaction parameters are often approximated from their solution-phase values, which are reported in 3D units (i.e., on a per-volume basis). The membrane thickness approximation is useful to interconvert the 2- and 3D units. Note that the units used for each of the parameters have to be consistent and interconvertible. We considered membrane thickness in dm and surface area in dm^2. By simply multiplying these values, we directly get the volume in dm^3, which is equivalent to liters.

The ratio of V_{cyt} to V_{mem} (here, ~182) is an important consideration when modeling membrane reactions because it provides a measure of the acceleration of membrane reactions versus cytosolic reactions (13). This acceleration is purely a concentration effect, which arises from membrane reactions being restricted to a small volume enveloping the membrane, whereas cytosolic reactions are distributed throughout the large cytosolic volume.

8.2. Concentrations

Understanding the relationship between concentration and population number is of paramount importance in signal transduction models. A realistic model is one, where the concentrations of the molecules and complexes stay within the biologically acceptable ranges. These ranges can be quite wide or narrow, depending on the type of molecule. Even within a biological context, concentrations of specific molecular species can change dramatically during signaling, often over an order of magnitude. To ensure biological relevance, the modeler must provide starting concentrations that reflect the acceptable concentration ranges.

Concentration is traditionally defined as the number of moles of a molecule (or complex) per unit volume of the container (expressed in liters). A mole equals 6.02214×10^{23} individual entities.

$$\text{Concentration (M)}\ C = \frac{N}{N_{Avo}\, V},$$

where N is number of entities (no unit), N_{Avo} is Avogadro number $= 6.02214 \times 10^{23}$/mol, V is volume (L).

To provide a sense of scale, one can keep in mind that in a volume of 1 pl, 1 nM concentration equals 602 molecules, but 1 μM equals 602,214 molecules.

Similarly, for the same number of molecules, the concentration can be large or small depending on the volume of the compartment. For example, 10^4 molecules per cell is 16.6 nM in the cytosol of volume 1 pl, but 3.02 μM in the membrane compartment of volume 5.5 fl.

Depending on the biology being modeled, different molecule types in the model can have drastically different concentration ranges. A global protein profiling study of yeast cells (49) shows a log-normal distribution with some proteins being as few as 50 molecules per cell and some as high as a million per cell. Most proteins are found at levels of many hundreds to a few thousands of molecules per cell. Eukaryotic cells are typically larger than yeast cells, but they also exhibit high variability in size and shape.

Typical concentration/population ranges for signal transduction models (from low to high) include the following:

1–10 molecules	Genes
10–100 molecules	mRNA transcripts
nM concentrations	Typical signal transduction proteins
μM concentrations	Typical secreted molecules
mM concentrations	Free triphosphate nucleotides (ATP and GTP), Ca^{2+}

Another effect of extremely small volumes (or extremely high concentrations) is molecular crowding. This can cause a breakdown of the well-mixed infinite dilution assumption, especially for bulky proteins. Crowding tends to reduce reaction probabilities and increase spatial heterogeneity because the reaction probability becomes more dependent on local conditions. Crowding can also affect assumptions made about other parameters, such as rate constants and equilibrium constants. The problem of identifying crowding effects in kinetic models is an active research area (50, 51). For a BNGL model, the modeler should be aware of the caveat that significant crowding can affect the rate constants that apply inside cells in comparison to those that are measured using purified proteins or cell extracts.

8.3. Equilibrium Constants

Consider a typical reversible bimolecular reaction. The ratio of the forward and reverse rate constants is referred to as the *equilibrium association constant* K_a and the reciprocal is referred to as the *equilibrium dissociation constant* K_d.

$$A + B \underset{k_r}{\overset{k_f}{\rightleftharpoons}} AB.$$

$$K_a = \frac{1}{K_d} = \frac{k_f}{k_r}.$$

At equilibrium, the flux in both directions is equal.

$$k_f[A]_{eq}[B]_{eq} = k_r[AB]_{eq} \Rightarrow K_a = \frac{1}{K_d} = \frac{k_f}{k_r} = \frac{[AB]_{eq}}{[A]_{eq}[B]_{eq}}.$$

Thus, the equilibrium constant is the ratio of the concentration of products and reactants at equilibrium. Suppose at equilibrium, bound and unbound concentrations of reactant A are equal (i.e., half of A is bound and half of A is free). This would restrict the value of unbound concentration of reactant B.

$$[A]_{eq} = [AB]_{eq} \Rightarrow K_d = [B]_{eq}.$$

Similarly,

$$[B]_{eq} = [AB]_{eq} \Rightarrow K_d = [A]_{eq}.$$

Thus, the dissociation equilibrium constant is also the concentration of one reactant for which the other reactant is half-bound at equilibrium. The dissociation equilibrium constant is also that concentration near which the changes in concentration of the bound complex are the largest.

Because of the above relations, the equilibrium constant (in either definition) is a powerful way to quantify the strength of a reversible association. The dissociation equilibrium constant has units of concentration and is often referred to as the *binding affinity*. The lower the value of K_d, the higher the strength of the interaction. Reversible protein–protein and protein–small molecule interactions have very wide ranges of affinities:

$K_d \approx$ (fM)	Extremely strong protein–inhibitor interactions
$K_d \approx$ (pM)	Typical strong interactions
$K_d \approx$ (nM)	Typical moderate interactions
$K_d \approx$ (μM)	Typical weak interactions
$K_d \approx$ (mM)	Extremely weak nonspecific interactions

Most functional protein–protein interactions in signal transduction are in the nM–μM range. Equilibrium constants are some of the most widely available types of experimental data and are often the first target of biophysical experiments if an interaction is hypothesized between two proteins. When using available literature values, the modeler must be aware of experimental caveats, such as use of truncated or chimeric proteins, non-physiological experimental conditions, etc.

8.4. Rate Constants

BNG requires individual microscopic rate constants for each interaction, i.e., *per-site asymmetric* rate constants. The conversion from macroscopic to microscopic rate constants is explained in the Subheading 5.1. In this section, we focus on how to arrive at the macroscopic constants themselves.

Often, these parameters are not explicitly determined or available. They are usually available indirectly in the form of equilibrium constants or Michaelis–Menten fits. Due to the unconstrained nature of the relationships between them, assumptions must often be made when modeling explicit reactions.

Consider the bimolecular association reaction described earlier.

$$A + B \underset{k_r}{\overset{k_f}{\rightleftharpoons}} AB.$$

The "forward" rate constant k_f, i.e., the bimolecular rate constant of association, is often estimated by theoretical derivation or Brownian dynamics simulations (reviewed in ref. 52). To simplify model parameterization in the face of insufficient data, a weak assumption can be employed on bimolecular association rate constants: similar sized molecules have similar diffusivities and therefore similar association rate constants under well-mixed conditions. The diffusion coefficient for small molecules in water places an upper limit on the possible diffusion coefficients for proteins and, consequently, the association rate constants. Typical ranges of values for the association rate constant (high to low) include the following (52):

10^{10} M^{-1}s^{-1}	Diffusion limit in water
10^{9}–10^{10} M^{-1}s^{-1}	Small molecule interactions
10^{7}–10^{9} M^{-1}s^{-1}	Protein–protein interactions involving long-range electrostatics
10^{6}–10^{7} M^{-1}s^{-1}	Typical protein–protein interactions
10^{3}–10^{6} M^{-1}s^{-1}	Interactions of bulky slowly diffusing proteins

Dissociation rate constants (k_r) are much harder to theoretically derive or estimate using simple assumptions. More often than not, they need to be experimentally verified. Dissociation rate constants can also vary over a much wider range than association rate constants for similarly sized molecules. If equilibrium constants are the only available data, a consistent approach to estimating dissociation rates is to assume a fixed association rate constant and compute the dissociation rate constant:

$$K_a = \frac{1}{K_d} = \frac{k_f}{k_r} \Rightarrow k_r = \frac{k_f}{K_a} = k_f K_d.$$

In the case of enzymatic catalysis, interactions are typically reported using the implicit enzyme parameters k_{cat} and K_M which are estimated under controlled conditions. However, as explained in the Subheading 5.3, in large and dense biological networks, it is preferable to model the reactions explicitly using the rate parameters k_f, k_r, and k_{cat}. The three explicit rate parameters can be derived from the two implicit enzyme parameters by making an additional assumption. To be consistent with our previous approach, the forward rate constant k_f can be fixed and the reverse rate estimated:

$$K_M = \frac{k_r + k_{cat}}{k_f} \Rightarrow k_r = K_M k_f - k_{cat}.$$

When determined experimentally, the dissociation rate k_r (U/s) is often expressed in terms of the stability of the formed complex. A typical quantification is the half-life $\tau_{1/2}$ which denotes the time taken (s) for half the complexes to dissociate or the exponential decay constant τ_e (s) which denotes the time taken for the concentration of the complex to decay to $1/e$ of the original value (Euler number $e = 2.71828$). The dissociation rate is inversely proportional to these decay constants.

$$k_r = \frac{\ln 2}{\tau_{1/2}} = \frac{1}{\tau_e}.$$

9. Downstream Signaling from the Membrane

In Subheading 6, we created and simulated a detailed model of receptor–ligand interactions. In this section, we broadly treat the rest of the signal transduction paradigm (see Fig. 8).

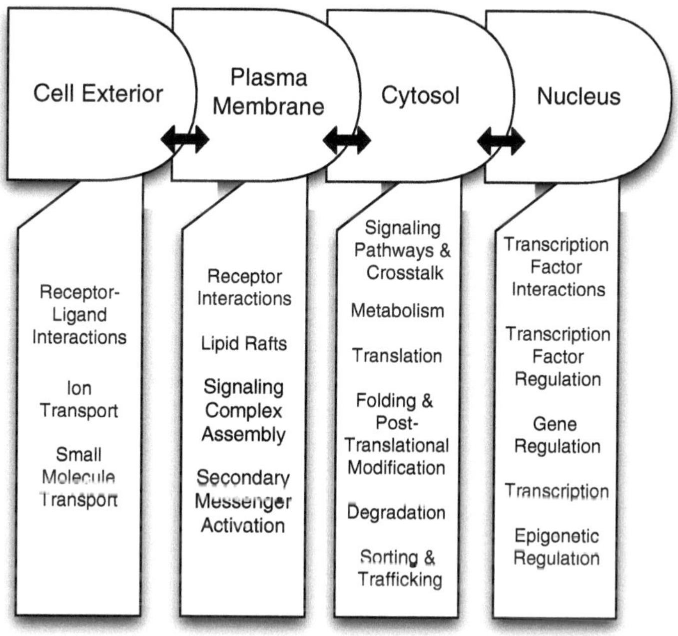

Fig. 8. General paradigm of signal transduction. At each compartmentalization of the cell, numerous processes of various types occur. These processes interact with each other and also with processes in adjacent compartments through means of transport. In general, a signal from the exterior is detected at the membrane, transduced through the cytoplasm, and concludes in gene regulation. This changes the transcription profile, subsequently affecting the protein distribution and can lead to a phenotypic change of state for the entire cell.

Signaling at the membrane causes the recruitment of molecules into an active signaling complex which possesses or modifies catalytic activities to change the distribution of messenger molecules in the cytosol and membrane. These messengers subsequently recruit and initiate signaling cascades. The signaling cascades interact with each other through shared molecules and usually terminate with the activation/attenuation of transcription factors in the nucleus. These transcription factors initiate/terminate synthesis of new molecules which can then modify the network structure to adapt to the incoming stimulus.

Here, we focus only on the EGFR (ErbB1) and the signals recruited by its homo-dimerization. One should keep in mind that ErbB1 is only one member of the growth factor-activated ErbB kinases and that homo- and hetero-dimerization result in a typically dense and intricate network that is still under experimental scrutiny. We have attempted to capture many of the essential mechanisms here that can be easily extended or modified.

9.1. Assembling the Signaling Complex

The assembly of the signaling complex at the membrane is a critical step linking the ligand binding that happens outside the cell to the signaling inside the cell. The EGFR is activated by ligand-induced enhancement of dimerization that is modeled in Subheading 6. The dimer formation introduces a conformational change that activates the tyrosine kinase domains on the receptor. These kinase domains then phosphorylate specific tyrosines on the receptor.

These phosphorylated tyrosine sites then form specific docking sites for several proteins that contain SH2, SH3, and PTB domains. Some recruited proteins, such as Grb2, Gab1, and Shc1, function as adaptors and scaffolds to facilitate recruitment of other proteins to the signaling complex. The receptor tyrosine kinase can also activate binding sites on these adaptors and scaffolds. Among the recruited proteins are several enzymes that perform one of the following two functions: (a) activate/deactivate other binding sites in the complex (typically through phosphorylation/dephosphorylation (53)) or (b) being themselves activated by the recruitment process, and change the distributions of messenger molecules in the cytosol and membrane.

While utilizing biochemical literature to formalize the recruitment interactions, it is important to distinguish sequential (processive) and parallel (distributive) interactions (54). *Sequential* interactions are those where every interaction is a prerequisite for the next interaction. *Parallel* interactions are those that are independent of each other, occurring on different regions of the same molecule. The typical phenomenological description of the recruitment process is often misleading, making it seem sequential, when in fact some interactions can be independent and parallel.

For example, a typical phenomenological statement might be "Phosphotyrosines on the receptor recruit Grb2 which in turn recruits Sos1." This makes it seem as if the Sos1–Grb2 interaction is dependent on the Grb2–phosphotyrosine interaction. A closer look at the structure and function of Grb2 shows that this implied linearization is false and that the Grb2–Sos1 and Grb2–phosphotyrosine interactions are independent of each other. The statement is then an observation of just one outcome of these two parallel interactions, which is Grb2 bound to both Sos1 and a receptor phosphotyrosine. It ignores the other two outcomes, where Grb2 is bound to only one or the other.

The older modeling literature is replete with such subjective interpretations and rule-based modeling was developed so that such arbitrary choices can be avoided (54). Structural information on different molecules is available in several databases, such as PDB (55) and UniProt (56), that can be exploited to model the interactions rigorously and accurately. An intimate knowledge of the structural nature of the interactions helps in determining which interactions are sequential and which are not.

A good way of distinguishing a sequential binding interaction is to ask the question: "Does a new binding site need to be created/activated for this interaction to occur?" If the answer is yes, then the interaction is sequential to the previous interaction. If the answer is no, then it is independent of the previous interaction. An intimate knowledge of the structural aspects of the proteins is very useful in this aspect. In the absence of conclusive evidence, it is better to assume that interactions are distributive/parallel rather than processive/sequential.

Knowledge Base:

> The binding of EGF ligand to EGFR receptor is cooperative with the homo-dimerization of EGFR. Ligand binding enhances dimerization, but ligand-binding sites on the dimer are negatively cooperative.

The modeling of this information is treated in the Subheading 6.

Knowledge Base:

> Dimerized EGFR autophosphorylates on several tyrosines in its amino acid sequence, including Y998, Y1016, Y1092, Y1110, Y1138, Y1172, and Y1197 (57). 2030

We model the phosphorylation with identical kinetics, although we could potentially use a different rate constant for each phosphorylation.

```
R(dim!+,Y998~0)   ->R(dim!+,Y998~p)    k_ph
R(dim!+,Y1016~0)  ->R(dim!+,Y1016~p)  k_ph
R(dim!+,Y1092~0)  ->R(dim!+,Y1092~p)  k_ph
R(dim!+,Y1110~0)  ->R(dim!+,Y1110~p)  k_ph
R(dim!+,Y1138~0)  ->R(dim!+,Y1138~p)  k_ph
R(dim!+,Y1172~0)  ->R(dim!+,Y1172~p)  k_ph
R(dim!+,Y1197~0)  ->R(dim!+,Y1197~p)  k_ph
```

Knowledge Base:

> Dimerized EGFR kinases are also responsible for activation of phosphorylation sites on Gab1 and Shc1 if they are present in the same complex. These include Y627 on Gab1 (58), tandem YXXP and YXXM motifs on Gab1 (58), and Y317 on Shc1 (59).

We can use a dot operator to show that these molecules are in the same complex when they are modified, even though there can be several ways in which the molecules could be recruited.

```
R(dim!+).Gab1(YXXP~0) -> R(dim!+).Gab1(YXXP~p) k_ph
R(dim!+).Gab1(YXXM~0) -> R(dim!+).Gab1(YXXM~p) k_ph
R(dim!+).Gab1(Y627~0) -> R(dim!+).Gab1(Y627~p) k_ph
R(dim!+).Shc1(Y317~0) -> R(dim!+).Shc1(Y317~p) k_ph
```

Knowledge Base:

> If the molecule Shp2 is recruited to the complex, it opposes the phosphorylation activity of the receptor tyrosine kinases (60).

In the absence of a dephosphorylation event, the phosphorylation sites will be saturated to phosphorylation even with a tiny amount of dimerization, as explained in Subheading 5.2. Shp2 recruited to the complex can dephosphorylate any of the kinase substrates:

```
Shp2().R(Y998~p) -> Shp2().R(Y998~0)        k_deph
Shp2().R(Y1016~p) -> Shp2().R(Y1016~0)      k_deph
Shp2().R(Y1110~p) -> Shp2().R(Y1110~0)      k_deph
Shp2().R(Y1138~p) -> Shp2().R(Y1138~0)      k_deph
Shp2().R(Y1172~p) -> Shp2().R(Y1172~0)      k_deph
Shp2().R(Y1197~p) -> Shp2().R(Y1197~0)      k_deph
Shp2().Gab1(YXXP~p) -> Shp2().Gab1(YXXP~0) k_deph
Shp2().Gab1(YXXM~p) -> Shp2().Gab1(YXXM~0) k_deph
Shp2().Gab1(Y627~p) -> Shp2().Gab1(Y627~0) k_deph
Shp2().Shc1(Y317~p) -> Shp2().Shc1(Y317~0) k_deph
```

Knowledge Base:

> The receptor phosphotyrosines recruit proteins in the following manner: Y1110, Y1138, Y1172, and Y1197 bind both Grb2 and Shcl. Y1092 binds only Grb2. Y1016 binds only Shp2. Y998 binds both Shp2 and Shcl (57). All binding interactions are reversible.

The interaction with the receptor is localized to the SH2 domain on Grb2. The N-terminal regions of Shcl, Shp2, and Rasa1 all contain SH2/SH3/PTB domains in some combination that enables binding to specific phosphotyrosines. They are modeled as generic "n" domains, referring to their N-terminal position.

```
# Binding of Y1110, Y1138, Y1172 and Y1197
R(Y1110~p) + Grb2(sh2) <-> R(Y1110~p!1).Grb2(sh2!1)
  kf_pY_grb2,kr_pY_grb2
R(Y1138~p) + Grb2(sh2) <-> R(Y1138~p!1).Grb2(sh2!1)
  kf_pY_grb2,kr_pY_grb2
R(Y1172~p) + Grb2(sh2) <-> R(Y1172~p!1).Grb2(sh2!1)
  kf_pY_grb2,kr_pY_grb2
R(Y1197~p) + Grb2(sh2) <-> R(Y1197~p!1).Grb2(sh2!1)
  kf_pY_grb2,kr_pY_grb2
R(Y1110~p) + Shc1(n) <-> R(Y1110~p!1).Shc1(n!1)
  kf_pY_shc1,kr_pY_shc1
R(Y1138~p) + Shc1(n) <-> R(Y1138~p!1).Shc1(n!1)
  kf_pY_shc1,kr_pY_shc1
R(Y1172~p) + Shc1(n) <-> R(Y1172~p!1).Shc1(n!1)
  kf_pY_shc1,kr_pY_shc1
R(Y1197~p) + Shc1(n) <-> R(Y1197~p!1).Shc1(n!1)
  kf_pY_shc1,kr_pY_shc1

# Binding of Y1092
R(Y1092~p) + Grb2(sh2) <-> R(Y1092~p!1).Grb2(sh2!1)
  kf_pY_grb2,kr_pY_grb2

# Binding of Y1016
R(Y1016~p) + Shp2(n) <-> R(Y1016~p!1).Shp2(n!1)
  kf_pY_shp2,kr_pY_shp2

# Binding of Y998
R(Y998~p) + Shc1(n) <-> R(Y998~p!1).Shc1(n!1)
  kf_pY_shc1,kr_pY_shc1
R(Y998~p) + Shp2(n) <-> R(Y998~p!1).Shp2(n!1)
  kf_pY_shp2,kr_pY_shp2
```

A contact map with defined symbols for components, interactions, and influences is useful in summarizing these reaction rules. The conventions followed for the contact maps in this tutorial are

Contact Map	Meaning
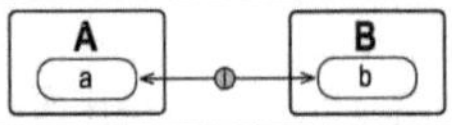	A has component a, B has component b Components a and b may form a bond
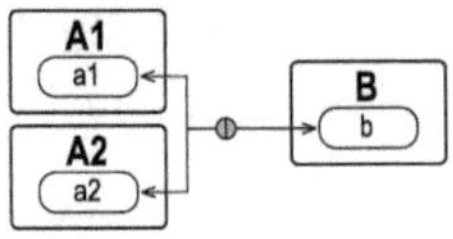	Component b in B may form a bond with either a1 in A1 or a2 in A2
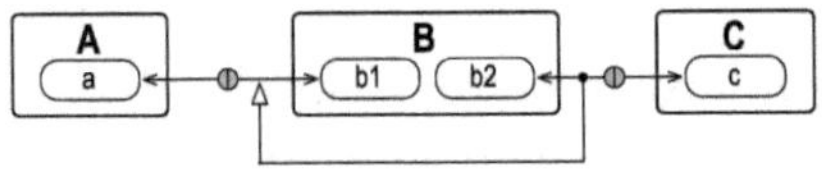	The b2-c bond has an influence on the a-b1 bond
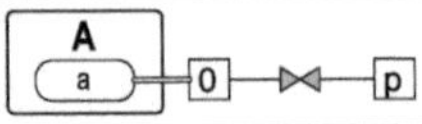	A has component a, with internal states 0 and p
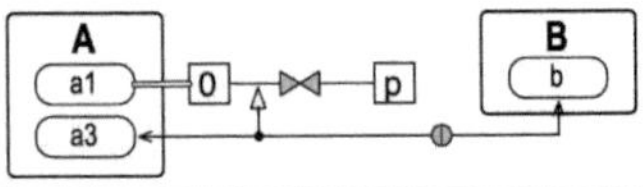	The a3-b bond is necessary for the conversion: A(a1~0) -> A(a1~p)
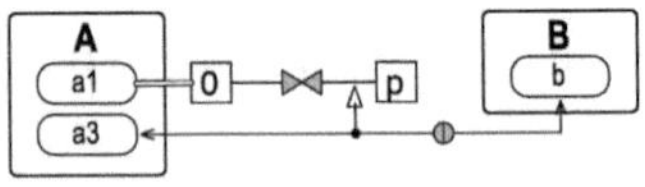	The a3-b bond is necessary for the conversion: A(a1~p) -> A(a1~0)
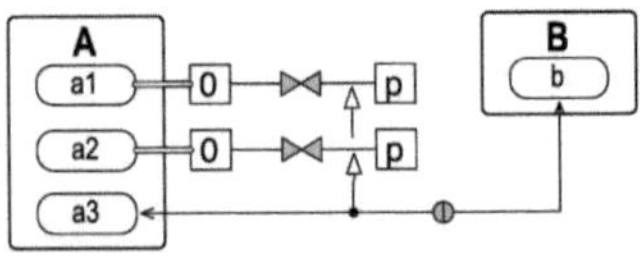	The a3-b bond is necessary for the conversions: A(a2~p) -> A(a2~0) A(a3~p) -> A(a3~0)
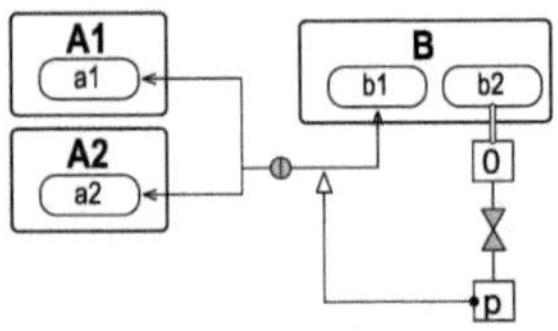	b1 may form a bond with either a1 or a2 b2~p has an influence on both a1-b1 and a2-b1 bonds
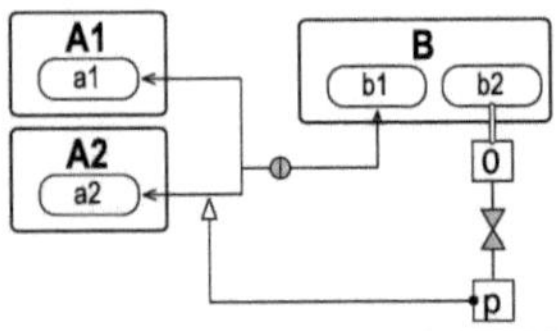	b1 may form a bond with either a1 or a2 b2~p has an influence on the a2-b1 bond, but not on the a1-b1 bond

Fig. 9. Contact map explanation. *Square boxes* are molecules and *curved boxes* are components. The *bond symbol* is a *crossed out circle* with *arrows* on both sides and represents a potential reversible bond between components. The different permutations of bonds can be expressed concisely by combining the *arrows* of the *bond symbol*. Influences are shown by *unidirectional* or *bidirectional arrows* with *white solid arrowheads*. Unidirectional influences always begin on a *solid dot*. The *gray bowtie symbol* indicates a reversible state change. An influence *arrow* terminating on one side of the bowtie is a catalyst to convert the state to the other side of the bowtie. Influences can terminate on bond interactions or on other influences.

shown in Fig. 9. A contact map of the kinase and phosphatase activities in the signaling complex and the primary binding interactions with the receptor are shown in Fig. 10.

The molecules recruited to the primary phosphotyrosines on the receptor have complex interactions among themselves. Grb2 and Shc1 are adaptor proteins that bind each other as well as other proteins. Gab1, a protein recruited via Grb2, is a scaffold protein with multiple binding sites.

Knowledge Base:

> Grb2 has three domains SH2, SH3N, and SH3C. In addition to receptor phosphotyrosines binding, the SH2 domain is capable of binding phosphorylated Y317 of Shc1 (59). The SH3N domain recruits Sos1 through the N-terminal domain of Sos1 (61). The SH3C domain recruits Gab1 through a proline-rich region on Gab1 (58). Certain serines on Sos1 (62) and certain serines and threonines on Gab1 (58) need to be unphosphorylated for the Grb2 binding to be effective.

```
Grb2(sh2) + Shc1(Y317~p) <->
Grb2(sh2!1).Shc1(Y317~p!1) kf_grb2shc1,kr_grb2shc1
Grb2(sh3n) + Sos1(S~0,n) <-> Grb2(sh3n!1).Sos1(S~0,n!1)
  kf_grb2sos1,kr_grb2sos1
Grb2(sh3c) + Gab1(ST~0,pro)<->Grb2(sh3c!1).Gab1
  (ST~0,pro!1) kf_grb2gab1,kr_grb2gab1
```

Knowledge Base:

> The Gab1 scaffold protein contains tandem YXXP motifs that bind the N-terminal domain of Rasa1 on phosphorylation. Similarly, a phosphorylated tandem YXXM motif binds the regulatory subunit p85 of PI3K. Phosphorylated Y627 recruits the Shp2 phosphatase. All of these binding events require that certain serines and threonines on Gab1 remain unphosphorylated (58).

```
Gab1(ST~0,YXXP~p)+Rasa1(n)<->Gab1(ST~0,YXXP~p!1).
  Rasa1(n!1) kf_gab1rasa1,kr_gab1rasa1
Gab1(ST~0,YXXM~p)+PI3K(p85)<->Gab1(ST~0,YXXP~p!1).
  PI3K(p85!1) kf_gab1pi3k,kr_gab1pi3k
Gab1(ST~0,Y627~p)+Shp2(n)<->Gab1(ST~0,YXXP~p!1).
  Shp2(n!1) kf_gab1shp2,kr_gab1shp2
```

The secondary binding interactions between the molecules recruited to the receptor are summarized in Fig. 11. The contact

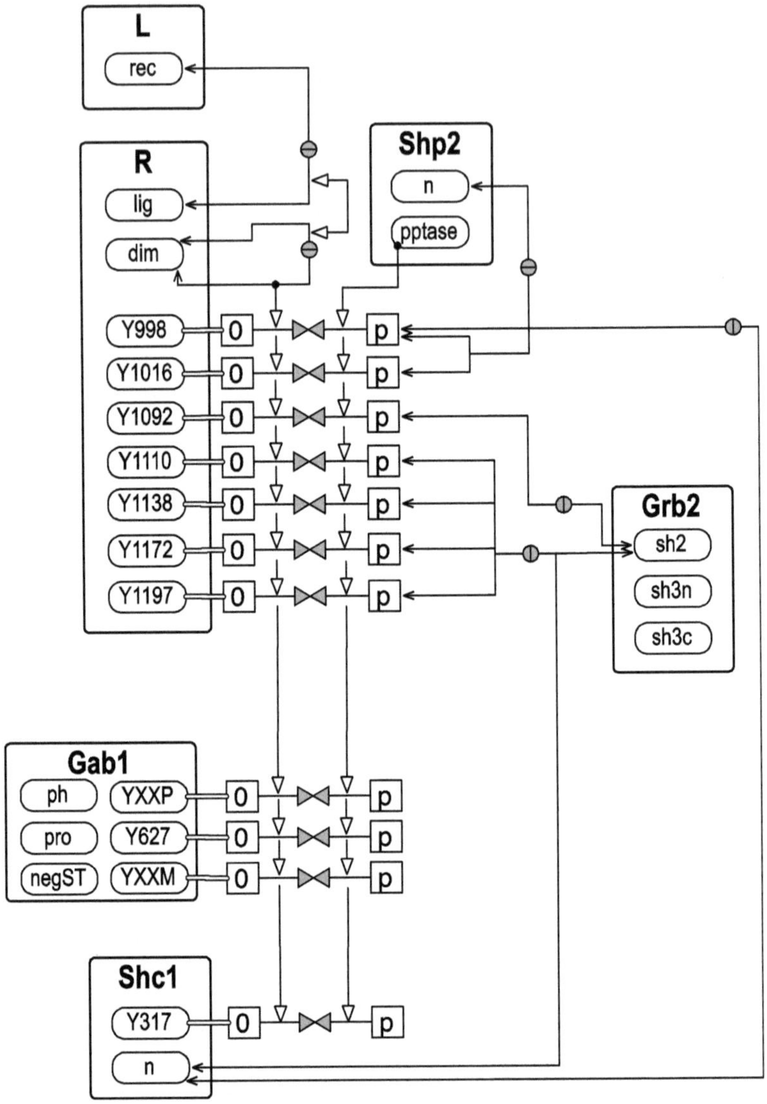

Fig. 10. Signal complex assembly, receptor interactions. This encapsulates the direct interactions of the receptor with other molecules. The receptor can bind the ligand, dimerize with another receptor, and the dimerized state activates the kinase activity of the receptor, resulting in phosphorylation of the many tyrosines on its tail. The kinase activity is limited to not only its own tyrosines, but also tyrosines on other molecules that may be in the complex, such as Gab1 and Shc1. The phosphotyrosines can bind and recruit other molecules. One of the recruits is a phosphatase Shp2 that dephosphorylates the tyrosines in the same complex.

map in Fig. 11 is to be considered complementary to the contact map in Fig. 10 and not exclusive.

9.2. Secondary Messenger Activation

As mentioned earlier, certain catalytic activities are activated on recruitment to the signaling complex. These catalytic activities rapidly change the distributions of certain effector molecules in

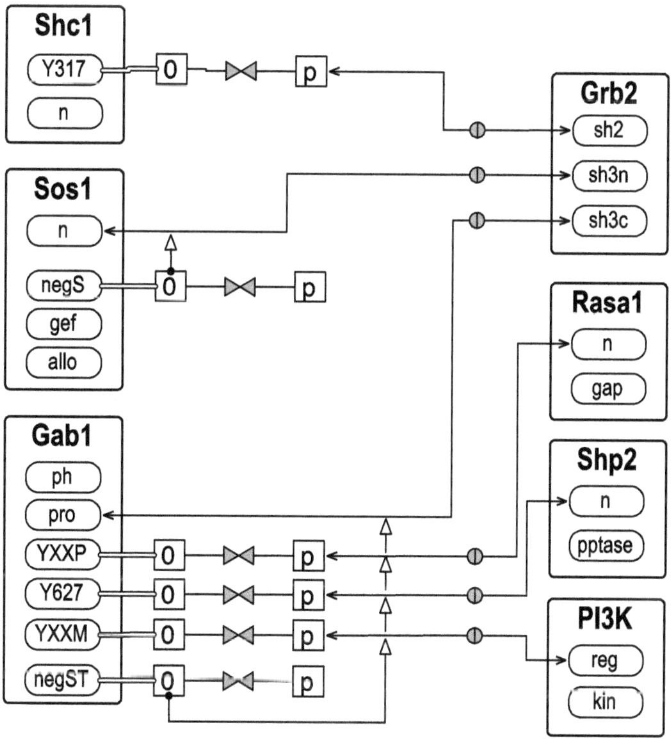

Fig. 11. Signal complex assembly, recruit interactions. The recruited molecules in the EGFR themselves have complicated binding properties with respect to each other. Certain serines on Sos1 and serines and threonines on Gab1 have a negative effect on the binding properties of Sos1 and Gab1, and this is utilized to create a feedback loop by molecules downstream of the signal.

the plasma membrane and cytosol. This manifests as a large phenotypic switch causing wholesale activation of corresponding cytosolic signaling pathways. In this section, we consider the activation of Ras, a GTPase switch, and PIP3, a phosphoinositide messenger.

9.2.1. PIP3 Activation

Phosphoinositides are small lipid molecules that function as secondary messengers in several eukaryotic pathways. Phosphoinositides can be interconverted by adding and removing phosphates onto the 3', 4', and 5' positions on the inositol moiety. The ratio and spatial distribution of the different phosphoinositides are exploited for signaling purposes. In growth factor receptor activation, the interconversion of phosphatidylinositol (4,5)-bisphosphate, also known as PIP2, and phosphatidylinositol (3,4,5)-trisphosphate is especially critical. PIP3 is maintained at very low levels by the activity of cytosolic PTEN phosphatase. PIP3 levels are drastically increased by recruitment of PI3K to the membrane. PIP3, thus, serves as a secondary messenger for activating several PH domain-containing proteins, an important one being Akt, which triggers the Akt/mTor pathway critical for cell proliferation and differentiation. The information here can be found in ref. 63. The rules describing change of state have the appropriate component underlined for emphasis.

Knowledge Base:

> PIP3 is maintained at low levels by conversion to PIP2 by PTEN phosphatase.

Here, we model PIP3 as a phosphoinositide with a 3' site targeted by the phosphatase.

```
PTEN(c2) + PI(3p~p) <-> PTEN(c2!1).PI(3p~0!1)
  kf_pip3_pten,kr_pip3_pten
PTEN(c2!1).PI(3p~0!1) -> PTEN(c2) + PI(3p~p) kcat_pten
R().PI3K(p110) + PI(3p~0) <->
```

Knowledge Base:

> The p110 domain of PI3K catalyzes conversion of PIP2 to PIP3 when recruited to the membrane.

```
  R().PI3K(p110!1).PI(3p~0!1) kf_pip2_pi3k,kr_pip2_pi3k
PI3K(p110!1).PI(3p~0!1) -> PI3K(p110) + PI(3p~p)
  kcat_pi3k
Gab1(ph) + PI(3p~p) <-> Gab1(ph!1).PI(3p~p!1)
  kf_pip3_gab1,kr_pip3_gab1
```

Knowledge Base:

> PIP3 recruits PH domain-containing proteins like Gab1 (58) and Akt to the membrane.

```
Akt(ph) + PI(3p~p) <-> Akt(ph!1).PI(3p~p!1)
  kf_pip3_akt, kr_pip3_akt
```

```
Gab1(ph!1,pro).PI(3p~p!1) + R().Grb2(sh3c)->
Gab1(ph,pro!2).Grb2(sh3c!2) + PI(3p~p) \
```

Knowledge Base:

> PIP3-recruited Gab1 at the membrane can be recruited to the signaling complex easily.

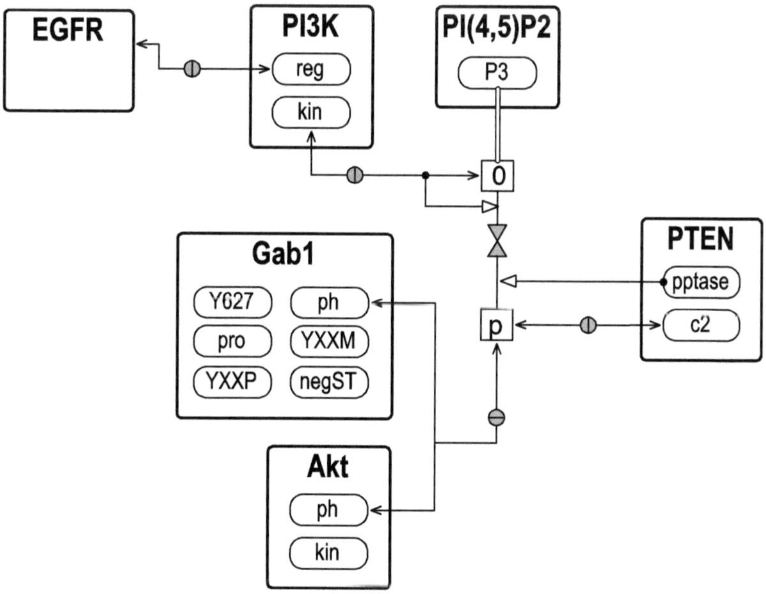

Fig. 12. PIP3 activation. PI(3,4,5)P2 or PIP3 is an important secondary messenger whose levels are kept at low basal levels by the phosphatase PTEN which catalyzes it to PI(4,5)P2 or PIP2. The third position on the inositol moiety is modeled as a separate component which can either be phosphorylated or unphosphorylated. PI3K is a kinase which can convert PIP2 to PIP3, but it is activated only by recruitment to the membrane by recruitment to the EGFR-signaling complex. PIP3 can then activate many other signaling cascades by recruitment, an important one being the kinase Akt. It can also strengthen the existing signals since Gab1 possesses a PH domain that can bind PIP3.

```
k_gab1grb2_mem
Akt(ph!+,kin) + mTor(S2448~0) <->
Akt(ph!+,kin!1).mTor(S2448~0!1) \
  kf_akt_mtor,kr_akt_mtor
Akt(ph!+,kin!1).mTor(S2448~0!1) -> Akt(ph!+,kin) +
  mTor(S2448~p) kcat_akt
```

Knowledge Base:

> PIP3-recruited Akt is an active kinase with downstream substrates, such as mTOR.

The PIP3 activation and downstream interactions modeled here are illustrated in Fig. 12.

9.2.2. Ras Activation

Ras is a small GTPase protein that is tethered to the membrane and binds with high affinity to free guanidine nucleotides (GTP and GDP) in the cytosol. Ras, in its basal state, slowly catalyzes the dephosphorylation of GTP into GDP. In the absence of ligand-induced signal, the majority of Ras molecules are bound to GDP.

When bound by a guanidine nucleotide exchange factor (GEF), Ras releases the GDP nucleotide and binds a free GTP nucleotide. Since free GTP is in excess over free GDP under normal conditions, we can assume that the binding event after GDP release is always with a GTP. The Ras population switches from a GDP-bound majority to a GTP-bound majority.

Ras-GTP, by binding to other molecules, initiates many signaling cascades, including the MAPK cascade. The intrinsic GTPase activity can be enhanced by a GTPase-activating protein (GAP) to provide the opposite effect, i.e., to quickly revert a Ras-GTP population to Ras-GDP. EGFR recruits both Sos1 (a GEF) and Rasa1 (a GAP) to the membrane through adaptors and scaffolds.

Ras has three subtypes: HRas, NRas, and KRas with many common interactions among them. Here, we model HRas as an example. Instead of modeling the nucleotide as a separate molecule, we model it as simply a state on an HRas component. This is because the binding of GDP/GTP to HRas is very strong and does not reverse spontaneously. Any free HRas quickly binds to excess free nucleotide. The nucleotide-binding site is underlined for emphasis whenever a GTP–GDP or GDP–GTP transition occurs.

The following information is taken from ref. 61. The rules describing change of state have the appropriate component underlined for emphasis.

Knowledge Base:

HRas has a protein-binding site and a nucleotide-binding site. The nucleotide-binding site can be assumed to bind GTP or GDP strongly. HRas has innate GTPase activity.

```
HRas(pbs,nbs~gtp) -> HRas(pbs,nbs~gdp) k_hras_gtpase
```

Knowledge Base:

Receptor-bound Rasa1 is a GAP, i.e., it enhances the GTPase activity of HRas.

```
R().Rasa1(gap)+HRas(pbs,nbs~gtp) ->
R().Rasa1(gap!1).HRas(pbs!1,nbs~gtp) kf_rasa1_hras
Rasa1(gap!1).HRas(pbs!1,nbs~gtp) -> Rasa1(gap) +
HRas(pbs,nbs~gtp) kr_rasa1_hras
Rasa1(gap!1).HRas(pbs!1,nbs~gtp) -> Rasa1(gap) +
HRas(pbs,nbs~gdp) kcat_rasa1
```

Knowledge Base:

> Sos1 recruited to the membrane can bind HRas at two places, an allosteric site and a catalytic site. Sos1 possesses GEF activity, i.e., it catalyzes the release of GDP from the HRas bound at the catalytic site. The GEF activity is enhanced by the presence of another HRas bound at the allosteric site. When GDP is released, GTP is assumed to bind immediately since it is in excess over GDP.

```
# HRas-GDP binding at GEF site
R().Sos1(gef)+HRas(pbs,nbs~gdp)->R().Sos1(gef!1).
  HRas(pbs!1,nbs~gdp) kf_sos1gef_hras Sos1(gef!1).
  HRas(pbs!1,nbs~gdp)-> Sos1(gef) + HRas(pbs,nbs~gdp)
  kr_sos1gef_hras

# HRas (both GTP and GDP) binding at allosteric site
R().Sos1(allo) + HRas(pbs) -> R().Sos1(allo!1).HRas
  (pbs!1) kf_sos1allo_hras
Sos1(allo!1).HRas(pbs!1) -> Sos1(allo) + HRas(pbs)
  kr_sos1allo_hras

# GEF activity in presence/absence of GTP at allo-site
HRas(pbs!1,nbs~gdp).Sos1(gef!1,allo)->  HRas(pbs,
nbs~gtp)+Sos1(gef,allo) kcat_sos1_1
HRas(pbs!1,nbs~gdp).Sos1(gef!1,allo!+)->HRas(pbs,
nbs~gtp)+Sos1(gef,allo!+) kcat_sos1_2
```

The modeled HRas activation is illustrated in Fig. 13.

9.3. Cytosolic Signaling—MAPK Cascade

The Ras proteins, on activation by GTP binding, go on to activate several pathways. One of the most widely studied Ras-activated pathways is the classical mitogen-activated protein kinase (MAPK) cascade. MAPKs are kinases that are heavily involved in growth-related transcription factor regulation. Previously, the paradigm of activation was that MAPKs were phosphorylated by MAP2Ks and MAP2Ks were phosphorylated by MAP3Ks, which in turn were activated by binding of Ras-GTP. However, as evidence accumulated, it became fairly obvious that the situation is not so simple. The presence of multiple subtypes, oligomerization, scaffolds, feedback mechanisms, and cell-specific conditions confound the study of this complex pathway.

As mentioned earlier, there are three subtypes of Ras proteins: HRas, KRas, and MRas. MAP3Ks are of three types too: Raf1, Raf2, and Raf3. They are subject to complex regulation from other signaling pathways as well as regulation by each other. Mek1 and Mek2 are MAP2K subtypes. MAPKs themselves have multiple families, such as the Erk (Erk1 and Erk2) and Jnk. Erk2

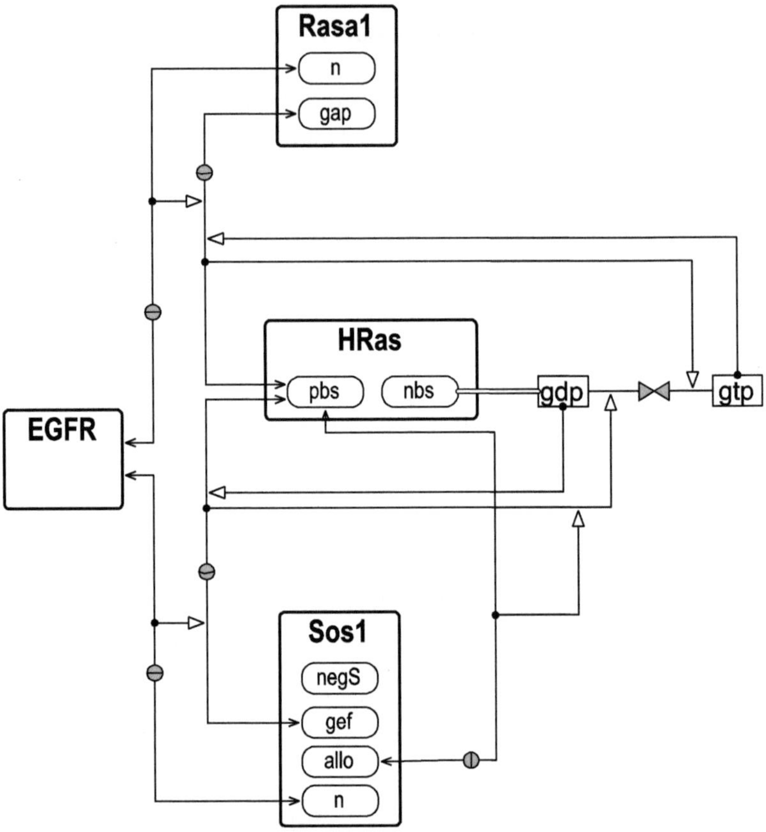

Fig. 13. Ras activation. Using HRas as an example, the complex interactions involved in Ras activation are shown. Rasa1 and Sos1 are recruited to the receptor in multiple ways (see Figs. 11 and 12). HRas is distributed on the membrane and comes into contact with Sos1 and Rasa1. Both Sos1 and Rasa1 have antagonistic activity towards HRas. Sos1 enables the quick release of bound HRas-GDP, letting HRas bind ubiquitous GTP and become activated. Rasa1 enhances the GTPase activity of HRas, speeding up the conversion of HRas-GTP to HRas-GDP.

can cause negative feedback by deactivating Gab1 and Sos1, early participants in Ras activation. The primary function of the MAPKs seems to be in the nucleus though, causing activation of growth-related transcription factors such as those of the AP1 family.

With these caveats, the model presented here omits many of the known details of the MAPK cascade and is provided to illustrate how the basic biochemistry can be encoded. The rules describing change of state have the appropriate component underlined for emphasis.

Knowledge Base:

HRas-GTP binds Raf1 and causes a conformational change (64). Raf1 is now an active kinase that can phosphorylate Mek1 (65) on S218 and S222. The phosphorylation is antagonized by the phosphatase PP2A (66).

```
# HRas-GTP binds Raf1
HRas(nbs~gtp,pbs) + Raf1(rbd) ->
HRas(nbs~gtp,pbs!1).Raf1(rbd!1) kf_hras_raf1
HRas(pbs!1).Raf1(rbd!1) -> HRas(pbs!1) + Raf1(rbd!1)
  kr_hras_raf1

# Ras-bound Raf1 is a kinase of Mek1
Raf1(rbd!+,kin) + Mek1(S218~0) ->
Raf1(rbd!+,kin!1).Mek1(S218~0!1) kf_raf1_mek1
Raf1(rbd!+,kin) + Mek1(S222~0) ->
Raf1(rbd!+,kin!1).Mek1(S222~0!1) kf_raf1_mek1
Raf1(kin!1).Mek1(S218~0!1) -> Raf1(kin!1) + Mek1
  (S218~0) kr_raf1_mek1
Raf1(kin!1).Mek1(S222~0!1) -> Raf1(kin!1) + Mek1
  (S222~0) kr_raf1_mek1
Raf1(kin!1).Mek1(S218~0!1) -> Raf1(kin) + Mek1
  (S218~p) kcat_raf1
Raf1(kin!1).Mek1(S222~0!1) -> Raf1(kin) + Mek1
  (S222~p) kcat_raf1

# PP2A is a phosphatase of Mek1
PP2A(pptase) + Mek1(S218~p) ->
PP2A(pptase!1).Mek1(S218~p!1) kf_pp2a_mek1
PP2A(pptase) + Mek1(S222~p) ->
PP2A(pptase!1).Mek1(S222~p!1) kf_pp2a_mek1
PP2A(pptase!1).Mek1(S218~p!1) -> PP2A(pptase) +
Mek1(S218~p) kr_pp2a_mek1
PP2A(pptase!1).Mek1(S222~p!1) -> PP2A(pptase) +
Mek1(S222~p) kr_pp2a_mek1
PP2A(pptase!1).Mek1(S218~p!1) -> PP2A(pptase) +
Mek1(S218~0) kcat_pp2a
PP2A(pptase!1).Mek1(S222~p!1) -> PP2A(pptase) +
Mek1(S222~0) kcat_pp2a
```

Knowledge Base:

> Biphosphorylated Mek 1 is an active kinase that can phosphorylate Erk2 (65) on T185 and Y187. The phosphorylation is antagonized by the phosphatase Dusp1 (68)

```
# Biphosphorylated Mek1 is a kinase of Erk2
Mek1(S218~p,S222~p,kin) + Erk2(T185~0)->
Mek1(S218~p,S222~p,kin!1).Er2(T185~0!1) \
  kf_mek1_erk2
Mek1(S218~p,S222~p,kin) + Erk2(Y187~0)->
Mek1(S218~p,S222~p,kin!1).Er2(Y187~0!1) \
  kf_mek1_erk2
```

```
Mek1(kin!1).Erk2(T185~0!1) -> Mek1(kin!1) + Erk2
  (T185~0) kr_mek1_erk2
Mek1(kin!1).Erk2(Y187~0!1) -> Mek1(kin!1) + Erk2
  (Y187~0) kr_mek1_erk2
Mek1(kin!1).Erk2(T185~0!1) -> Mek1(kin) + Erk2
  (T185~p) kcat_mek1
Mek1(kin!1).Erk2(Y187~0!1) -> Mek1(kin) + Erk2
  (Y187~p) kcat_mek1

# Dusp1 is a phosphatase of Erk2
Dusp1(pptase) + Erk2(T185~p) ->
Dusp1(pptase!1).Erk2(T185~p!1) kf_dusp1_erk2
Dusp1(pptase) + Erk2(Y187~p) ->
Dusp1(pptase!1).Erk2(Y187~p!1) kf_dusp1_erk2
Dusp1(pptase!1).Erk2(T185~p!1) -> Dusp1(pptase) +
Erk2(T185~p) kr_dusp1_erk2
Dusp1(pptase!1).Erk2(Y187~p!1) -> Dusp1(pptase) +
Erk2(Y187~p) kr_dusp1_erk2
Dusp1(pptase!1).Erk2(T185~p!1) -> Dusp1(pptase) +
Erk2(T185~0) kcat_dusp1
Dusp1(pptase!1).Erk2(Y187~p!1) -> Dusp1(pptase) +
Erk2(Y187~0) kcat_dusp1
```

Signal transduction networks exhibit significant *positive and negative feedback*, i.e., downstream effectors doubling back and modifying (enhancing or attenuating) the properties of upstream signal proteins. This enables ultra-sensitivity in regulation and adaptive responses to incoming signals. Interesting dynamic phenomena, such as oscillations and travelling waves, are also observed in some cases due to such feedback (73).

Knowledge Base:

> Biphosphorylated Erk2 can dimerize with other Erk2 (phosphorylated or unphosphorylated). The dimer is catalytically active on many substrates (53), including negative regulatory sites on Gab1 (58) and Sos1 (67).

```
Erk2(T185~p,Y187~p,dim)+Erk(dim) ->
Erk2(T185~p,Y187~p,dim!1)+Erk2(dim!1) kf_erk2_dim
Erk2(dim!1).Erk(dim!1) -> Erk2(dim) + Erk(dim)
  kr_erk2_dim
Erk2(dim!+,kin) + Gab1(ST~0) ->
Erk2(dim!+,kin!1).Gab1(ST~0!1) kf_erk2_gab1
Erk2(kin!1).Gab1(ST~0!1) -> Erk2(kin) +
Gab1(ST~0) kr_erk2_gab1
Erk2(kin!1).Gab1(ST~0!1) -> Erk2(kin) +
Gab1(ST~p) kcat_erk2_gab1
```

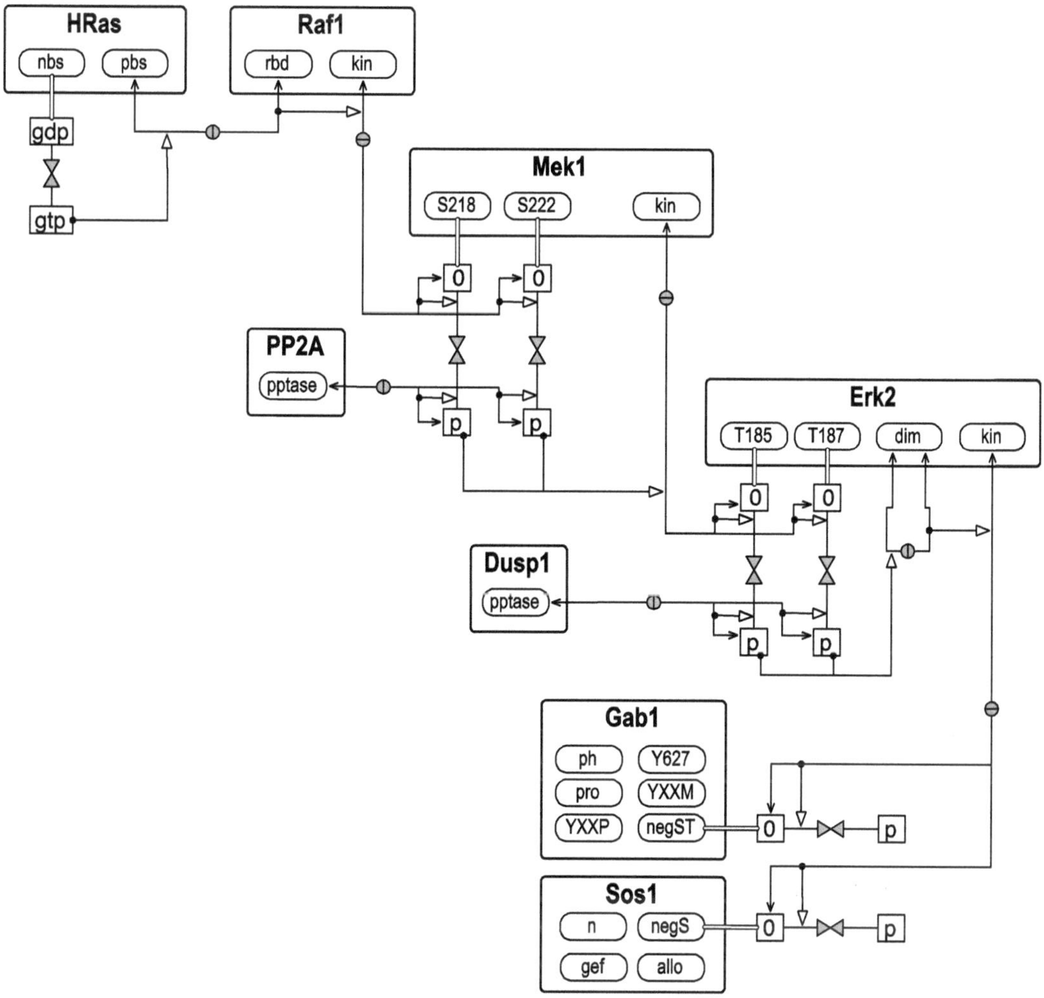

Fig. 14. MAPK cascade. The canonical MAPK cascade is shown for illustration purposes. In reality, the pathway is much more complicated. Here, the example G-protein HRas activates Raf1 by binding, when HRas is in GTP-bound state. The HRas-GTP-bound Raf1 undergoes a conformational change and becomes an active kinase. It binds to and phosphorylates serines on Mek1 and this is opposed by an example phosphatase PP2A. Phosphorylated Mek1 is an active kinase and phosphorylates Erk2 on its threonines, and this is opposed by a dual-specificity phosphatase, Dusp1, shown as an example. Phosphorylated Erk2 can dimerize and this positively influences its kinase activity. Erk2 has several system-wide substrates, two important ones being Gab1 and Sos1 which allow negative feedback control over the transduced signal.

```
Erk2(dim!+,kin) + Sos1(S~0) -> Erk2(dim!+,kin!1).
  Sos1(S~0!1) kf_erk_sos1
Erk2(kin!1).Sos1(S~0!1) -> Erk2(kin) + Sos1(S~0)
  kr_erk_sos1
Erk2(kin!1).Sos1(S~0!1) -> Erk2(kin) + Sos1(S~p)
  kcat_erk2_sos1
```

The modeled rules are illustrated in Fig. 14.

9.4. Gene Regulation

Extracellular signals have their furthest reaching effects when the transduced signals culminate in the nucleus. This is usually accomplished by changing the distribution of transcription factors in the nucleus causing repression or enhancement of mRNA transcript production from multiple genes.

The central dogma of biology, i.e., transcription followed by translation, can be explicitly implemented using rules. Both involve creation of new molecules (mRNA transcripts and proteins).

```
Gene(a) -> Gene(a) + Transcript(a) k_transcription
Transcript(a) -> Transcript(a) + Protein(a)
  k_translation
```

If stochastic fluctuations (due to genes turning on and off) are important to the system, then it is important to restrict the number of molecules of the `Gene` to the actual copy number to replicate real behavior.

A simplifying assumption can be made, where all the transcription, translation, and post-synthetic modification steps can be compressed into a single step. This amounts to assuming that the delay caused by sequence of steps in creating the product is negligible when compared to the rates at which the protein is produced (see ref. 48 supplementary material). The effective reaction would be:

```
Gene(a) -> Gene(a) + Protein(a) k_syn
```

Protein synthesis can also be made dependent on transcription-factor binding.

```
Gene(tf) + TF(gene) <-> Gene(tf!0).TF(gene!0)
  kf_gene_tf,kr_gene_tf
Gene(tf!+) -> Gene(tf!+) + Protein(a) k_syn
```

Often, the binding can be assumed to be in quasi-equilibrium on the timescale on which genes are expressed so that the expression rate becomes a function of the transcription factor concentration. An example is shown here using a Hill rate law (see Subheading 5.3.2 for details).

```
TF()+Gene() -> TF()+Gene()+Protein() Hill(k,K,n)
```

In the rest of this section, we implement reaction rules pertaining to gene regulation by a cytosolic signal. The rules will be modeled in the compartmental context, retaining the same compartmental hierarchy in Subheading 7.4 and Fig. 5. Specifically, we model the transduction of signal from two parallel MAPK pathways in the form of Erk2 and Jnk1 into the nucleus and activation of the AP1 transcription factor. The information presented here is reviewed in refs. 69–71.

Knowledge Base:

> Biphosphorylated Erk2 and Jnk1 can reversibly homodimerize in both cytosolic and nuclear compartments. It is sufficient that one partner is phosphorylated.

```
Erk2(T185~p,Y187~p,dim)+Erk2(dim) ->\
 Erk2(S218~p,S222~p,dim!1)+Erk2(dim!1)\
 kf_erk2_dim
Erk2(dim!1).Erk2(dim!1) -> Erk2(dim) +Erk2(dim)\
 kr_erk2_dim
Jnk1(T183~p,Y185~p,dim)+ Jnk1(dim) ->\
 Jnk1(T183~p,Y185~p,dim!1)+ Jnk1(dim!1)\
 kf_jnk1_dim
Jnk1(dim!1).Jnk1(dim!1) -> Jnk1(dim) + Jnk1(dim)\
 kr_jnk1_dim
```

Since homo-dimerization occurs in both cytosolic and nuclear compartments, it is sufficient to model them with universal reaction rules.

Knowledge Base:

> Both Erk2 and Jnk1 monomers passively diffuse into the nucleus. Erk2 and Jnk1 dimers also translocate to the nucleus at a faster rate.

Since the nucleus and cytosol are bridged by the nuclear membrane, we use bridged-volume molecule transport rules. We also need to allocate separate microscopic rates for monomers and dimers.

```
# Erk2 and Jnk1 monomer transport
Erk2(T185,Y187,dim,kin)@Cyt <-> Erk2(T185,Y187,\
 dim,kin)@Nuc k_tr_erk2_m,k_tr_erk2_m
Jnk1(T183,Y185,dim,kin)@Cyt <-> Jnk1(T183,Y185,\
 dim,kin)@Nuc k_tr_jnk1_m,k_tr_jnk1_m

# Erk2 and Jnk1 dimer transport
Erk2(T185,Y187,dim!1,kin)@Cyt.Erk2(T185,Y187,\
  dim!1,kin)@Cyt<->Erk2(T185,Y187,dim!1,kin)@Nuc.\
Erk2(T185,Y187,dim!1,kin)@Nuc k_tr_erk2_d,\
 k_tr_erk2_d
Jnk1(T183,Y185,dim!1,kin)@Cyt.Jnk1(T183,Y185,\
  dim!1,kin)@Cyt<->Jnk1(T183,Y185,dim!1,kin)@Nuc.\
Jnk1(T183,Y185,dim!1,kin)@Nuc k_tr_jnk1_d,\
 k_tr_jnk1_d
```

Note that the other components on Erk2 and Jnk1 need to be in unbound state to allow transport.

Knowledge Base:

> Fos and Jun are transcription factors synthesized and maintained at a certain level in the nucleus.

```
FosGene()@Nuc -> FosGene()@Nuc + Fos(y~0)@Nuc
  k_syn_fos
JunGene()@Nuc -> JunGene()@Nuc + Jun(y~0)@Nuc
  k_syn_jun
Fos()@Nuc -> 0 k_del_fos
Jun()@Nuc -> 0 k_del_jun
```

Instead of simply creating a preexisting number of Fos and Jun molecules, we now have a dynamic equilibrium of synthesis/degradation events that leads to a steady-state number of Fos and Jun.

Knowledge Base:

> Activated Erk2 and Jnk1 dimers can phosphorylate Fos and Jun, respectively.

```
Erk2(dim!+,fos) + Fos(y~0)<-> Erk2(dim!+,fos!0).\
  Fos(y~0!0) kf_erk_fos,kr_erk_fos
Erk2(dim!+,fos!0).Fos(y~0!0)-> Erk2(dim!+,fos) +\
  Fos(y~p) k_ph_erk
Jnk1(dim!+,kin) + Jun(y~0)<-> Jnk1(dim!+,kin!0).\
  Jun(y~0!0) kf_jnk_kin,kr_jnk_kin
Jnk1(dim!+,kin!0).Jun(y~0!0)-> Jnk1(dim!+,kin) +\
  Jun (y~p) k_ph_jnk
```

Since there are no direct or indirect transport mechanisms for Fos and Jun, universal rules are sufficient to model these interactions. Activated Fos and Jun can dimerize to form the transcription factor AP1, which shows binding affinity for other transcription factors and certain promoter DNA sequences. The binding sites are shared across both Fos and Jun subunits. Since the Fos–Jun dimer seems to behave as an independent molecule type with binding sites, we create a new molecule type called AP1.

Knowledge Base:

> Activated Fos and Jun can hetero-dimerize to form AP1 (Fos–Jun). AP1 can be degraded.

```
Fos(y~p) + Jun(y~p) <-> AP1(g,nfat) kf_kin_kin,
 kr_kin_kin
AP1()@Nuc -> 0 k_del_ap1
```

Knowledge Base:

AP1 can bind to promoter sequences and activate target genes on its own or in combination with other transcription factors (NFAT). AP1-bound genes enable synthesis of target proteins in the cytosol.

```
AP1(nfat) + NFAT(ap1) <-> AP1(nfat!0).NFAT(ap1!0)\
 kf_ap1_nfat,kr_ap1_nfat
AP1(g) + TargetGene(prm) <-> AP1(g!0).TargetGene\
 (prm!0) kf_ap1_gene,kr_ap1_gene
TargetGene(prm!+)@Nuc -> TargetGene(prm!+)@Nuc +\
 TargetProtein@Cyt k_syn_ap1
```

The gene regulation model is illustrated in Fig. 15.

10. Good Modeling Practice

This section is intended as a summary of the critical points from the previous sections, which taken together provide guidelines for the creation of comprehensible, intuitive and biologically realistic models using BNGL.

- Name molecules after important indivisible entities in the model.
- Name components after biological substructures (domains, motifs, amino acid sequence position, etc.), their binding partner, or their designated function.
- The fewer the transformations in a reaction rule, the more realistic it is.
- Unimolecular and bimolecular reaction orders are realistic. Higher orders are less so.
- The rate constant used should be the *asymmetric, per-site rate constant*, which would require scaling by volume and Avogadro number in BNG and only by Avogadro number in cBNG.
- When writing reaction rules, omit components that do not influence the rate of the reaction.
- When two (or more) reactions have the same reaction center but different kinetics, use multiple reaction rules to model them, unless the difference is due to symmetry and multiplicity effects.

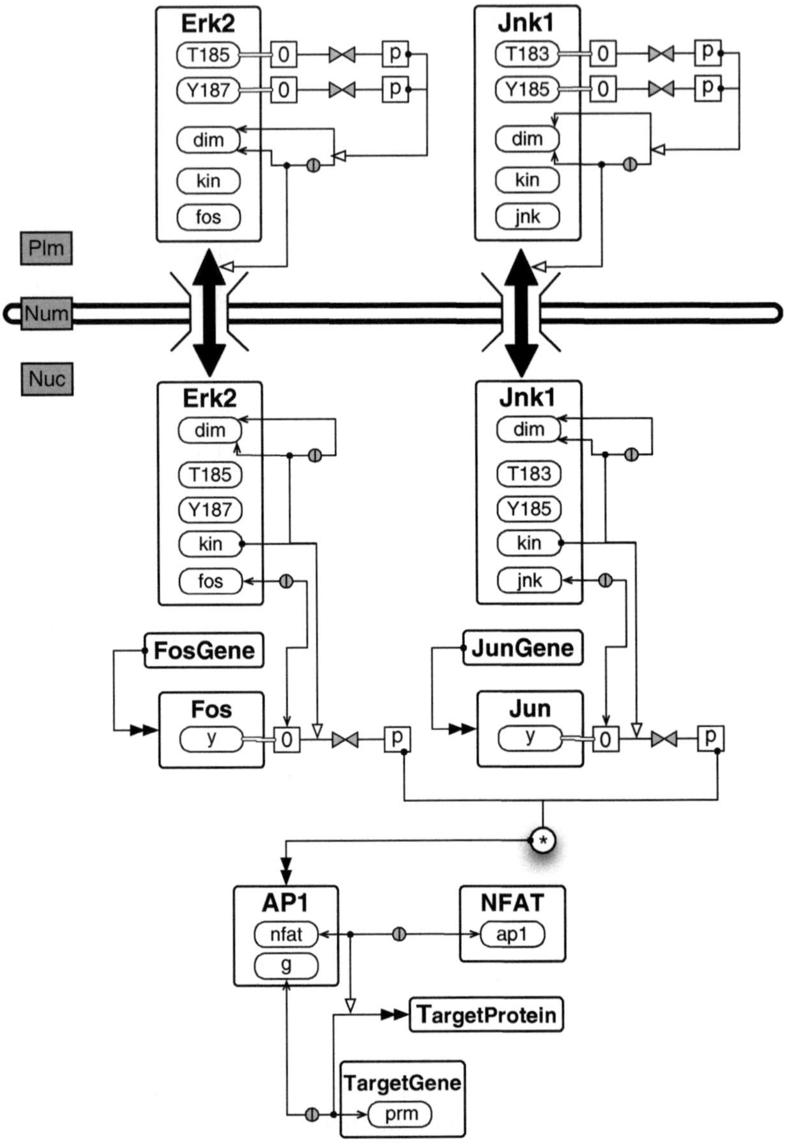

Fig. 15. AP1 activation. Erk2 and Jnk1 are MAP kinases that are activated through many pathways, especially growth factor signals. Both Erk2 and Jnk1 can translocate to the nucleus, with the transport being enhanced by dimerization. In the nucleus, they can activate Fos and Jun transcription factors by phosphorylation. Fos and Jun are actively synthesized (as shown by the *black double arrowhead*) and maintained at equilibrium. Activated Fos and Jun can bind to each other to form the AP1 transcription factor. Since AP1 has many distinct activities on its own, it is modeled as a separate molecule. The *circle* with the *star* indicates that one molecule of Fos and one molecule of Jun are destroyed to synthesize one molecule of AP1. AP1 can bind to other transcription factors, such as NFAT, and the complex can initiate transcription at several target genes. Here, we have also subsumed transcription, translation, and posttranslational modification into a single process.

- Cooperative interactions must obey detailed balance.
- Use wildcards in reaction rules and observables to finely tailor matching conditions in patterns.
- When synthesizing a new species in a reaction rule, provide the full species specification.
- Most transformations are reversible in a realistic model. Unless there is specific information on irreversibility, there should exist rules that:
  - Dephosphorylate all phosphorylation sites.
  - Break all bonds that can be formed.
  - Degrade all molecules that can be synthesized.
- Deletion rules delete the whole species by default. Use `DeleteMolecules` keyword to delete only the respective molecules and not the whole species.
- When transporting molecules in cBNG, check to see if you are transporting molecules or species.
- Avoid rate law approximations. Prefer explicit mechanisms.
- Use structural knowledge to infer how interactions affect one another. In the absence of information assume they are independent (i.e. distributive or parallel).
- Model interactions as sequential (i.e. processive) only if there is reasonable structural information that such is the case.
- If small concentration changes can cause drastic effects, use a stochastic simulator.
- If the network is too big, use a network-free simulator.
- To simulate a typical biological experiment, always equilibrate before perturbing.
- Avoid having to generate the same network repeatedly, especially with large networks.

Acknowledgments

We gratefully acknowledge support from NIH grant GM76570, NSF grant CCF-0829788, and NSF Expeditions in Computing Grant 0926181. We would also like to thank Arshi Arora, Leonard A. Harris, William S. Hlavacek, and Justin S. Hogg for their helpful comments on the manuscript.

References

1. Kholodenko BN, Hancock JF, Kolch W (2010) Signalling ballet in space and time. Nat Rev Mol Cell Biol 11:414–426
2. Kholodenko BN (2006) Cell-signalling dynamics in time and space. Nat Rev Mol Cell Biol 7:165–176
3. Scott JD, Pawson T (2009) Cell signaling in space and time: where proteins come together and when they're apart. Science (New York, NY) 326:1220–1224
4. Natarajan M, Lin K-M, Hsueh RC, Sternweis PC, Ranganathan R (2006) A global analysis of cross-talk in a mammalian cellular signalling network. Nat Cell Biol 8:571–580
5. Hecker M, Lambeck S, Toepfer S, van Someren E, Guthke R (2009) Gene regulatory network inference: data integration in dynamic models – a review. Biosystems 96:86–103
6. Hlavacek WS, Faeder JR (2009) The complexity of cell signaling and the need for a new mechanics. Sci Signal 2:46
7. Hlavacek W, Faeder J, Blinov M (2006) Rules for modeling signal-transduction systems. Sci STKE 344:6
8. Bachman JA, Sorger P (2011) New approaches to modeling complex biochemistry. Nat Methods 8:130–131
9. Faeder J, Blinov M, Hlavacek W (2009) Rule-based modeling of biochemical systems with BioNetGen. Methods Mol Biol 500:113–167
10. Blinov ML, Faeder JR, Goldstein B, Hlavacek WS (2004) BioNetGen: software for rule-based modeling of signal transduction based on the interactions of molecular domains. Bioinformatics 20:3289–3291
11. Danos V, Laneve C (2004) Formal molecular biology. Theor Comput Sci 325:69–110
12. Feret J, Danos V, Krivine J, Harmer R, Fontana W (2009) Internal coarse-graining of molecular systems. Proc Natl Acad Sci USA 106:6453–6458
13. Haugh JM, Lauffenburger DA (1997) Physical modulation of intracellular signaling processes by locational regulation. Biophys J 72:2014–2031
14. Michaelis L, Menten M (1913) Die kinetik der invertinwirkung. Biochem Z 49:333–369
15. Briggs G, Haldane J (1925) A note on the kinetics of enzyme action. Biochem J 19:338
16. Cornish-Bowden A (2004) Fundamentals of enzyme kinetics, 3rd edn. Portland Press, London
17. Chen WW, Niepel M, Sorger PK (2010) Classic and contemporary approaches to modeling biochemical reactions. Gene Dev 24:1861–1875
18. Tracy TS, Hummel MA (2004) Modeling kinetic data from in vitro drug metabolism enzyme experiments. Drug Metab Rev 36:231–242
19. Sabouri-Ghomi M, Ciliberto A, Kar S, Novak B, Tyson JJ (2008) Antagonism and bistability in protein interaction networks. J Theor Biol 250:209–218
20. Sneddon MW, Faeder JR, Emonet T (2011) Efficient modeling, simulation and coarse-graining of biological complexity with NFsim. Nat Methods 8:177–183
21. Chen WW, Schoeberl B, Jasper PJ, Niepel M, Nielsen UB, Lauffenburger DA, Sorger PK (2009) Input-output behavior of ErbB signaling pathways as revealed by a mass action model trained against dynamic data. Mol Syst Biol 5:239
22. Barik D, Baumann WT, Paul MR, Novak B, Tyson JJ (2010) A model of yeast cell-cycle regulation based on multisite phosphorylation. Mol Syst Biol 6:405
23. Hänggi P (2002) Stochastic resonance in biology how noise can enhance detection of weak signals and help improve biological information processing. Chemphyschem 3:285–290
24. Gillespie D (1976) A general method for numerically simulating the stochastic time evolution of coupled chemical reactions. J Comput Phys 22:403–434
25. Yang J, Monine MI, Faeder JR, Hlavacek WS (2008) Kinetic Monte Carlo method for rule-based modeling of biochemical networks. Phys Rev E 78:031910
26. Borisov N, Chistopolsky A, Faeder J, Kholodenko B (2008) Domain-oriented reduction of rule-based network models. IET Syst Biol 2:342–351
27. Conzelmann H, Saez-Rodriguez J, Sauter T, Kholodenko BN, Gilles ED (2006) A domain-oriented approach to the reduction of combinatorial complexity in signal transduction networks. BMC Bioinformatics 7:34
28. Lok L, Brent R (2005) Automatic generation of cellular reaction networks with Moleculizer 1.0. Nat Biotechnol 23:131–136
29. Faeder JR, Blinov M, Goldstein B, Hlavacek W (2005) Rule-based modeling of biochemical networks. Complexity 10:22–41
30. Danos V, Feret J, Fontana W, Krivine J (2007) Scalable simulation of cellular signaling networks. Lect Notes Comput Sci 4807:139–157

31. Le Novére N, Shimizu TS (2001) STOCHSIM: modelling of stochastic biomolecular processes. Bioinformatics 17:575–576
32. Colvin J, Monine MI, Faeder JR, Hlavacek WS, Hoff DDV, Posner RG (2009) Simulation of large-scale rule-based models. Bioinformatics 25:910–917
33. Colvin J, Monine MI, Gutenkunst RN, Hlavacek WS, Hoff DDV, Posner RG (2010) RuleMonkey: software for stochastic simulation of rule-based models. BMC Bioinformatics 11:404
34. Walker F, Orchard SG, Jorissen RN, Hall NE, Zhang H-H, Hoyne PA, Adams TE, Johns TG, Ward C, Garrett TPJ, Zhu H-J, Nerrie M, Scott AM, Nice EC, Burgess AW (2004) CR1/CR2 interactions modulate the functions of the cell surface epidermal growth factor receptor. J Biol Chem 279:22387–22398
35. Jorissen RN, Walker F, Pouliot N, Garrett TPJ, Ward CW, Burgess AW (2003) Epidermal growth factor receptor: mechanisms of activation and signalling. Exp Cell Res 284:31–53
36. Zhang X, Gureasko J, Shen K, Cole PA, Kuriyan J (2006) An allosteric mechanism for activation of the kinase domain of epidermal growth factor receptor. Cell 125:1137–1149
37. Tao R, Maruyama I (2008) All EGF (ErbB) receptors have preformed homo-and heterodimeric structures in living cells. J Cell Sci 121:3207–3217
38. Kholodenko BN, Demin OV, Moehren G, Hoek JB (1999) Quantification of short term signaling by the epidermal growth factor receptor. J Biol Chem 274:30169–30181
39. Schoeberl B, Eichler-Jonsson C, Gilles ED, Müller G (2002) Computational modeling of the dynamics of the MAP kinase cascade activated by surface and internalized EGF receptors. Nat Biotechnol 20:370–375
40. Wofsy C, Goldstein B, Lund K, Wiley HS (1992) Implications of epidermal growth factor (EGF) induced egf receptor aggregation. Biophys J 63:98–110
41. Ollivier JF, Shahrezaei V, Swain PS (2010) Scalable rule-based modelling of allosteric proteins and biochemical networks. PLoS Comput Biol 6:e1000975
42. Macdonald JL, Pike LJ (2008) Heterogeneity in EGF-binding affinities arises from negative cooperativity in an aggregating system. Proc Natl Acad Sci USA 105:112–117
43. Alvarado D, Klein DE, Lemmon MA (2010) Structural basis for negative cooperativity in growth factor binding to an EGF receptor. Cell 142:568–579
44. Harris LA, Hogg JS, Faeder JR (2009) Compartmental rule-based modeling of biochemical systems. Proceedings of the 2009 winter simulation conference, 13–16 Dec 2009, Austin TX, USA
45. Saltelli A, Ratto M, Tarantola S, Campolongo F (2005) Sensitivity analysis for chemical models. Chem Rev 105:2811–2827
46. Seewöster T, Lehmann J (1997) Cell size distribution as a parameter for the predetermination of exponential growth during repeated batch cultivation of CHO cells. Biotechnol Bioeng 55:793–797
47. Swanson JA, Lee M, Knapp PE (1991) Cellular dimensions affecting the nucleocytoplasmic volume ratio. J Cell Biol 115:941–948
48. Lipniacki T, Puszynski K, Paszek P, Brasier AR, Kimmel M (2007) Single TNFalpha trimers mediating NF-kappaB activation: stochastic robustness of NF-kappaB signaling. BMC Bioinformatics 8:376
49. Ghaemmaghami S, Huh W-K, Bower K, Howson RW, Belle A, Dephoure N, O'Shea EK, Weissman JS (2003) Global analysis of protein expression in yeast. Nature 425:737–741
50. Lee B, LeDuc PR, Schwartz R (2008) Stochastic off-lattice modeling of molecular self-assembly in crowded environments by Green's function reaction dynamics. Phys Rev E Stat 78:031911
51. Kim JS, Yethiraj A (2010) Crowding effects on association reactions at membranes. Biophys J 98:951–958
52. Gabdoulline RR, Wade RC (2002) Biomolecular diffusional association. Curr Opin Struct Biol 12:204–213
53. Alberts B, Bray D, Lewis J, Raff M, Roberts K, Watson JD (1994) Molecular biology of the cell, 3rd edn. Garland Publishing, New York
54. Hlavacek WS, Faeder JR, Blinov ML, Perelson AS, Goldstein B (2003) The complexity of complexes in signal transduction. Biotechnol Bioeng 84:783–794
55. Berman HM, Westbrook J, Feng Z, Gilliland G, Bhat TN, Weissig H, Shindyalov IN, Bourne PE (2000) The protein data bank. Nucleic Acids Res 28:235–242
56. Consortium U (2010) The universal protein resource (UniProt) in 2010. Nucleic Acids Res 38:D142–D148
57. Schulze WX, Deng L, Mann M (2005) Phosphotyrosine interactome of the ErbB-receptor kinase family. Mol Syst Biol 1:2005.0008
58. Gu H, Neel BG (2003) The "Gab" in signal transduction. Trends Cell Biol 13:122–130

59. Ravichandran KS (2001) Signaling via Shc family adapter proteins. Oncogene 20:6322–6330
60. Neel B, Gu H, Pao L (2003) The 'Shp'ing news: SH2 domain-containing tyrosine phosphatases in cell signaling. Trends Biochem Sci 28:284–293
61. Das J, Ho M, Zikherman J, Govern C, Yang M, Weiss A, Chakraborty AK, Roose JP (2009) Digital signaling and hysteresis characterize ras activation in lymphoid cells. Cell 136:337–351
62. Langlois WJ, Sasaoka T, Saltiel AR, Olefsky JM (1995) Negative feedback regulation and desensitization of insulin- and epidermal growth factor-stimulated p21ras activation. J Biol Chem 270:25320–25323
63. Seminario M-C, Precht P, Bunnell SC, Warren SE, Morris CM, Taub D, Wange RL (2004) PTEN permits acute increases in D3-phosphoinositide levels following TCR stimulation but inhibits distal signaling events by reducing the basal activity of Akt. Eur J Immunol 34:3165–3175
64. Terai K, Matsuda M (2005) Ras binding opens c-Raf to expose the docking site for mitogen-activated protein kinase kinase. EMBO Rep 6:251–255
65. Huang W, Kessler DS, Erikson RL (1995) Biochemical and biological analysis of Mek1 phosphorylation site mutants. Mol Biol Cell 6:237–245
66. Grethe S, Pцrn-Ares MI (2006) p38 MAPK regulates phosphorylation of Bad via PP2A-dependent suppression of the MEK1/2-ERK1/2 survival pathway in TNF-alpha induced endothelial apoptosis. Cell Signal 18:531–540
67. Kamioka Y, Yasuda S, Fujita Y, Aoki K, Matsuda M (2010) Multiple decisive phosphorylation sites for the negative feedback regulation of SOS1 via ERK. J Biol Chem 285:33540–33548
68. Caunt CJ, Rivers CA, Conway-Campbell BL, Norman MR, McArdle CA (2008) Epidermal growth factor receptor and protein kinase C signaling to ERK2: spatiotemporal regulation of ERK2 by dual specificity phosphatases. J Biol Chem 283:6241–6252
69. Cobb MH, Goldsmith EJ (2000) Dimerization in MAP-kinase signaling. Trends Biochem Sci 25:7–9
70. Chang L, Karin M (2001) Mammalian MAP kinase signalling cascades. Nature 410:37–40
71. Shaulian E, Karin M (2002) AP-1 as a regulator of cell life and death. Nat Cell Biol 4: E131–E136
72. Weiss J (1997) The Hill equation revisited: uses and misuses. FASEB J 11:835–841
73. Sauro HM, Kholodenko BN (2004) Quantitative analysis of signaling networks. Prog Biophys Mol Biol 86:5–43

Chapter 10

Computational Modeling of Signal Transduction Networks: A Pedagogical Exposition

Ashok Prasad

Abstract

We give a pedagogical introduction to computational modeling of signal transduction networks, starting from explaining the representations of chemical reactions by differential equations via the law of mass action. We discuss elementary biochemical reactions such as Michaelis–Menten enzyme kinetics and cooperative binding, and show how these allow the representation of large networks as systems of differential equations. We discuss the importance of looking for simpler or reduced models, such as network motifs or dynamical motifs within the larger network, and describe methods to obtain qualitative behavior by bifurcation analysis, using freely available continuation software. We then discuss stochastic kinetics and show how to implement easy-to-use methods of rule-based modeling for stochastic simulations. We finally suggest some methods for comprehensive parameter sensitivity analysis, and discuss the insights that it could yield. Examples, including code to try out, are provided based on a paper that modeled Ras kinetics in thymocytes.

Key words: Computational modeling, Systems biology, Signal transduction, Stochastic simulations, Bifurcations, Ras activation, Biochemical networks

1. Introduction

Biological organisms can be regarded as spatially extended networks of biochemical and biophysical interactions, encased in a self-organizing physical structure. The physical structures, from the organization of cell membranes to the distribution of tissues, are themselves maintained via active processes by the very same networks of interactions. The interactions that make up the network span at least 8 orders of magnitude in length scale, starting from interactions between molecules to interactions between higher order structures. A very large subset of biological networks are networks of biochemical reactions between biomolecules; however, physical interactions through mechanical forces,

Xuedong Liu and Meredith D. Betterton (eds.), *Computational Modeling of Signaling Networks*, Methods in Molecular Biology, vol. 880, DOI 10.1007/978-1-61779-833-7_10,

entropic forces (especially osmosis), and electrical forces also play a role. From this point of view, life is an emergent property of these complex networks and systems biology the science devoted to their study.

According to one popular definition, systems biology is concerned with the acquisition and analysis of large and complex biological datasets. This definition is at best only partially true. At a conceptual level systems biology arose out of the recognition that biological phenomenon are emergent properties of complex networks. Acquiring information about these networks is of course a necessary first step, but what is even more important is making the conceptual leap from information to explanation of biological phenomena. There is still a huge amount to be learnt about the different biological networks that make up a single cell, even for bacterial cells. However, even at the current stage of knowledge, biological explanations require integrating many pathways into an explanatory whole. Such a task demands mathematical analysis, since the nonlinearity of many interactions and complexity of even simple networks usually implies that intuition is an unreliable guide.

Analysis of biologically important network and network modules is already a huge field. In this pedagogical exposition, we cannot do justice to all of it. We will therefore concentrate on the analysis of signal transduction networks. We will also use a mathematical model for Ras activation in T cells (1, 2) as an illustrative example of modeling.

2. Modeling Molecular Networks in the Cell

Very broadly speaking there are three overlapping biochemical networks in cells: the metabolic networks, the gene transcription networks, and the signal transduction networks. Metabolic networks are concerned with the synthesis of ATP, the primary energy source of the cell, and the synthesis of the other building blocks of the cell such as amino acids and lipids. Gene transcriptional networks are the networks of the proteins that control gene transcription and translation. Signal transduction networks link the cell to its environment, and enable the cell to process information and take decisions depending upon the changes in the environment and the specific signals received. Signal transduction networks are therefore key for cellular decision making for both single-celled organisms such as bacteria as well as cells that constitute multicellular organisms.

All cells perform sensory functions, such as ability to sense chemical gradients, forces, changes in osmotic pressure, and changes in electrostatic fields. Sensory cells, that is, cells that are specialized to perform some particular sensory function, such as record a photon of light like rod cells in the retina (3), or respond to minute changes in peptide affinities like T cells (4), perform these functions with particular sensitivity. The exquisite sensitivity that these cells possess are a product of the signal-processing capabilities of the signal transduction networks of these cells (5). One of the broad aims of computational models has therefore been to understand these information-processing capabilities with molecular detail.

2.1. Identifying the Question

The first step in mathematical modeling of signal transduction networks is formulation of the question. Mathematical modeling typically involves a number of assumptions at every step. Care needs to be taken that the question being asked justifies the assumptions being made. Usually there is a biological question of interest that requires explanation or molecular insight. The first step therefore is formulation of the biological question as a molecular question. As an example we take the model constructed in ref. (1) to explain the dependence of positive and negative selection of T cells on the affinity of the TCR-pMHC interaction (6). Selection of a single T cell is a long process with a timescale of days (7). However, ref. (1) assumes, based on some empirical evidence, that provided the signaling environment is not drastically changed (i.e., the source of stimulation is suddenly removed), the selection decision may be taken in minutes, and Ras activation is the molecular marker for this selection decision (8, 9). The model in ref. (1) reconstructs the key part of the network that leads from the stimulus to this molecular marker for the selection decision. The question then becomes how this part of the network processes information from the signal and how that is translated into the specific state of the marker chosen. To answer this question, the set of biochemical reactions so identified is then simulated by representing it as a system of differential equations or stochastic equations.

2.2. The Law of Mass Action

Mathematical modeling of biochemical reactions is based on the law of mass action. Put simply this law states that any chemical reaction proceeds at a rate that is proportional to the products of the concentrations of the reactants, raised to the power of the *molecularity* of the reactant. The rate here is defined as the time rate of change of concentration. Mass action assumes that in the volume under question, the reactants are well mixed. Under these assumptions, this law is clearly based on the assumption of a constant probability of encounter and for any encounter, a constant probability of it being converted into a reaction. Under these assumptions then for a bimolecular reaction

$$A + B \xrightarrow{k} C, \qquad \text{(i)}$$

the law of mass action implies that the rate of change of the product C is given by,

$$\frac{d[C]}{dt} = k[A][B], \tag{1}$$

where the square brackets denote concentrations. The proportionality constant k is called the kinetic rate constant and in general it depends upon the characteristics of molecular collisions in the medium and the average internal state of the reactant molecules. If the chemical reaction is reversible, $A + B \rightleftharpoons C$, then the product C is increased by the reaction of A with B but diminished by the decay of C. The rate of change of the concentration of C is then given by

$$\frac{d[C]}{dt} = k_{on}[A][B] - k_{off}C \tag{2}$$

where k_{on} and k_{off} are the rates of the forward and reverse reactions, respectively. The two reaction rate constants have different units. The forward rate constant for a second-order reaction like the one above has the units $[\mathrm{Conc}]^{-1}[\mathrm{time}]^{-1}$ while the backward rate constant, like all first order reactions, has the units $[\mathrm{time}]^{-1}$.

In this simple form, the law of mass action holds for *elementary reactions*, i.e., reactions with direct physical interactions between the reactants. However, many reactions can take place by more complex mechanisms. Enzymatic reactions are one example in which the rate of reaction does not follow the mass-action law. These reactions may be described by special rate laws, such as the Michaelis–Menten rate law. Reactions involving multistep cooperative binding are often simplified as single-step reactions that follow a rate law given by a Hill function. In many situations, the rate law is given by some power law, i.e., the rates depend upon products of concentrations to different powers that may not equal the molecularity of the reactions. Such rate laws are called generalized mass-action rate laws. Some studies have shown that reactions taking place in effectively lower dimensions or those taking place in excessively crowded environments may follow more complex rate laws such as generalized mass action (10) instead of simple mass action.

For some complex reaction mechanisms computational simulation is often simplified if the reaction system can be broken down into elementary reaction steps that follow mass-action kinetics. The Michaelis–Menten scheme for enzymatic kinetics is precisely such a scheme that decomposes an enzymatic reaction into elementary reaction steps that follow mass-action. In this case the necessity for invoking rate laws that may depend upon certain assumptions vanishes, at the cost of adding more parameters to the simulation.

2.3. Steady States

Many mathematical analyses are carried out under the assumption that the reaction system will reach an equilibrium state or a steady-state. A steady-state is reached when all time derivates become zero, i.e. concentrations have stopped changing with time. Note that this does not necessarily mean that chemical reactions are not taking place. For the reversible reaction,

$$A + B \underset{k_{-1}}{\overset{k_1}{\rightleftharpoons}} C \tag{ii}$$

equilibrium means that the rates of conversion of the forward and backward reaction have equalized. However for the irreversible reaction, Reaction (i), the only steady state is reached after all the A has converted into B. For the reversible reaction, Reaction (ii), the steady-state concentrations of the reactants and products must obey $k_{\text{on}}[A][B] = k_{\text{off}}[C]$, or,

$$K_{eq} = \frac{k_{on}}{k_{off}} \frac{[C]}{[A][B]}. \tag{3}$$

The value of the ratio K_{eq} is a thermodynamic constant, in general given by,

$$K_{eq} = \exp\left(-\frac{\Delta G}{RT}\right) \tag{4}$$

where ΔG is the change in the Gibbs free energy due to the reaction, T is the temperature and R is the gas constant.

The steady state assumption needs to be made with care. One of the important aspects to take into account are the timescales of the processes taking place. For example, if a reaction network contains both fast and slow reactions, then one could safely assume that the fast reactions equilibrate. Signal transduction processes typically are much faster than gene transcription for mammalian cells. In T cell activation, for example, early signaling processes could take place in seconds while gene transcription takes hours and days. A network that links signaling events to protein translation is separable into fast and slow processes, and the assumption that the fast reactions equilibrate to slowly changing protein concentrations is reasonable. A network consisting of only early signaling events may not reach equilibrium in the timescale of the experiments that the models are based on. In such cases, it is the dynamic pattern of changes in protein concentrations with time rather than steady-state concentrations that are relevant.

Note that even when a separation of timescales exists, other factors could prevent a true steady state from being reached. For example, active transport mechanisms could exist that move activated molecules to other cellular compartments where they are trapped, as has been suggested for TGFβ signaling (11).

2.4. Elementary Reaction Kinetics

Most chemical reactions in biology can be divided into three elementary types: simple binding and unbinding, cooperative binding, and enzymatic modifications.

2.4.1. Binding–Unbinding

Consider a simple reaction scheme in which two molecules, A and B, reversibly bind to each other. At the length scales of the molecules and the amino acids of the proteins taking part in the reactions, even simple binding reactions are in reality not-so-simple. This is especially true for many binding reactions where an allosteric change takes place after binding. Allosteric conformational changes typically expose some residues on another part of the protein's surface that allows binding reactions and enzymatic modifications to take place. For example, ligand–receptor binding on the cell surface can lead to a conformational change in the cytoplasmic tail of the receptor molecule, allowing other downstream molecules to bind. A relevant question here is what happens if the ligand dissociates after the downstream molecule has bound to the exposed residues. Thermodynamic considerations suggest that receptor binding to the downstream protein is probably destabilized by this dissociation that would be reflected in the use of a faster dissociation rate. Things can be more complicated when what is involved is an enzymatic modification, such as a phosphorylation of the cytoplasmic tail. Thinking through these details is often necessary even when specifying binding–unbinding reactions in the network. Practical modeling of biochemical reactions usually requires an act of judgement on the choice of some reaction scheme over another, and clarity on the basis of the choice is always desirable.

2.4.2. Cooperative Binding

Another common class of biological reactions is cooperative binding. Consider for example a molecule P that requires two molecules of S to bind sequentially for activation. The reaction scheme can be written as

$$P + S \overset{K_1}{\rightleftharpoons} P_1 \tag{iii}$$

$$P_1 + S \overset{K_2}{\rightleftharpoons} P_2 \tag{iv}$$

The steady-state assumption along with mass conservation leads to the following expression for the steady-state concentrations:

$$P_2^{\star} = \frac{K_1 K_2 P_0 S^{\star 2} - K_1 K_2 P_1^{\star} S^{\star 2}}{1 + K_1 K_2 S^{\star 2}}, \tag{5}$$

where the starred terms refer to the steady-state concentrations and $P_0 = P + P_1 + P_2$. For positive cooperativity, i.e., when $K_2 > K_1$, the intermediate state P_1 can be assumed negligible yielding

$$\frac{P_2^{\star}}{P_0} = \frac{K_1 K_2 S^{\star 2}}{1 + K_1 K_2 S^{\star 2}} \tag{6}$$

for the steady-state concentration $P_2^\star$. Note that this is equivalent to assuming the single step equation:

$$P + 2S \overset{K_1 K_2}{\rightleftharpoons} P_2. \tag{v}$$

The functional form

$$f(x) = \frac{Kx^n}{1 + x^n}, \tag{7}$$

where K is a constant, is called a Hill function and is an obvious generalization to the case of n-binding sites where intermediate states can be neglected. In mathematical biology, Hill functions are often used as effective rate laws that capture cooperative binding, and the Hill coefficient n is a measure of the cooperativity or the sigmoidality of the process. A similar rate law is often used in models of gene transcription for the dependence of the rate of transcription of a protein on the concentration of the repressors in many synthetic biology experiments and models (12). This form is also based on the assumption that repressor multimerization follows a generalized version of the scheme given in Reaction (v).

2.4.3. Enzymatic Kinetics

The simplest representation of an enzymatic reaction as a series of elementary reaction steps is the Michaelis–Menten scheme. This scheme represents the enzymatic reaction as occurring through two elementary steps, a reversible binding step in which the enzyme binds to its substrate and an irreversible modification step:

$$E + S \underset{k_{off}}{\overset{k_{on}}{\rightleftharpoons}} ES \overset{k_{cat}}{\rightarrow} P + E \tag{vi}$$

Under the assumption that the complex ES is in equilibrium (sometimes called quasiequilibrium), we can easily derive the following expression for the reaction rate:

$$\frac{dP}{dt} = \frac{k_2[E]_0[S]}{K_M + [S]} \tag{8}$$

where the constant K_M is given by

$$K_M = \frac{k_{-1} + k_2}{k_1}. \tag{9}$$

In order to avoid making the quasi-equilibrium assumption, an alternative to using the Michaelis–Menten rate equation, Eq. 8, is to use the reaction scheme given in Reaction No. (vi) based on simple mass action.

2.5. Stochastic Modeling

In the previous section, we have dealt with describing basic methods for modeling cellular processes by constructing a network of biochemical reactions and then simulating them on a computer. We mainly focused, however, on deterministic modeling using differential equations. In this section, we briefly describe some common methods for stochastic simulations.

2.5.1. Stochastic Chemical Reactions

The representation of chemical reactions by systems of differential equations involves two important assumptions. The first is that the time rate of change of the concentrations of the species can be treated as a continuous function. The second assumption is that the time rate of change is a deterministic process. In reality both assumptions are strictly speaking incorrect. Chemical concentrations of species change by integer amounts; hence, concentration changes cannot be continuous. Furthermore, chemical reactions are stochastic rather than deterministic, i.e., they take place by a random process.

The continuous and deterministic assumption actually works quite well provided the concentrations are high enough so that (a) the continuous approximation is reasonable for the concentrations and (b) the typical size of fluctuations is small relative to the number of molecules, i.e., $N^{-1/2} \ll 1$. However, even in such cases, the stochastic nature of the chemical reaction process can manifest itself. For example, close to a saddle-node bifurcation a bimodal population of the concentration of an activated protein is a signal of that stochasticity. In situations when the number of molecules is much lower, fluctuations cannot be ignored and in fact may even play an important role in cellular decision making (13, 14).

It is therefore often necessary and desirable to perform both stochastic and deterministic simulations for a given biochemical system. Stochastic simulations are actually quite easy to perform and for small systems not computationally expensive. The standard algorithm for stochastic simulations is the Gillespie algorithm and its later variants that are described in a number of excellent papers and reviews (15). Moreover, a number of excellent software packages that can perform stochastic simulations of the biochemical network exist. Some of these, like the Simbiology®; toolbox in Matlab®; and the VirtualCell®; simulation environment, can do both deterministic and stochastic simulations.

2.6. Example: Kinetics of Thymocyte Selection

Let us take as an example a model constructed to explain the results of an experiment on thymocyte positive and negative selection (6), which probed the relationship between the selection decision in the thymus and the affinity between the TCR and the pMHC complex. The experiment observed that the percentage of selected thymocytes increases as a gradual function of affinity, reaches a maximal plateau (positive selection) and then drops sharply to zero after the affinity crosses a threshold (negative selection). The model based on this paper (1) noted that a major qualitative distinction between positively selected and negatively selected thymocytes appeared to be the kinetics of activation of the G-protein Ras and its downstream target Erk. In negatively selected thymocytes, both Ras and Erk appeared to be strongly activated with a sharp, digital-looking response. In positive selection, however, Ras and Erk activation was weaker and did not show a sharp temporal response. These considerations led to the choice of Ras as the molecular marker of the selection decision during early signaling events.

The modelers then constructed a minimal model of Ras activation based on integrating a large number of experimental papers on T cell biology (1). Since ref. (6) reported a monotonic relationship between thymocyte affinity and the activation of the tyrosine kinase, ZAP70, the level of activated ZAP70 was chosen as the input and activation of Ras as the output of the model. The intermediate steps are as follows. ZAP70 phosphorylates the adapter protein LAT on several tyrosines sites. LAT assembles a signaling complex in a synergistic fashion, including recruitment of PLCγ1, a complex consisting of Gads, SLP76, Itk, and Vav, and a complex consisting of Grb2 and SOS. PLCγ1 is activated by Itk and then cleaves PIP2 on the cell membrane, producing DAG and IP3. The latter activates calcium channels, leading to a burst of calcium signaling that activates RasGRP, a cytosolic Ras guanine-exchange factor. Activated RasGRP is recruited to the cell membrane by DAG, and activates Ras. SOS, another Ras guanine-exchange factor, is recruited to the cell membrane by Grb2 (and in fact is often found complexed with Grb2), and also activates Ras.

2.7. Network Topology, Motifs, and Modules

Once the question of interest is clarified, and the biochemical reactions involved in the signaling process written down, it is useful to take a global look at the structure of the network so formed. This could be useful for multiple reasons.

1. Qualitative behavior of the biochemical network can sometimes be predicted from the topology of the network, that is, from the structure of interconnections that the network possesses. In principle, once the topology of the network has been delineated, it should be possible to then divide parameter space into different regions that correspond to different qualitative behavior of the network, i.e., different phenotypes (16). In practice this is a very difficult undertaking except for small networks. In some limited cases, it may be possible to obtain some information about the steady states of the system based on the topology of the network alone (17, 18). These possibilities will not be discussed further, and the interested reader is referred to the references cited above.
2. However, the other reason to have a global view of the network structure is to look for motifs and modules in the network.

A motif is a small subnetwork of a specific type that recurs frequently in biological networks (19). The presence of these motifs can throw a lot of light on the dynamic properties of the networks. Some of the most important motifs in gene transcription networks include auto-regulation and feedforward regulation (20). A module of a network is the term used for a subnetwork that for certain purposes can be studied in isolation from the rest of the network. Usually this is only possible for a portion of the network with no feedback, i.e., when the output of the subnetwork does not influence the input. While in general this is not true for biochemical networks in a cell, it is quite possible that either weak or slow feedback can be a valid basis for the assumption of modularity. Note that a motif may or may not be a module.

Examination of modules and motifs present in networks helps in building simple effective models that help gain intuition about the network dynamics. For example, if it is possible that some module, say a positive feedback loop or negative regulation of a negative regulator, dominates the response of the molecular markers under consideration. It should then be possible to gain insight into the entire system by examining the behavior of that particular module or motif. Let us go back to the example of activation of the Ras protein in T cells to illustrate this point.

In the example that we are considering, we see the following steps taking place. (1) Multisite phosphorylation of the adapter protein LAT. (2) Assembly of a signaling complex on LAT. (3) Activation of Ras by RasGRP. (4) Activation of Ras by SOS. Analysis of the biochemistry of the last two steps revealed a key difference in the expected kinetics. SOS was a guanine-exchange factor with an inbuilt positive feedback, while RasGRP did not show any feedback kinetics. Therefore, this network contained a positive feedback motif, which will be analyzed in detail in the next section.

2.8. Bifurcation Analysis

Work done by the Kuriyan lab showed that there was a possible positive feedback loop in Ras activation due to the enzyme SOS. SOS had an allosteric Ras-binding site and Ras binding at this site increased the overall catalytic rate of Ras activation about fivefold if RasGDP bound but about 75-fold if RasGTP bound. Since SOS converts RasGDP to RasGTP, thereby making more RasGTP available to bind at the allosteric site, there is positive feedback in Ras activation (21, 22).

Before simulating this entire network we begin by trying to understand the positive feedback module in Ras activation (23). Using the Michaelis–Menten form for the rate equations for enzymatic reactions, and using SOS concentration for the input signal, we can write down the following simplified model:

$$S + RD \underset{k_{11}}{\overset{k_1}{\rightleftharpoons}} SRD \tag{vii}$$

$$S + RT \underset{k_{21}}{\overset{k_2}{\rightleftharpoons}} SRT \tag{viii}$$

$$SRD + RD \xrightarrow{K_{3m}, k_3} SRD + RT \tag{ix}$$

$$SRT + RD \xrightarrow{K_{4m}, k_4} SRT + RT \tag{x}$$

$$RT + GAP \xrightarrow{K_{5m}, k_5} RD + GAP \tag{xi}$$

The corresponding differential equations and conservation laws are

$$\frac{d[S]}{dt} = -k_1[S][RD] + k_{11}[SRD] - k_2[S][RT] + k_{21}[SRT] \tag{10}$$

$$\frac{dSRT}{dt} = k_2[S][RT] - k_{21}[SRT] \tag{11}$$

$$\begin{aligned} \frac{dRT}{dt} = & -k_2[S][RT] + k_{21}[SRT] + \frac{k_3[RD][SRD]}{K_{3m} + [RD]} \\ & + \frac{k_4[RD][SRT]}{K_{4m} + [RD]} - \frac{k_5[GAP][RT]}{K_{5m} + [RT]} \end{aligned} \tag{12}$$

$$\alpha = [S] + [SRD] + [SRT] \tag{13}$$

$$\beta = [RD] + [RT] + [SRD] + [SRD]. \tag{14}$$

To understand the qualitative behavior of the system it is useful to make the steady-state analysis by setting all the time derivates in Eqs. 10–14 to zero. Solving the resulting set of polynomial equations then yields a vector of steady-state protein concentrations, also called a *fixed point*. Since the algebraic equations are nonlinear, a given set of parameter values can sometimes give us more than one vector of concentrations that satisfy the equations. These multiple fixed points are very interesting because they tell us that the system may possibly exist in two or more different steady states under the same circumstances. However, this is only possible when the fixed points are stable. An unstable fixed point is like a regular solid cone balanced on its narrow tip; symmetry ensures that there exists one position where there are no forces making the cone topple over. However, tiny displacements from this position can make the cone tip over, making this balancing act near impossible for a narrow tip. The same cone resting on its flat broad base is in a stable equilibrium since it can be tilted quite a bit and it will still fall back on its base (24, 25).

The fixed points of a systems of equations depend upon kinetic parameters and initial concentrations, and when these are varied, the structure of fixed points changes. For example, changing the initial concentration of a protein which behaves like the input of a module beyond some threshold may lead to the system making a transition from having a single stable fixed point to two stable fixed points. Such transitions are called bifurcations (25), and indicate a qualitative change in the behavior of the system. This qualitative change in the context of cell biology could represent a change of cell state, or a cellular decision. Basic one-dimensional bifurcation analysis, i.e., with respect to changes in one parameter, requires the calculation of fixed points and their stability as the parameter of interest is varied. This is not possible to do analytically for the system of equations arising from Eqs. 10 to 14, but a similar simpler system can be understood using semi-analytical arguments since the basic process comprises of an autocatalytic activation with a saturable deactivation (2).

It is usually more useful to carry out the analysis numerically, and a few easy to use software packages that can perform bifurcation analysis exist. The most well-known program is AUTO, and though using AUTO directly is nontrivial, some useful packages that can integrate systems of equations and also provide front ends to AUTO, such as Xppaut (26) and Oscill8 (27) exist. These front ends employ an intuitive syntax that allows the differential equation system given above to be written in a text file which is then run by the software. An example text file containing the code for Oscill8, which is quite user friendly, is shown in Box 1.

The file contains all the information the software needs to simulate the system of differential equations. In order to perform bifurcation analysis, we need to vary the total amount of SOS, which is set by α, from $\alpha = 5$ which is the value in the file to $\alpha = 80$. The results of the bifurcation analysis performed using Oscill8 are shown in Fig. 1.

Figure 1 shows that at low SOS concentration there is hardly any production of RasGTP. As SOS concentration increases, the system undergoes a saddle-node bifurcation, and acquires three fixed points. However, only two of these, the lowest and the highest, are stable. The middle fixed point, plotted in the lighter color, is unstable. As SOS concentration increases further, the system goes through another saddle-node bifurcation, and becomes monostable again, with RasGTP at the maximal level of activation.

Bifurcation diagrams like this immediately lead to the following conclusions: (1) the transition from a low RasGTP state to a high RasGTP state must be very sharp, almost digital; (2) the system will show hysteresis, i.e., it follows the lower path when we go from low SOS to high SOS, but will follow the upper path when we go from high SOS to low SOS; and (3) in the bistable regime, in between the two bifurcation points, there is likely to be a bimodal distribution of RasGTP because of intrinsic noise in the chemical reactions.

Box 1
The code for the differential equation system in Eqs. (10-14) for use in Oscill8. This code generated the bifurcation diagram in Fig. (1)

```
sosrd=alpha-S-SRT
RD=beta-RT-SRD-SRT
K3m=(k3c+k31)/k3
K4m=(k4c+k41)/k4
K5m=(k5c+k51)/k5
dS/dt=-k1*S*RD+k11*SRD-k2*S*RT+k21*SRT
dSRT/dt=k2*S*RT-k21*SRT
dRT/dt=-k2*S*RT+k21*SRT+(k4c*RD*SRD/(K4m+RD))+
(k3c*RD*SRT/(K3m+RD))-k5c*GAP*RT/(K5m+RT)
param alpha=5
param beta=300
param k1=.02
param k11=3
param k2=.025
param k21=.4
param k3c=0.04
param k3=0.02
param k31=.1
param k4c=0.003
param k4=0.02
param k41=1
param k5c=0.1
param k5=0.05
param k51=1
param GAP=10
init SRT=0
init RT=0
```

Bifurcation analysis of the idealized model of Ras activation yields some qualitative insight into the dynamics of Ras activation with SOS. But it is not yet established that the full network has these properties and the role of RasGRP is also not clear. Therefore it is necessary now to put together the various pieces and go back to the larger network.

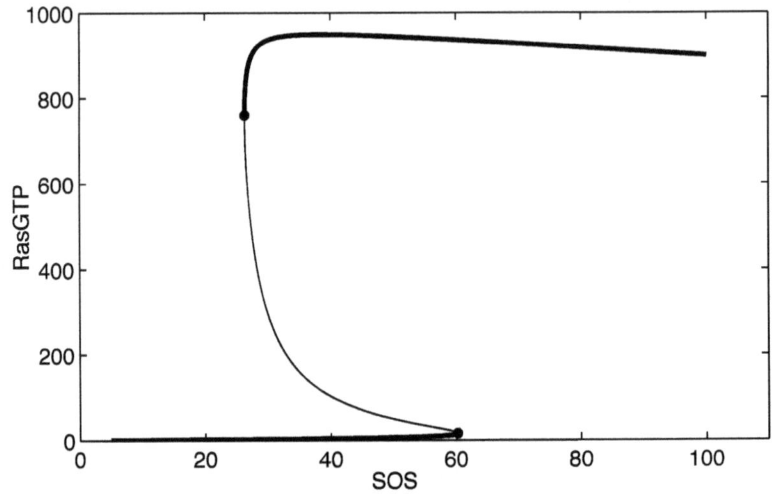

Fig. 1. One parameter bifurcation diagram obtained by varying the concentration of SOS by using the software Oscill8. Each point is a fixed point of RasGTP concentration corresponding to the SOS concentration shown in the x-axis. The *lighter line* denotes unstable fixed points; the *darker lines* are stable fixed points. The bifurcation points are marked.

2.9. Systems of Biochemical Reactions

The standard method of representing a system of chemical reactions that facilitates its deterministic simulation as a system of differential equations is by the stoichiometric matrix. The stoichiometric matrix can be constructed in the following manner. Construct a matrix $\{S\}$ with each row corresponding to each chemical reaction, and each column corresponding to every chemical species. For every element in the matrix, s_{ij} we ask if the species j participates in the chemical reaction i as either a reactant or product. If it does, then s_{ij} = number of molecules of j that are created in reaction i, with a minus sign for reactants and a positive sign for products. If a species does not participate in the reaction, the corresponding entry in the matrix is 0. Any specific reaction, say the k-th reaction, can therefore be written as

$$\sum_{j=1}^{N_s} s_{kj} C_j = 0, \tag{15}$$

where s_{kj} is the corresponding element of the stoichiometric matrix and $\{C\}$ is a vector of chemical species that follows the same numbering as for the stoichiometric matrix, and N_s is the number of species. We can define the *molecularity* of reactants in each reaction by using a simple command that picks out only the negative elements in the matrix $\{S\}$, i.e.,

$$\begin{aligned} m_{ij} &= | s_{ij} | \quad \text{if} \quad s_{ij} < 0 \\ &= 0 \quad \text{otherwise.} \end{aligned} \tag{16}$$

If every reaction in the network follows mass-action kinetics, then the reaction rate for each reaction is given simply by

$$r_i = k_i \prod_{j=1}^{N_s} C_j^{m_{ij}}, \qquad (17)$$

where k_i is the kinetic rate constant. In other words the reaction rate is the product of the concentrations of the reactants raised to the power of their molecularity. The chemical reaction network can now be written

$$\frac{dC_j}{dt} = \sum_{i=1}^{N_r} s_{ij} r_i, \qquad (18)$$

or in vector notation after introducing the vector of reaction rates, $\{\boldsymbol{R}\} = \{r_i\}$,

$$\frac{dC}{dt} = \boldsymbol{SR}. \qquad (19)$$

If the chemical reactions do not follow simple mass-action kinetics, the appropriate changes need to be made to the vector of reactions rates, $\{\boldsymbol{R}\}$.

The above steps can be easily encoded into a script in numerical software such as Matlab®; that produces the network of differential equations starting from the stoichiometric matrix, which then can then be solved using an inbuilt solver. The Matlab®; toolbox called Simbiology®; can perform all of these steps (and more) starting from a list of chemical reactions in an intuitive text format. While in principle the procedure we have outlined above is all what is required for differential equations, in practice one often needs to deal with another hurdle, called combinatorial complexity, in which the number of possible chemical reactions increases exponentially.

2.10. Combinatorial Complexity

The Ras activation network also provides an example of combinatorial complexity. The first step in the reaction network of ref. (1) was the phosphorylation of the protein LAT (Linker for Activation of T cells) by the tyrosine kinase, ZAP70. LAT has 9 or 10 tyrosines that are phosphorylated by ZAP70, but extensive mutation analysis has determined that LAT with just 3 tyrosine sites is close to wild type in many signaling characteristics (28). Now the first of these tyrosine sites, Y132, can be either unphosphorylated and either empty or bound with ZAP70, phosphorylated and empty or bound to PLCγ1 or a generic phosphatase. Both the remaining two tyrosines share the unphosphorylated states and the phosphatase-bound state, and in addition can bind either Grb2-SOS or Gads. Grb2-SOS has two binding sites for Ras, one of which can bind both RasGTP and RasGDP. Counting all the possibilities, we conclude that Y132 can exist in 5 states and Y191 and Y226 can exist in 10 states each. The total possible number of LAT-containing complexes therefore is found by multiplying all the possibilities for each state together, that is, $5 \times 10 \times 10 = 500$.

Addition of just one more tyrosine site with an additional 10 possible states increases the number of allowable LAT complexes to 5,000. This explosion in the possible number of states of multisite binding proteins is named combinatorial complexity.

Note that there are already many simplifying assumptions in the calculation for the number of states of LAT, such as treating Grb2 and SOS as constitutively bound and ignoring both complexes assembled by Gads as well as complexes consisting of multiple LAT molecules. Without these simplifying assumptions, the number of states would have been even larger. If LAT has 500 states, it must have an even larger set of reactions to reach these states. This number can be easily estimated by adding together reactions of each type. For example, reactions that involve Grb2-SOS binding and unbinding to either Y191 or Y226 on LAT can occur a total of $2 \times 5 \times 10 \times 1 = 100$ ways. It can be seen that the total number of reactions that even the simplified model of LAT takes part in would number in the thousands! Note that the number of reactions of each type is still quite small, for example Grb2-SOS binding to LAT Y191 and to Y226 are two different types of reactions but because of numerous LAT states they can take place in 100 ways. However, in the absence of cooperativity, these reactions involve just two forward rate constants and two unbinding rate constants.

Combinatorial complexity can be ignored by assuming away the multisite characteristic of the adapter protein, i.e., treating it as a single-site protein. However this has two possible pitfalls. Firstly, for important adapter proteins such as LAT, or p53, the effects of mutations of specific sites may be of great interest. This requires a more realistic representation of the molecule. Secondly, the presence of multiple phosphorylation sites may itself be playing a role in signal transduction (29, 1). Ignoring multiple sites carries the risk of ignoring the possible roles that these sites may be playing in signal transduction. Thus, as with all simplifying assumptions in systems biology, this one too should be made with care.

If combinatorial complexity is not to be ignored, how does one deal with it? The recommended method is not to try to list the reactions by hand, which is both tedious and error prone, but to use software for this task. This comes under the rubric of what has been described as rule-based simulations.

2.11. Rule-Based Simulations

Rule-based simulation software allows the formulation of rules using a transparent syntax (at least in theory) that provides the user a means to specify molecules, their allowed states and classes of allowed reactions. On completing the description of the reaction system in terms of rules, the software iteratively expands the network and lists all reactions between the molecules that satisfy the rules. Examples of such software packages are BioNetGen (30), SSC (31). Both not only expand the rules and construct the chemical reaction network, they also perform efficient stochastic simulations of the reactions.

Box 2
SSC code for generating the reactions for phosphorylating LAT by Zap70. The software also creates an executable for simulating the reactions. The parameter values needs to be passed to the executable during simulation in a separate file

```
——-Specify Geometry This is for well mixed simulations—-
region PM
box width 1 height 1 depth 1
subvolume edge 1
———Specify Species and initial concentrations—————
new LAT(p1="off",p2="off",p3="off") at L0
new zap at Z0
——-Reactions—————
rxn P:LAT(p1="off",bs1#) Q:zap(bs#) at kzapon1 → P.bs1#Q.bs
rxn LAT(p1="off",bs1#1) zap(bs#1) at kzapoff1 → break 1
rxn P:LAT(p1="off",bs1#1) zap(bs#1) at kzapcat → P.p1="on" ; break 1
rxn P:LAT(p2="off",bs2#) Q:zap(bs#) at kzapon2 → P.bs2#Q.bs
rxn LAT(p2="off",bs2#1) zap(bs#1) at kzapoff2 → break 1
rxn P:LAT(p2="off",bs2#1) zap(bs#1) at kzapcat → P.p2="on"; break 1
rxn P:LAT(p3="off",bs3#) Q:zap(bs#) at kzapon3 → P.bs3#Q.bs
rxn LAT(p3="off",bs3#1) zap(bs#1) at kzapoff3 → break 1
rxn P:LAT(p3="off",bs3#1) zap(bs#1) at kzapcat → P.p3="on"; break 1
————what to record. For everything use record all——-
record "pY1pY2pY3" LAT(p1="on",p2="on",p3="on")
```

BioNetGen is now part of the simulation environment called VirtualCell (32), which can also perform differential equation and partial differential equation simulations in realistic geometries.

SSC performs very efficient stochastic simulations and is also capable of performing spatially explicit simulations. In Box 2, we show a simple example of SSC syntax to encode all the possible reactions that lead to the phosphorylation of LAT. SSC requires that all rules should be written as a text file with the extension.rxn. The file should begin by specifying the geometry and the reacting species and their concentrations. Since spatially explicit simulations are not being discussed here, the geometry specified is a cubic box with subvolume edge equal to the size of the box, i.e., there are no subvolumes. Species with different states, i.e., phosphorylation states, must be declared with the state explicitly noted; binding sites are declared in the reaction statements. The reaction statements start with rxn on each line followed by the rule for that reaction. For example,

rxn P:LAT(p1=”off”,bs1#) Q:zap(bs#) at kzapon1 → P.bs1#Q.bs

is an instruction to the software to look for all patterns with the LAT molecule having its state p1=”off” and its binding site bs1 empty and for zap molecules with binding sites bs empty and connect the two binding sites with a rate kzapon1. P and Q are labels for these patterns which need not be unique to a reaction. The reaction

rxn LAT(p1="off",bs1#1) zap(bs#1) at kzapoff1 → break 1

instructs the software to look for a pattern where the binding site bs1 of LAT is connected to the binding site bs of zap (hence, they have the same associated number) and break that connection with a rate kzapoff1. When this code is run, SSC constructs the set of all reactions consistent with these rules, and also makes an executable file that can do an efficient stochastic simulation of the reaction. Readers interested in using SSC or BioNetGen can download them and try them out using the tutorials and manuals that come with these programs and that can be found on their websites.

The code corresponding to the model of ref. (1) is too long to be reproduced here and is available from the author upon request. Though ref. (1) used a manually constructed system of biochemical equations for the main results, this code is equivalent, has fewer assumptions, and can reconstruct all the results of that paper. In brief, the simulations show that:

1. The signal strength determines how much of LAT is fully phosphorylated. The concentration of fully phosphorylated LAT is a sigmoidal function of signal strength due to multiple phosphorylation sites.
2. Fully phosphorylated LAT assembles a stable complex of PLCγ1, Gads, and Grb2-SOS because of cooperative interactions between these molecules.
3. Partially phosphorylated LAT can activate the RasGRP pathway but cannot activate the SOS pathway of Ras activation efficiently.
4. Weak signals therefore activate Ras through RasGRP while strong signals activate both pathways. RasGRP serves to prime the SOS pathway by making RasGTP available for it.

2.11.1. Stochastic Simulations as a Cellular Population

Every time evaluation of a stochastic simulation produces a different temporal path. Therefore to compare the results of an experiment using a bulk measurement like western blots with stochastic

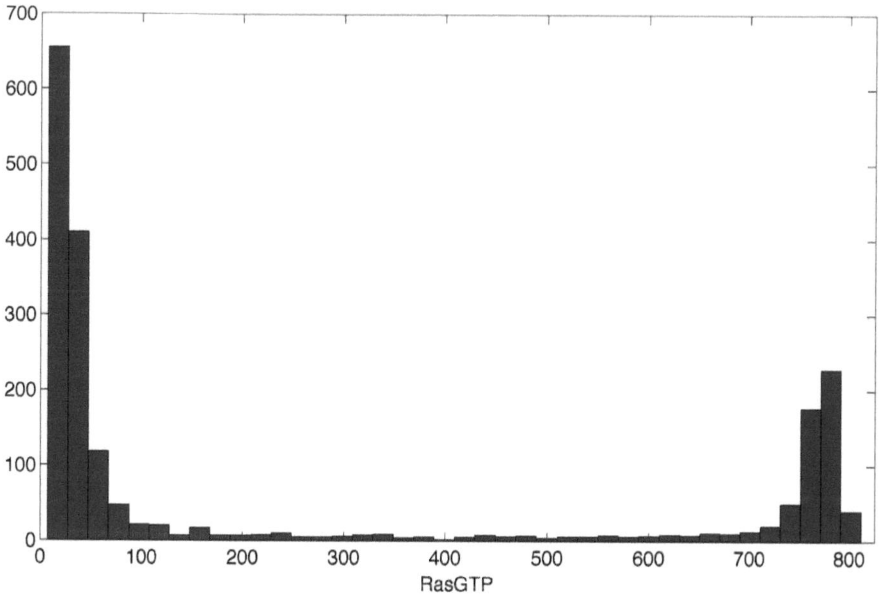

Fig. 2. Histogram of Ras activation levels from 2000 stochastic simulations of the SSC code written to illustrate the modeling of ref. (1). Each simulation ran for 3,000 s and the average level of RasGTP of the last 50 s was calculated and binned to produce this histogram.

simulations, the results of several simulations need to be averaged together. However, if the average of the simulations represents the bulk measurement, an individual simulation can be thought of as representing a single cell. While a single simulation cannot be compared to any single cell, the statistical properties of the simulation data can certainly be compared with the statistical properties of data from a population of cells. Note that real cells live in a noisy environment and are somewhat different to begin with. Real cells therefore have both extrinsic noise and intrinsic noise, where the former can be regarded as environmental noise and the latter as noise emanating from the random nature of chemical processes. However, the simulation only has intrinsic noise, though extrinsic noise can be artificially simulated by, for example, varying initial concentrations of reacting proteins.

In the case of Ras activation, thousands of stochastic simulations can be run, and a histogram of Ras concentration at some timepoint of choice plotted. For both low levels and high levels of antigenic strength, defined as low and high concentrations of Zap70, the histograms are unimodal distributions, but for intermediate strengths, which correspond to the bistable regime of Fig. 1 the histogram is bimodal, as can be seen in Fig. 2. At the signal strength of the simulation shown in Fig. 2 most cells have not activated and stay at the lower fixed point, but some of them have been kicked to the upper fixed point by the inherent intrinsic noise of chemical processes.

Simulations that activate the cell through the RasGRP pathway alone could also have been carried out. This can be achieved by increasing concentration of DAG in the simulation and bypassing the early signal transduction network. The histograms of Ras activation in this case show unimodality at all strengths. The last exercise also corresponds to an actual experiment: T cells can be activated by giving them PMA, a synthetic DAG analog. Both predictions were confirmed by flow cytometry experiments on active Erk (1).

2.12. Parameter Sensitivity

We have so far glossed over the problems of parameters in modeling biological networks. This is probably the biggest hurdle before the development of quantitative models for at least what is well known in cell biology. In the current example, several parameters, including some of the most important such as enzymatic constants of SOS were in fact measured in vitro. Though this yields only equilibrium constants, it does serve to constraint the kinetic rate constants to some extent. Other parameters needed to be assumed. This is unfortunately quite typical of computational modeling especially in mammalian cells. Apart from kinetic parameters, concentrations are also generally not known though some rough measurements do exist in the literature, though often not for the cell type of interest. Therefore even after exercising a great deal of care in sifting the literature for measurements and making inspired guesses based on similar proteins in other cell types, the next necessary computational step becomes extensive parameter sensitivity analysis.

Ideally a molecular biology paper that proposes a molecular mechanism for some cellular phenomenon should attempt to both model that phenomenon and measure key quantities that are required for that model. Since this is unfortunately not yet standard operating procedure, partly because of the experimental difficulties involved, parameter sensitivity analysis is necessary to ensure that the molecular mechanism proposed is robust. In a sense, the task of parameter sensitivity analysis is to understand part of the design space of the biochemical network. If it can be shown that certain combinations of the parameters are important rather than single parameters, that reduces the dimensionality of the problem significantly and the design space can be comprehensively investigated.

There are two kinds of parameter sensitivity that can be calculated. Local parameter sensitivity s_j is defined as the proportional response of the output Υ to a small perturbation of the parameter k_j, i.e., it is an elasticity with respect to the parameter,

$$s_j = \frac{\Delta \Upsilon}{\Upsilon} / \frac{\Delta k_j}{k_j}. \quad (20)$$

Note that s_j is a function of time. While local parameter sensitivity is useful, it does not really explore parameter space. For the latter, it is necessary to make large discrete jumps in the values of the parameters. For a realistic model with tens of parameters, this often implies running the model again and again with parameters differing by orders of magnitudes and examining the results.

It is also necessary to perform multidimensional parameter sensitivity analysis in order to understand the effects of changing all the parameters or a subset of the parameters at once. The most useful way of carrying this out involves picking the set of parameters from a distribution. For example, ref. (1) picked parameters from a normal distribution with mean equal to the base parameter value and standard deviation chosen as a proportion of the mean, and re-ran the model for every different set.

The final aim of parameter sensitivity analysis is to help in determining the constraints on the model. Ideally these should take the form of inequality constraints on the parameters that specify the area of validity. At the current state of knowledge, these constraints are expressed as qualitative knowledge; but the hope is that with the expansion of biological modeling techniques we should be able to understand the domain of validity of the models with greater precision.

3. Conclusion: Using Modeling for Biological Explanation

Parameter sensitivity simulations are also like experiments; they predict the result of introducing a perturbation of the system. In fact they correspond directly to gain-of-function, loss-of-function, and knockout mutations. If the model has proved robust, then the next logical step is the generation of predictions. The advantage of computational models is that these predictions can be made with a fair amount of precision even at the qualitative level. In the final analysis, the ability of a model to survive the testing of its qualitative predictions, either implicitly or explicitly, is a measure of its success.

What then does modeling add to biological explanation? Firstly the analysis of small modules often leads to new explanations and phenomenon being discovered, which would not have been possible without mathematical modeling. Secondly, the model provides a stronger basis for arguing that a proposed molecular scheme does in fact do what it is supposed to do. More importantly in some ways, it also indicates how and when the proposed scheme may fail, or what its weaknesses are. That is one reason why careful construction of models and extensive parameter sensitivity analysis are so essential. Finally, as our ability to model develops, these models bear the promise of helping develop a quantitative and predictive theory of biological phenomenon.

References

1. Prasad A, Zikherman J, Das J, Roose JP, Weiss A, Chakraborty AK (2009) Origin of the sharp boundary that discriminates positive and negative selection of thymocytes. Proc Natl Acad Sci U S A 106(2):528–533
2. Das J, Ho M, Zikherman J, Govern C, Yang M, Weiss A, Chakraborty AK, Roose JP (2009) Digital signaling and hysteresis characterize ras activation in lymphoid cells. Cell 136 (2):337–351
3. Barlow HB, Levick WR, Yoon M (1971) Responses to single quanta of light in retinal ganglion cells of the cat. Vision Res Suppl 3:87–101
4. MM Davis, Krogsgaard M, Huse M, Huppa J, Lillemeier BF, Li QJ (2007) T cells as a self-referential, sensory organ. Annu Rev Immunol 25:681–695
5. Wylie DC, Das J, Chakraborty AK (2007) Sensitivity of t cells to antigen and antagonism emerges from differential regulation of the same molecular signaling module. Proc Natl Acad Sci U S A 104(13):5533–5538
6. Daniels MA, Teixeiro E, Gill J, Hausmann B, Roubaty D, Holmberg K, Werlen G, Hollander GA, Gascoigne NR, Palmer E (2006) Thymic selection threshold defined by compartmentalization of ras/mapk signalling. Nature 444(7120):724–729
7. McCaughtry TM, Wilken MS, Hogquist KA (2007) Thymic emigration revisited. J Exp Med 204(11):2513–2520
8. Starr TK, Jameson SC, Hogquist KA (2003) Positive and negative selection of t cells. Annu Rev Immunol 21:139–176
9. Werlen G, Hausmann B, Naeher D, Palmer E (2003) Signaling life and death in the thymus: timing is everything. Science 299(5614): 1859–1863
10. Savageau MA (1998) Development of fractal kinetic theory for enzyme-catalysed reactions and implications for the design of biochemical pathways. Biosystems 47(1–2):9–36
11. Schmierer B, Hill CS (2005) Kinetic analysis of smad nucleocytoplasmic shuttling reveals a mechanism for transforming growth factor beta-dependent nuclear accumulation of smads. Mol Cell Biol 25(22):9845–9858
12. Gardner TS, Cantor CR, Collins JJ (2000) Construction of a genetic toggle switch in escherichia coli. Nature 403(6767):339–342
13. Artyomov MN, Das J, Kardar M, Chakraborty AK (2007) Purely stochastic binary decisions in cell signaling models without underlying deterministic bistabilities. Proc Natl Acad Sci USA 104(48):18958–18963
14. To T-L, Maheshri N (2010) Noise can induce bimodality in positive transcriptional feedback loops without bistability. Science 327(5969):1142–1145
15. Gillespie DT (2007) Stochastic simulation of chemical kinetics. Annu Rev Phys Chem 58:35–55
16. Savageau MA, Coelho PMBM, Fasani RA, Tolla DA, Salvador A (2009) Phenotypes and tolerances in the design space of biochemical systems. Proc Natl Acad Sci USA 106(16):6435–6440
17. Craciun G, Tang Y, Feinberg M (2006) Understanding bistability in complex enzyme-driven reaction networks. Proc Natl Acad Sci USA 103(23):8697–8702
18. Angeli D, Ferrell JE, Sontag ED (2004) Detection of multistability, bifurcations, and hysteresis in a large class of biological positive-feedback systems. Proc Natl Acad Sci USA 101(7):1822–1827
19. Alon U (2007) Network motifs: theory and experimental approaches. Nat Rev Genet 8(6):450–461
20. Milo R, Shen-Orr S, Itzkovitz S, Kashtan N, Chklovskii D, Alon U (2002) Network motifs: simple building blocks of complex networks. Science, 298(5594):824–827
21. Sondermann H, Soisson SM, Boykevisch S, Yang SS, Bar-Sagi D, Kuriyan J (2004) Structural analysis of autoinhibition in the ras activator son of sevenless. Cell 119(3):393–405
22. Freedman TS, Sondermann H, Friedland GD, Kortemme T, Bar-Sagi D, Marqusee S, Kuriyan J (2006) A ras-induced conformational switch in the ras activator son of sevenless. Proc Natl Acad Sci U S A 103(45):16692–16697
23. Das J, Kardar M, Chakraborty AK (2009) Positive feedback regulation results in spatial clustering and fast spreading of active signaling molecules on a cell membrane. J Chem Phys 130(24):245102
24. Murray JD (1989) Mathematical biology. Springer, Berlin
25. Strogatz SH (1994) Nonlinear dynamics and chaos: with applications to physics, biology, chemistry and engineering. Perseus Books, Reading, Mass
26. Ermentrout B (2002). Simulating, analyzing, and animating dynamical systems: a guide to XPPAUT for researchers and students. SIAM, Philadelphia

27. http://oscill8.sourceforge.net/ Last Accessed April 3 2012
28. Zhu M, Janssen E, Zhang W (2003) Minimal requirement of tyrosine residues of linker for activation of t cells in tcr signaling and thymocyte development. J Immunol 170(1):325–333
29. Thomson M, Gunawardena J (2009) Unlimited multistability in multisite phosphorylation systems. Nature 460(7252):274–277
30. Hlavacek WS, Faeder JR, Blinov ML, Posner RG, Hucka M, Fontana W (2006) Rules for modeling signal-transduction systems. Sci STKE 2006(344):re6
31. Lis M, Artyomov MN, Devadas S, Chakraborty AK (2009) Efficient stochastic simulation of reaction-diffusion processes via direct compilation. Bioinformatics 25(17):2289–2291
32. http://www.nrcam.uchc.edu/index.html Last Accessed April 3 2012

Chapter 11

Modeling Spatiotemporal Dynamics of Bacterial Populations

Hao Song and Lingchong You

Abstract

Quantitative modeling of spatiotemporal dynamics of cells facilitates understanding and engineering of biological systems. Using a synthetic bacterial ecosystem as a workbench, we present the approach to mathematically simulate the spatiotemporal population dynamics of the ecosystem. A description of ecosystem's genetic construction and model development is firstly given. Parameter estimation and computational approach for the derived partial differential equations (PDEs) are then given. Spatiotemporal pattern formation is computed by numerically solving the PDE model. Biodiversity of the ecosystem and its impacts by cellular seeding distance and motility are computed according to the cell distribution patterns.

Key words: Partial differential equations, Spatiotemporal dynamics, Finite element method, Comsol Multiphysics, Synthetic ecosystem

1. Introduction

Complex regulatory networks of interacting intracellular and extracellular signals underlie cell responses to environmental variations and determine cell fate (1). Such complex networks usually manifest nonlinear dynamics, and cannot be intuitively understood. Mathematical modeling is, therefore, a powerful tool not only helping to elucidate the underlying regulatory mechanisms, but also predicting possible cell behaviors at diverse environments, as well as identifying emergent properties of biological systems (2).

One simple modeling strategy is to assume that the cellular components are well mixed, that is, the concentrations of the participants are uniformly distributed everywhere (3). Ordinary differential equations are used in this case to quantitatively describe the temporal dynamics of the species' concentrations. This strategy,

Xuedong Liu and Meredith D. Betterton (eds.), *Computational Modeling of Signaling Networks*,
Methods in Molecular Biology, vol. 880, DOI 10.1007/978-1-61779-833-7_11,

although highly simplified, has provided many nonintuitive insights into the nonlinear regulatory dynamics of gene circuits and cell signaling (4), for example bistable switch in cell cycle entry (5), long-term memory formation in neurons (6, 7), and oscillatory gene regulation in circadian rhythms (8).

However, in many biological processes such as cellular chemotaxis (9, 10), cell polarization (11), and tissue development (12–14), the spatial gradients and heterogeneity of intracellular and extracellular signaling molecules are dominant and underlie their spatiotemporal behaviors. Therefore, in order to understand their design principles, both the temporal evolution and spatial distribution of the signaling molecules should be coherently accounted for (3, 15–17). In these biological processes, the intracellular dynamics of a single cell is often coupled with extracellular cell–cell communications to coordinate the orchestrated evolution of the overall population dynamics (18). In such cases, partial differential equations (PDEs) should be employed to describe the spatiotemporal dynamics of the system as a whole.

1.1. Modeling Spatiotemporal Dynamics by PDE

The PDE model is extensively used to quantitatively describe how the quantities of interest in a biological process change with time and how their spatial distributions evolve via reaction-transport processes. To construct a PDE model, the interacting relationships between species of a biological system should be well defined first of all. Then, the spatial domain, that is the boundary of the system of interest, should be constrained. Based on the reaction mechanism and spatial domain specification, we can use the mass balance rule to build the PDE model. The temporal rate of change of a species in a particular spatial location is determined by three coupled components: (1) all processes that produce or synthesize the species; (2) all processes that consume or degrade the species; and (3) the net species movement rate into and out of that specific location by physical transport. These mass balance processes can be translated mathematically into PDE to describe how the species' concentrations change in time and space (19, 20):

$$\frac{\partial c_i(x,y,z,t)}{\partial t} = \sum_{j=1}^{m} v_{i,\text{production}} - \sum_{k=1}^{n} v_{k,\text{consumption}} + v_{\text{transport}}, \quad (1)$$

where the concentration of species-ith, that is $c_i(x,y,z,t)$, is a function of time (t) and spatial coordinates (x,y,z). The left-hand side of Eq. 1 is the partial derivative of the $c_i(x,y,z,t)$ with respect to time. The first two terms of the right-hand side of Eq. 1 state that the rate of change of c_i at a spatial location is the sum of all reaction rates that produce the species-ith minus all reaction rates that consume the species-ith at that specific location. The last term of Eq. 1, $v_{\text{transport}}$, is the net flux rate of the species-ith that move in and out of the spatial location (x,y,z). Species' diffusion and convection are two main transport mechanisms that are usually considered to contribute to $v_{\text{transport}}$, which is mathematically represented by Bird et al. (21):

$$v_{\text{transport}} = D_i \nabla^2 c_i - v \cdot \nabla c. \quad (2)$$

The first term of the right-hand side of Eq. 2 describes the diffusion transport of the species-*i*th with a diffusivity D_i, and $\nabla^2 c_i = \frac{\partial^2 c_i}{\partial x^2} + \frac{\partial^2 c_i}{\partial y^2} + \frac{\partial^2 c_i}{\partial z^2}$ in standard Cartesian coordinate system. The second term of the right-hand side of Eq. 2 describes the velocity-driven transport (convection) of the species-*i*th, and $v \cdot \nabla c_i = v_x(x, y, z) \frac{\partial c_i}{\partial x} + v_y(x, y, z) \frac{\partial c_i}{\partial y} + v_z(x, y, z) \frac{\partial c_i}{\partial z}$ in Cartesian coordinates. In this chapter, the convection transport of cells takes the form of chemotaxis, which is detailed in the following.

In general, the reaction rate expressions in Eq. 1 may be sophisticated and nonlinear; therefore, an analytical solution to the PDE of Eq. 1 cannot be obtained in most cases. Numerical solution to the PDE has to be taken to study the spatiotemporal dynamics of biological systems. In this chapter, we use a synthetic ecosystem (22) (Fig. 1) to illustrate the basic steps involved in the construction and solution of a PDE model. While this example deals with a bacterial system, the techniques are generally applicable to other intracellular or intercellular dynamics, where the spatial dynamics are important (20, 23).

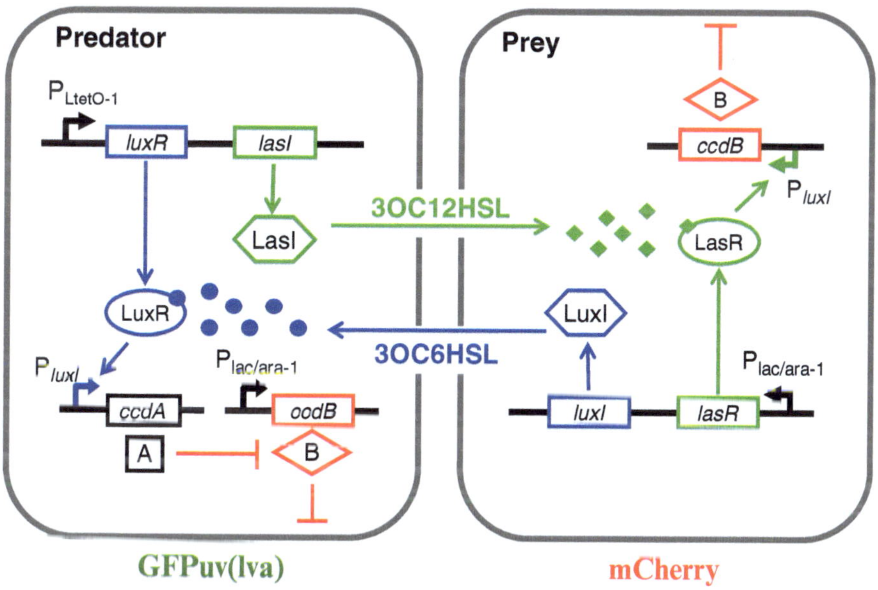

Fig. 1. Schematic gene circuits and detailed reaction mechanism of the synthetic predator–prey ecosystem. Gene circuits that compromise the synthetic predator–prey ecosystem. The system consists of two types of engineered *Escherichia coli* bacteria controlling each other's survival and death via two different QS signals. *Outer boxes* represent cell walls. *Arrows* represent activation or production; *blunt arrows* represent inhibition or killing. CcdB is controlled by $P_{lac/ara-1}$ in the predator and P_{luxI} in the prey. CcdA is controlled by P_{luxI}. The lux system is shown in blue and the las system in *green*. The QS genes in the predator are controlled by $P_{LtetO-1}$, and those in the prey by $P_{lac/ara-1}$. *Filled circles* represent 3OC6HSL and *filled diamonds* represent 3OC12HSL. The predator–prey interaction is activated by IPTG. Adapted from refs. 22, 35 with permission.

2. Materials

2.1. Gene Circuit Construction of the Synthetic Ecosystem

As shown in Fig. 1, two *Escherichia coli* populations ("predator" and "prey") communicate and regulate each other's density through quorum sensing (QS), a mechanism by which many bacteria broadcast and respond to changes in their local population density (24). Two QS modules, LuxI/LuxR from *Vibrio fischeri* and LasI/LasR from *Pseudomonas aeruginosa*, are used to enable two-way communication between the predator and prey populations. When the prey density is low, the predator cells die due to constitutive expression of a suicide gene (*ccdB*). In prey cells, an acyl-homoserine lactone (AHL), 3OC6HSL, is synthesized by LuxI. As the prey density increases, 3OC6HSL accumulates in the culture, and at sufficiently high concentrations, it is bound by the transcriptional regulator LuxR in the predator cells leading to increased expression of an antidote gene (*ccdA*) to rescue the predators. The predator cells in turn produce another AHL, 3OC12HSL, by LasI that permeates into the prey cells where it is bound by LasR, which activates expression of the killer *ccdB* gene, inducing "predation." This system satisfies the broader definition of predation for a two-species ecosystem, where one species (prey) suffers from the growth of the second (predator) and the latter benefits from the growth of the former (25). The two populations also compete for shared resources in coculture. Thereby, the programmed predation and the competition for nutrients occur through chemical diffusion.

The gene circuits of the synthetic ecosystem have well-characterized genetic components and well-defined interactions, making it an ideal workbench for mathematical model development and experimental validation. Integration of modeling and experiment further deepens our understanding about how gene regulation, cell–cell communication, and cellular motility dictate spatiotemporal population dynamics of ecosystems. Such integrative methodology is generally applicable to study other biological processes.

2.2. Mathematical Modeling of the Synthetic Ecosystem

To mathematically describe how the predator and the prey cells interact in both temporal and spatial domains, we first construct a PDE model for the synthetic ecosystem. One of the mostly adopted numerical approach to solve PDE is finite element methods (26). There exist a few available simulation packages to implement the numerical methods to solve PDE, such as Virtual Cell (27) and Comsol Multiphysics® (28, 29). The COMSOL Multiphysics provides a powerful simulation environment that can streamline the PDE numerical solving process: defining the PDE geometry, specifying the PDE, automatic meshing, and numerical solving and post-processing of the computational results for analysis and visualization of spatiotemporal dynamics of the system.

3. Methods

Here, we use the synthetic ecosystem (Fig. 1) to demonstrate how to construct and solve the PDE model of the system and to study its spatiotemporal dynamics (see Note 1). We employ Comsol Multiphysics software package to numerically solve the PDE for the synthetic ecosystem. By solving the PDE model, we can obtain the knowledge about how does the combination of cellular motility and long-range interactions via diffusible chemicals affect biodiversity.

3.1. Development of PDE Model of the Synthetic Ecosystem

1. The dynamics of the synthetic *E. coli* predator–prey system (Fig. 1) was modeled by accounting for the key reactions during the functioning of this system. Several components were assumed at a quasi-steady state, including all mRNAs, transcriptional regulators, and killer proteins. This is equivalent to assuming that processes leading to changes in these species are at a much faster timescale than the rest of the processes. These simplifying assumptions do not significantly change the qualitative nature of the system dynamics.
2. Based on the simplification assumptions made above, a coupled PDE model was developed to describe the system spatiotemporal dynamics, including species interactions, chemical diffusion, cellular motility (diffusion and chemotaxis), and nutrient consumption.

   (a) Cell (predator-c_1, prey-c_2) diffusion, chemotaxis, growth, and death:

$$\frac{\partial c_1}{\partial t} = D_{p1}\nabla^2 c_1 - \alpha_1 \nabla\left[\frac{c_1}{(K_{\text{tax}} + s)^2}\nabla s\right] + k_{p1} c_1 \frac{s}{K_s + s} \times \left(\delta_1 + \delta_2 \frac{A_2}{K_2 + A_2}\right) - d_{p1} p_1 \tag{3}$$

$$\frac{\partial c_2}{\partial t} = D_{p2}\nabla^2 c_2 - \alpha_2 \nabla\left[\frac{c_2}{(K_{\text{tax}} + s)^2}\nabla s\right] + k_{p2} c_2 \frac{s}{K_s + s} \times \left(\frac{K_1}{K_1 + A_1}\right) - d_{p2} c_2, \tag{4}$$

   where Eqs. 3 and 4 describe the dynamics of the predator (c_1 [ng/ml]) and the prey (c_2 [ng/ml]) in the solid phase, respectively; the nutrients (s) were assumed to play the role of chemoattractants. The bacteria are diffused according to simple Fickian diffusion with diffusion coefficients D_p (D_{p1} for the predator, D_{p2} for the prey). A well-established mathematical expression for cell chemotaxis that fitted various experiments (30–33) was adopted in the model (the second term in the right-hand side of Eqs. 3 and 4).

(b) Diffusion, synthesis, and degradation of the QS signals, AHLs (A_1: 3OC12HSL synthesized by the predator; A_2: 3OC6HSL synthesized by the prey):

$$\frac{\partial A_1}{\partial t} = D_A \nabla^2 A_1 + k_{A1} c_1 - d_{A1} A_1 \tag{5}$$

$$\frac{\partial A_2}{\partial t} = D_A \nabla^2 A_2 + k_{A2} c_2 - d_{A2} A_2 \tag{6}$$

(c) Diffusion, consumption of the nutrients (denoted by s, also playing as the chemoattractants):

$$\frac{\partial s}{\partial t} = D_s \nabla^2 s - \lambda_1 k_{p1} c_1 \frac{s}{K_s + s}\left(\delta_1 + \delta_2 \frac{A_2}{K_2 + A_2}\right) - \lambda_2 k_{p2} c_2 \frac{s}{K_s + s}\left(\frac{K_1}{K_1 + A_1}\right) \tag{7}$$

Therefore, the Eqs. 3–7 constitute the reaction-diffusion-chemotaxis PDE model that describes the spatiotemporal evolution of the synthetic ecosystem (Fig. 1).

3. The base set of parameter values of the model and their references are listed in Table 1. Several parameter values are directly taken from the literature or derived from literature data. For other parameters where lack quantitative information (e.g., those governing gene expression process), educated guesses that are biologically feasible and able to simulate our experimental findings were used (see Note 2).

4. The boundary conditions and initial conditions were defined according to the experimental setup of seeding the ecosystem in a Petri dish. In the 2-dimensional (2D) PDE model, a square domain ($x = [-4, 4]$ (cm), $y = [-4, 4]$ (cm)) was constrained for simulations, in consistent with the experimental setup. The non-flux boundary condition (∇x (at boundary) $= 0$, where $x = p_1, p_2, A1, A2$, and s in Eqs. 3–7) was used for all simulations.

5. The initial conditions vary according to the initial seeding of the predator and prey cells. For the case that the predator and prey cells were seeded in partitioned habitats (Figs. 2 and 3), for example, seeding distance $d = 1$ cm, the initial conditions are:

$$p_1 = \begin{Bmatrix} 10, (x = -0.5, y = 0) \\ 0, \text{ (elsewhere)} \end{Bmatrix}, p_2 = \begin{Bmatrix} 10, (x = 0.5, y = 0) \\ 0, \text{ (elsewhere)} \end{Bmatrix},$$

$A_1 = 0$; $A_2 = 0$, and $s = 10$.

Table 1
Basal parameter values for the PDE model Eqs. 3–7

Parameters	Description	Basal values	References
k_{p1}	Predator growth rate in M9	2.4 [h^{-1}]	Estimated
K_s	Concentration of substrate necessary to reach half-maximum of the growth rate	5 [μM]	(31)
d_{p1}	Predator death rate constant	0.01 [h^{-1}]	Estimated
K_2	Concentration of 3OC6HSL necessary to half-maximum of the pLuxI promoter	10 [nM]	(36)
δ_1	Basal growth rate coefficient of predator (at $A_2 = 0$)	0.5	Estimated
δ_2	Prey's rescuing effect upon predator	1.8	Estimated
D_{p1}	Predator diffusivity in 0.2% M9 agar	0.013 [cm^2/h]	(33)
α_1	Predator chemotaxis constant	0.001 [cm^2 μM/h]	(30–33, 37)
K_{tax}	Chemoattractant threshold concentration to induce chemotactic motility	1 [μM]	(33)
k_{p2}	Prey growth rate	1 [h^{-1}]	Estimated
d_{p2}	Prey death rate constant	0.01 [h^{-1}]	Estimated
K_1	Concentration of 3OC12HSL necessary to half-maximum of the pLuxI promoter	100 [nM]	(36)
D_{p2}	Prey diffusivity in 0.2% M9 agar	0.01 [cm^2/h]	(33)
α_2	Prey chemotaxis constant in 0.2% M9 agar	0.001 [cm^2 μM/h]	(30–33, 37)
k_{A1}	Synthesis rate of 3OC12HSL by predator	0.3 [nM/h ng ml]	Estimated
d_{A1}	Degradation rate constant of 3OC12HSL	0.02 [h^{-1}]	(38)
D_{A1}	Diffusivity of 3OC6HSL in 0.2% M9 agar	0.4 [cm^2/h]	Estimated
D_{A2}	Diffusivity of 3OC12HSL in 0.2% M9 agar	0.03 [cm^2/h]	Estimated
k_{A2}	Synthesis rate of 3OC6HSL by prey	0.125 [nM/h ng ml]	Estimated
d_{A2}	Degradation rate constant of 3OC6HSL	0.1 [h^{-1}]	(38)
D_s	Diffusivity of nutrients (or chemoattractants)	0.2 [cm^2/h]	Estimated
λ_1	Consumption rate constant of nutrients by predator	0.05 [μM ml/ng]	Estimated
λ_2	Consumption rate constant of nutrients by prey	0.05 [μM ml/ng]	Estimated

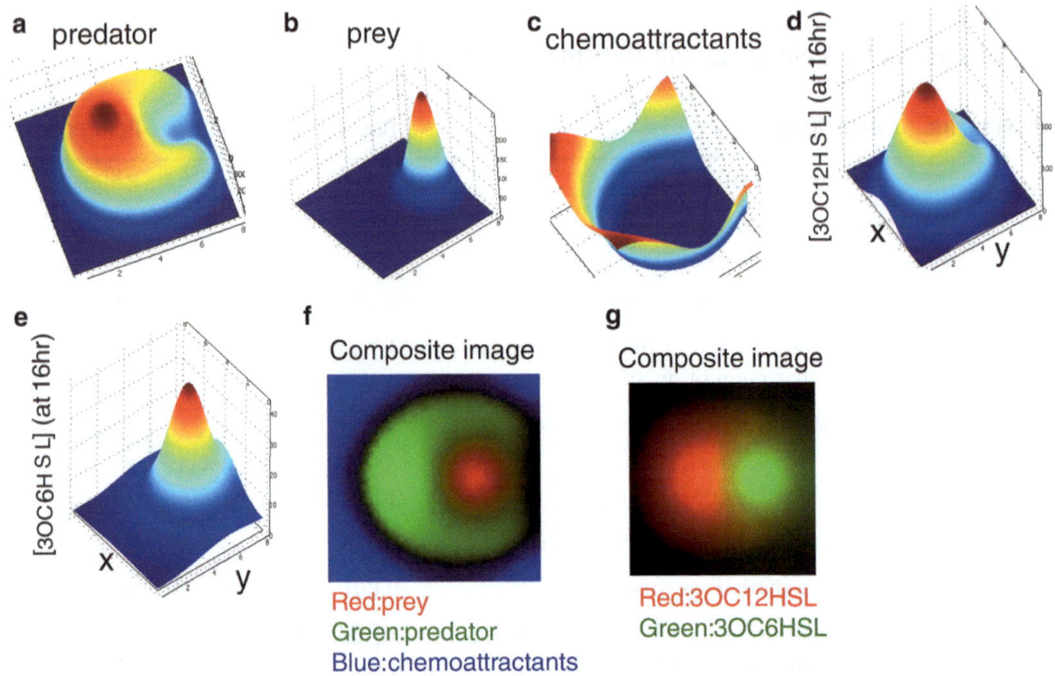

Fig. 2. Computations showed that (1) the affinity of the predator for growing is higher in the vicinity of the prey and that (2) the chemoattractants were exhausted in the wake of the populations' migration. (**a**) The simulated 3D distribution of the predator cell density (at $t = 20$ h). The predator had a higher cell density in the vicinity of the prey because the rescuing signal (3OC6HSL) concentration produced by the prey was higher in this region; thus, the rescuing of the predator was more significant in the vicinity of the prey than at a distance. (**b**) The 3D distribution of the prey cell density (at $t = 20$ h). (**c**) The corresponding 3D distribution of the chemoattractants concentration (at $t = 20$ h). The chemoattractants were almost exhausted in the path of dispersion of the predator and prey from the inoculation point. (**d**) The 3D distribution of the 3OC12HSL corresponding to the predator–prey pattern (in (**a**) and (**b**)). (**e**) The 3D distribution of the 3OC6HSL corresponding to the predator–prey pattern. (**f**) The composite image that demonstrates the distribution of the predator density (*green*), the prey density (*red*), and the chemoattractants' concentration (*blue*). (**g**) The composite images of the distribution of the two AHLs signals are shown in (**d**) and (**e**), where the *red* represents 3OC12HSL and the *green* represents 3OC6HSL. Adapted from ref. 35 with permission.

6. The 2D PDE model was numerically solved by the COMSOL Multiphysics software. In the solid phase, the spatial-averaged density of each population over the entire spatial domain was calculated by $< c_i > = \frac{\int_\Omega c_i(x,y,t)\mathrm{d}x\mathrm{d}y}{\int_\Omega \mathrm{d}x\mathrm{d}y}$, where $c_i(x,y)$ $(i = 1, 2)$ was the density distribution of the predator or the prey at a location-(x,y) in the 2D domain. The fraction of this population was $x_i = < c_i > /(< c_1 > + < c_2 >)$..
7. To quantify biodiversity in a mixed culture, the modified Simpson's biodiversity index (BI) was adopted (34): $\mathrm{BI} = 1 - \sum_{i=1}^{2} x_i^2 = 2x_1x_2$, where x_i is the fraction of the ith population in a coculture ($i = 1$ for the predator and 2 for the prey). In the

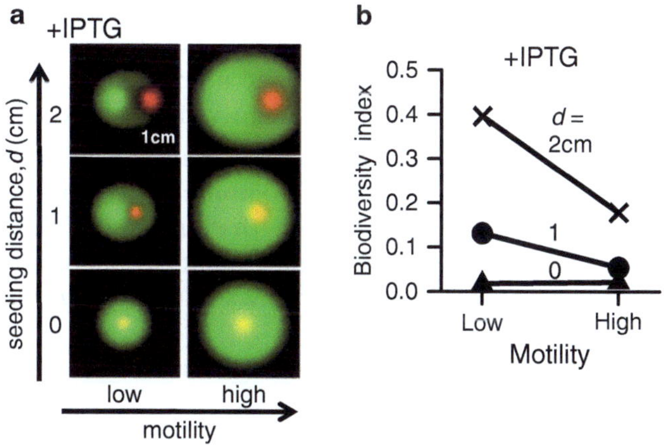

Fig. 3. Reduced motility promoted biodiversity if seeding habitats were partitioned. (**a**) Simulated predator–prey patterns at high and low motility with different seeding segregation distances. Image snapshots were taken at 20 h after the simulation initiation. (**b**) Modeling showed that decreasing motility increased the BI for a sufficient segregation distance ($d = 1$ or 2 cm), but not for $d = 0$. To reduce motility in modeling, we decreased the values of the cellular diffusivities (D_{p1}, D_{p2}) and chemotaxis constants (α_1, α_2) by fourfold from those of the high motility (base parameter values in Table 1 were used). The BIs were computed based on the predator–prey patterns at 20 h (**a**). Adapted from ref. 35 with permission.

solid phase culture, x_i was calculated by using the average densities over a spatial domain:

$$x_i = \int_\Omega p_i \mathrm{d}x\mathrm{d}y \Big/ \left(\int_\Omega p_1 \mathrm{d}x\mathrm{d}y + \int_\Omega p_2 \mathrm{d}x\mathrm{d}y \right).$$

The BI reaches a maximum for $x_1 = x_2 = 0.5$.

3.2. Numerical Result of the Spatiotemporal Dynamics of the Ecosystem

1. To simulate cellular pattern formation in partitioned habitats in a Petri dish, the predator and the prey were seeded separately at two focal points with varying distance in between, leaving the other locations with zero density. Simulations furnished the spatiotemporal pattern formation of the cells and communication signals: the predator (Fig. 2a), the prey (Fig. 2b), the chemoattractants (Fig. 2c), 3OC6HSL produced by the predator (Fig. 2d), and 3CO6HSL produced by the prey (Fig. 2e). The 2D composites were also obtained: Fig. 2f manifests the distribution of the predator, prey, and chemoattractants; and Fig. 2g shows the distribution of 3OC6HSL and 3OC12HSL.
2. The impact of seeding distance on the pattern formation and biodiversity was further examined. The predator and prey cell distribution patterns at different seeding distances and motility are shown in Fig. 3a. Increasing the segregation distance between the two populations would reduce the strength of long-range interactions, including killing and rescuing by the

diffusible AHL signals and competition for chemoattractants and nutrients. This reduction would lead to less killing, faster growth, and more chemotaxis of the prey, resulting in a larger prey's territory and density.

3. According to the cell distribution patterns in Fig. 3a, the BI is further computed and demonstrated in Fig. 3b. The model predicted that increasing the segregation distance between the seeding points would promote biodiversity when the populations were sufficiently segregated.

4. Notes

1. Mathematical modeling provides overall and quantitative understanding of the biological system, and predicts possible system behaviors under various conditions which are subject to experimental investigation. A few rounds of model development and experimental validation are usually required in the course of building up a predictive and comprehensive model.
2. In mathematical modeling of spatiotemporal population dynamics, the parameter values played an essential role in determining the systems dynamics. Therefore, parameter values should be carefully chosen. One should try to search relevant literatures and find as much as possible information on the parameter values that were experimentally measured. If not obtainable, one should estimate the feasible range of parameter values according to scientific experience. Whenever possible, one should measure or estimate parameter values using his or her own experimental data on their specific context. One golden rule in determining parameter values is to make the modeling predictions qualitatively and quantitatively match experimental findings at various conditions.

Acknowledgments

This study was supported by a Startup Grant (H.S.), AcRF TIER 1 Grant (H.S.), the National Institute of Health (5R01CA118486, L.Y.), a David and Lucile Packard Fellowship (L.Y.), and a DuPont Young Professor Award (L.Y.).

References

1. Hanahan D, Weinberg RA (2000) The hallmarks of cancer. Cell 100:57–70
2. Aldridge BB, Burke JM, Lauffenburger DA, Sorger PK (2006) Physicochemical modelling of cell signalling pathways. Nat Cell Biol 8:1195–1203
3. Kholodenko BN (2006) Cell-signalling dynamics in time and space. Nat Rev 7:165–176
4. Tyson JJ, Chen KC, Novak B (2003) Sniffers, buzzers, toggles and blinkers: dynamics of regulatory and signaling pathways in the cell. Curr Opin Cell Biol 15:221–231
5. Yao G, Lee TJ, Mori S, Nevins JR, You L (2008) A bistable Rb-E2F switch underlies the restriction point. Nat Cell Biol 10:476–482
6. Song H, Smolen P, Av-Ron E, Baxter DA, Byrne JH (2006) Bifurcation and singularity analysis of a molecular network for the induction of long-term memory. Biophys J 90:2309–2325
7. Bhalla US, Iyengar R (1999) Emergent properties of networks of biological signaling pathways. Science 283:381–387
8. Roenneberg T, Chua EJ, Bernardo R, Mendoza E (2008) Modelling biological rhythms. Curr Biol 18:R826–R835
9. Iglesias PA, Devreotes PN (2008) Navigating through models of chemotaxis. Curr Opin Cell Biol 20:35–40
10. Levchenko A, Iglesias PA (2002) Models of eukaryotic gradient sensing: application to chemotaxis of amoebae and neutrophils. Biophys J 82:50–63
11. Onsum MD, Rao CV (2009) Calling heads from tails: the role of mathematical modeling in understanding cell polarization. Curr Opin Cell Biol 21:74–81
12. Shvartsman SY, Coppey M, Berezhkovskii AM (2008) Dynamics of maternal morphogen gradients in Drosophila. Curr Opin Genet Dev 18:342–347
13. Umulis D, O'Connor MB, Othmer HG (2008) Robustness of embryonic spatial patterning in *Drosophila melanogaster*. Curr Top Dev Biol 81:65–111
14. Koch AJ, Meinhardt H (1994) Biological pattern formation: from basic mechanisms to complex structures. Rev Mod Phys 66:1481–1507
15. Kholodenko BN (2009) Spatially distributed cell signalling. FEBS Lett 583:4006–4012
16. Neves SR, Iyengar R (2009) Models of spatially restricted biochemical reaction systems. J Biol Chem 284:5445–5449
17. Neves SR, Tsokas P, Sarkar A, Grace EA, Rangamani P, Taubenfeld SM, Alberini CM, Schaff JC, Blitzer RD, Moraru II, Iyengar R (2008) Cell shape and negative links in regulatory motifs together control spatial information flow in signaling networks. Cell 133:666–680
18. Freeman M (2000) Feedback control of intercellular signalling in development. Nature 408:313–319
19. Edelstein-Keshet L (1988) Mathematical models in biology. McGraw-Hill Companies
20. Birtwistle MR, Kholodenko BN (2009) Endocytosis and signalling: a meeting with mathematics. Mol Oncol 3:308–320
21. Bird RB, Stewart WE, Lightfoot EN (2001) Transport phenomena, 2nd edn. Wiley
22. Balagadde FK, Song H, Ozaki J, Collins CH, Barnet M, Arnold FH, Quake SR, You L (2008) A synthetic *Escherichia coli* predator-prey ecosystem. Mol Syst Biol 4:187
23. Reeves GT, Muratov CB, Schupbach T, Shvartsman SY (2006) Quantitative models of developmental pattern formation. Dev Cell 11:289–300
24. Camilli A, Bassler BL (2006) Bacterial small-molecule signaling pathways. Science 311:1113–1116
25. Taylor RJ (1984) Predation. Chapman and Hall Ltd, New York, NY
26. Morton KW, Mayers DF (2005) Numerical solution of partial differential equations: an introduction, 2nd edn. Cambridge University Press, Cambridge
27. Slepchenko BM, Schaff JC, Macara I, Loew LM (2003) Quantitative cell biology with the Virtual Cell. Trends Cell Biol 13:570–576
28. van Schijndel J (2008) Integrated modeling using MatLab, Simulink and COMSOL: with heat, air and moisture applications for building physics and systems. VDM Verlag Dr. Müller
29. Zimmerman WBJ (2006) Multiphysics modeling with finite element methods. World Scientific Publishing Company
30. Ford RM, Lauffenburger DA (1991) Analysis of chemotactic bacterial distributions in population migration assays using a mathematical model applicable to steep or shallow attractant gradients. Bull Math Biol 53:721–749
31. Lauffenburger DA (1991) Quantative studies of bacterial chemotaxis and microbial population dynamics. Microb Ecol 22:175–185
32. Woodward DE, Tyson R, Myerscough MR, Murray JD, Budrene EO, Berg HC (1995) Spatio-temporal patterns generated by *Salmonella typhimurium*. Biophys J 68:2181–2189
33. Murray JD (2005) Mathematical biology, vol 2. Oxford University Press

34. Peet R (1974) The measurement of species diversity. Annu Rev Ecol Syst 5:285–307
35. Song H, Payne S, Gray M, You L (2009) Spatiotemporal modulation of biodiversity in a synthetic chemical-mediated ecosystem. Nat Chem Biol 5:929–935
36. Collins CH, Leadbetter JR, Arnold FH (2006) Dual selection enhances the signaling specificity of a variant of the quorum-sensing transcriptional activator LuxR. Nat Biotechnol 24:708–712
37. Keller EF, Segel LA (1971) Traveling bands of chemotactic bacteria: a theoretical analysis. J Theor Biol 30:235–248
38. Kaufmann GF, Sartorio R, Lee SH, Rogers CJ, Meijler MM, Moss JA, Clapham B, Brogan AP, Dickerson TJ, Janda KD (2005) Revisiting quorum sensing: discovery of additional chemical and biological functions for 3-oxo-N-acylhomoserine lactones. Proc Natl Acad Sci USA 102:309–314

Chapter 12

Discrete Dynamic Modeling of Signal Transduction Networks

Assieh Saadatpour and Réka Albert

Abstract

Newly available experimental data characterizing different processes involved in signaling pathways have provided the opportunity for network analysis and modeling of these interacting pathways. Current approaches in studying the dynamics of signaling networks fall into two major groups, namely, continuous and discrete models. The lack of kinetic information for biochemical interactions has limited the wide applicability of continuous models. To address this issue, discrete dynamic models, based on a qualitative description of a system's variables, have been applied for the analysis of biological systems with many unknown parameters. The purpose of this chapter is to give a detailed description of Boolean modeling, the simplest type of discrete dynamic modeling, and the ways in which it can be applied to analyze the dynamics of signaling networks. This is followed by practical examples of a Boolean dynamic framework applied to the modeling of the abscisic acid signal transduction network in plants as well as the T-cell survival signaling network in humans.

Key words: Boolean dynamic modeling, Synchronous method, Asynchronous method, Signal transduction

1. Introduction

Recent years have witnessed a growing interest in the study of signal transduction pathways due to their pivotal role in adapting to various environmental conditions. The process of sensing a signal in the extracellular environment and its subsequent transduction and reaching of targets is carried out through a cascade of interactions (1). A wealth of experimental data characterizing various aspects of signaling pathways has provided the basis for network reconstructions to visualize and better investigate the properties of these interacting pathways. Network reconstruction of cellular signaling pathways involves the representation of proteins, secondary messengers, and small molecules as nodes and the interactions among these components as edges. This graphical representation

Xuedong Liu and Meredith D. Betterton (eds.), *Computational Modeling of Signaling Networks*,
Methods in Molecular Biology, vol. 880, DOI 10.1007/978-1-61779-833-7_12,

of the system usually contains directed edges, where the direction of each edge implies the regulation of the target node by the source node. In addition, any edge can be characterized by a positive or negative sign denoting activation or inhibition, respectively.

The compilation of all available knowledge about a given signaling system into a coherent network representation facilitates studying the role of each component in the system in achieving an appropriate response to a particular stimulus. Furthermore, by virtue of modeling approaches, one can capture the properties of the system as a whole, such as the long-term behavior of the system in a prespecified condition or its dynamical behavior in response to various internal or external perturbations. The model's predictions can in turn lead to generation of testable hypotheses regarding the underlying principles governing the behavior of the signaling system.

Dynamic models describe the evolution of a system over time. In dynamic models, the nodes of the network are characterized by states (representing, e.g., concentration or activity), and the states of the nodes change in time according to the interactions encapsulated in the network. The modeling approaches tackling the temporal behavior of the signal transduction networks can be of either quantitative or qualitative nature. The quantitative dynamic models are usually recast as a set of differential equations to capture the continuous change over time of different components in the system (2). The variables in this system of equations represent the relative abundance of the components, whose temporal change may be highly dependent on the abundance of other components as well as on kinetic parameters. Although continuous models tend to be accurate, the paucity of kinetic information for individual biochemical interactions has limited the wide applicability of these models.

Discrete dynamic approaches provide an alternative to continuous models for the analysis of less characterized biological systems with many unknown parameters. In these models, the nodes' states are qualitative and the time variable is often considered to be discrete. Examples of discrete dynamic models include Boolean network models (3, 4), finite-state logical models (5–7), finite difference equations (8, 9), and Petri nets (10–12). In this chapter, we focus on Boolean networks which have been extensively employed in modeling biological systems. In Boolean networks, each node is described by only two qualitative states, denoted by 1 (ON) and 0 (OFF). These models were originally proposed by Kauffman (3) and Thomas(4) for the analysis of gene regulatory networks. In its simplest form, known as a synchronous Boolean model (13), the states of all the nodes of the network are updated simultaneously at multiples of a common time step. However, these models fail to take into account the variety of timescales in many molecular interaction systems, which may range from fractions of a second to hours (14). This triggered the development of asynchronous models (4) that provide more variability to the update order. In asynchronous models, the state of each node can be updated based on an individual timescale (15–17).

In this chapter, we explain the steps necessary to construct and analyze a Boolean dynamic model of a signaling network. We also describe successful applications of Boolean dynamic modeling on two signal transduction networks: the abscisic acid (ABA) signal transduction network in plants as well as T-cell survival signaling in humans.

2. Boolean Dynamic Modeling

Biological regulatory networks can be represented by a directed graph in which the nodes are different components of a system and the edges describe regulatory interactions among the components. In particular, the nodes of a signal transduction network may include proteins, mRNAs, and small molecules. The direction of each edge in the network follows the direction of mass transfer or information propagation from the upstream node to the downstream node. In Boolean models, each node of the network is characterized by an ON (1) or OFF (0) state corresponding to, for example, activity/inactivity of a transcription factor or above/below-threshold concentration of a molecule. The time variable is considered to be discrete, i.e., it takes nonnegative integer values. The future state of a node at each time step is determined by the current states of the nodes regulating it (input nodes) through a set of Boolean functions (rules). These Boolean functions, usually expressed via the logical operators AND, OR, and NOT, represent the regulatory relationships between the components. Based on the output of these logical functions, which is a 0 or 1 value, the state of a node either remains the same or changes to the opposite value (i.e., from 0 to 1 or vice versa). The state of the system at each time step is then expressed as a vector whose components represent the state of the nodes at that time step. By updating the nodes' states at each time step, the state of the system evolves over time and eventually settles into an "attractor" (long-time behavior). The attractors are divided into two categories: fixed points, in which the state of the system does not change (i.e., a single state), and complex attractors, where the system oscillates among a set of states.

Boolean dynamic modeling of signal transduction networks consists of the following main steps:

1. Reconstructing the network
2. Identifying Boolean functions

3. Implementing time
4. Analyzing the dynamics of the system
5. Testing the validity and robustness of the dynamic model
6. Analyzing the effect of node perturbations

In the following, we explain these steps in detail and illustrate them through an example.

2.1. Reconstructing the Network

The first step towards modeling signal transduction networks within a Boolean dynamic framework is to reconstruct the network by synthesizing all the relevant information about the system of interest. This is usually done through an extensive literature search. Experimental evidence can provide information about both the components of the system as well as the regulatory relationships among them (18). Examples of the experimental evidence leading to identification of the components of a signaling system (nodes of the network) include experiments demonstrating that the activity or concentration of a protein changes once the respective input signal is induced or when a known component of the signaling system is knocked out. Alternatively, experiments showing that the output of the signaling system is altered upon a mutation of a gene suggest that the gene product is involved in the process, and experiments indicating that the artificial over-expression of an intermediary node affects the output of the signal transduction process imply that the node can be considered as a candidate component. Examples of experimental evidence that provide information on the interactions between the components (edges of the network) include (1) biochemical evidence representing the direct interaction between two components, such as protein–protein interactions or transcription factor–gene interactions, or (2) pharmacological evidence, such as exogenous application of a chemical, alluding to an indirect relationship between two components.

After compiling all such information about the signal transduction process, one needs to synthesize this information into a consistent network. It is usually useful to summarize the collected evidence in a table that lists all the nodes that are incorporated into the signaling process as well as the regulatory relationships between the nodes. The example below illustrates this first step.

Example 1: Let us consider a signaling pathway with four components A, B, C, and D in which A is the network's input node (signal) and D is the output node. Information regarding the causal relationships is summarized in Table 1.

The network representation of this simple example is given in Fig. 1a, where the directed edge → or —| denotes activation or inhibition of the target node by the upstream node, respectively. The Boolean functions governing the nodes' states in this network

Table 1
Components and the causal relationships between them incorporated in Example 1

Components	Causal relationships
A	A activates B A inhibits C
B	B activates C B activates D
C	C inhibits B C activates D
D	–

a

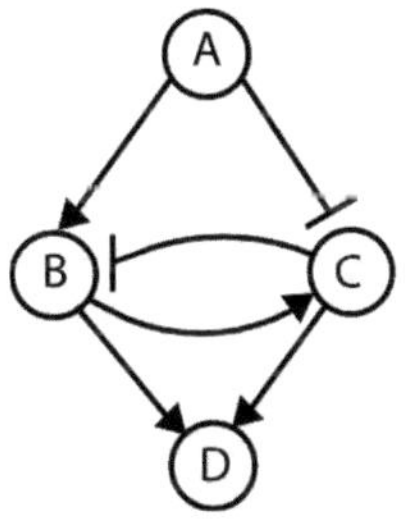

b

Node	Boolean rule
A	—
B	B* = A AND (NOT C)
C	C* = (NOT A) OR B
D	D* = B AND C

Fig. 1. Graphical representation and Boolean functions for the network given in Example 1. (**a**) Graphical representation of the network given in Example 1 with the causal relationships listed in Table 1. A is the network's input node, B and C are intermediary nodes, and D is the output node. The directed edge → or —| represents activation or inhibition of the downstream node by the upstream node, respectively. (b) Boolean rules governing the nodes' states in the network given in Example 1. *Asterisk* signifies the future state of a node, and for simplicity the state of each node is represented by its label.

are listed in Fig. 1b. We elaborate more on the Boolean rules in the next section.

When distilling experimental evidence into qualitative regulatory relationships, besides the causal component-to-component relationships such as "A activates B," some double-causal inferences such as "A activates the process through which B activates C" may also be obtained from both genetic and pharmacological experiments (18). In this case, if A is an enzyme catalyzing a reaction by which B is transformed into C, then the above statement corresponds to a direct interaction and we can assume that both A and B activate C. Otherwise, if the interaction between B and C is not direct, A can be assumed to activate a putative intermediary node of the B–C pathway. A rigorous implementation of this idea is offered by the software package NET-SYNTHESIS (19) (http://www.cs.

uic.edu/~dasgupta/network-synthesis) that synthesizes a signaling network consistent with all experimental observations. Inputting a text file containing causal or double-causal relationships to this software will output a simplified network representation along with a file containing the edges of the inferred network (19).

Additional information can also be incorporated in the network representation (15, 17). For example, if there is experimental evidence for the genetic interaction of two proteins but no information on their physical interaction, a putative intermediary node can be added between the two proteins. In addition, both nodes and edges of the network may be color coded to describe different functionalities or location of the components or to represent regulatory relationships between different types of biological entities. Conceptual nodes that represent a phenomenon rather than a physical component may be involved in the network, as well.

The network representation of a signal transduction process is a static description of the system. Graph theoretical analysis of this network using measures, such as node degree, distance, betweenness centrality, and clustering coefficient (20), can be performed to extract information about the structure of the underlying biological system. In order to understand the temporal behavior of a system, we need to construct a dynamic model for this static representation.

2.2. Identifying Boolean Functions

Once the network backbone is synthesized, the next step is to identify the Boolean functions governing the state changes of the nodes in the network. The Boolean function for a given node is determined by the nature of interactions between that node and the input nodes directly interacting with it and is formulated via the Boolean operators AND, OR, and NOT. When the given node has a single upstream node, the corresponding Boolean function has only one variable. More precisely, if a given node O has only one activator or inhibitor I, then its Boolean function can be written as one of the two following forms:

$$O^* = I, \tag{1}$$

$$O^* = \text{NOT } I, \tag{2}$$

where *asterisk* signifies the future state of node O, and for simplicity the state of node O is represented by its label. In the former case, the state of node O follows the state of node I while in the latter case, the state of node O is opposite to the state of node I. Based on this model, updating the state of node O in both cases will occur with a time delay, meaning that once the state of node I is determined, the state of node O will be determined accordingly after a gap time. In cases where the given node has multiple inputs, in addition to NOT, the AND or OR operators can be used to account for conditional or independent regulation, respectively. In conditional regulation, the co-expression of two (or more) inputs is required for activating the target node, whereas in independent regulation the activity of any of the inputs suffices to activate the target node. In order to decide if an interaction is a conditional or an independent regulation, one should refer to the relevant biological literature. Let us assume that in Example 1 the presence of A and the absence of C are required for activation of B; then, the Boolean function of B can be written as:

$$B^* = A \text{ AND } (\text{NOT } C). \tag{3}$$

Similarly, we assume that either the absence of A or the presence of B is necessary for activation of C, and activation of both B and C is required for D to be turned ON, thus leading to the following Boolean functions:

$$C^* = (\text{NOT } A) \text{ OR } B. \tag{4}$$

$$D^* = B \text{ AND } C. \tag{5}$$

In the case where a component is regulated by more than two regulators, its corresponding Boolean function can be a combination of AND, OR, and NOT operations, depending on the nature of biological interactions involved. In addition to the Boolean functions which provide a condensed representation of the regulatory effects on a specific compound, one can use the so-called truth tables to provide an exhaustive list of possible future states for the regulated compound for each specific combination of regulators' states. We note that the truth table for a node i has 2^{m_i} rows, where m_i is the number of regulators of node i.

2.3. Implementing Time

Having a reconstructed network along with the Boolean functions in hand, the next step is to specify the implementation of time. Let us denote the N nodes of a given network by $i = 1, 2, \ldots, N$. The future state of each node i at the time step $t + 1$, denoted by $X_i(t + 1)$, is determined by the current (or future) states of its input nodes and possibly its own current state. Different update methods are available for computing $X_i(t + 1)$. In the simplest formalism, called the synchronous model (13), the states of all the nodes of the network are updated simultaneously at multiples of a common time step (for simplicity, we consider this common time step to be 1). More precisely,

$$X_i(t+1) = F_i(X_1(t), \ldots, X_N(t)), \tag{6}$$

where F_i is the Boolean function corresponding to node i $(i = 1, 2, \ldots, N)$. Synchronous models assume similar timescales for all the processes possible in a given system. This assumption is often unrealistic in cellular biological processes due to the variety of timescales associated with, for example, transcription/translation and posttranslational mechanisms. For instance, the timescale of kinase/phosphatase reactions and protein conformational changes (both posttranslational mechanisms) is thousandths of a second, and the timescale of transcription is hundreds of seconds (14). To account for different timescales, asynchronous models may be used in which the state of each node is updated according to its own timescale. Here, we describe three different asynchronous algorithms, called random order asynchronous (21, 22), general asynchronous (22), and deterministic asynchronous (23).

In the random order model, at each time step (or round of update), the nodes' states are updated based on a random update order chosen from all possible permutations of the nodes. In this case, the state of node i at time step $t + 1$, i.e., $X_i(t + 1)$, is obtained according to the most recent updated state of its input nodes:

$$X_i(t+1) = F_i(X_1(\tau_{i1}), ..., X_N(\tau_{iN})), \quad (7)$$

where $\tau_{ij} \in \{t, t+1\}$, for any $i, j = 1, 2, ..., N$. When the position of input node j is after node i in the update order, $\tau_{ij} = t$ should be used; otherwise, $\tau_{ij} = t + 1$ should be used. Notably, with this update method, the states of all the nodes are updated exactly once during each round of update.

In the general asynchronous model, at each time step, a node is chosen at random and its state will be updated. Thus, in this approach, it is quite possible that a node chosen in the current time step will be selected again in the subsequent time step. Finally, in the deterministic asynchronous model, each node i is associated with a preselected time unit γ_i and its state is updated at multiples of that time unit, i.e., $t = k\gamma_i$, where k is any positive integer number (23):

$$X_i(t+1) = \begin{cases} F_i(X_1(t), ..., X_N(t)) & \text{if } t+1 = k\gamma_i, \quad k = 1, 2, ... \\ X_i(t) & \text{otherwise} \end{cases}. \quad (8)$$

If the time units γ_i are not known a priori, they can be selected randomly from a time interval. As in this method there is a deterministic updating order that depends on the time units, it is possible that the state of one node is updated several times while the other only once.

Several software libraries exist for modeling biological systems. An open-source python package called BooleanNet (24) is available online at http://code.google.com/p/booleannet/. The input to this software is a set of Boolean rules in a text file, and each of the updating schemes presented in this section can be implemented. Alternatively, a newly developed R package, called BoolNet (25), can be used for assembling and analyzing synchronous, general asynchronous, as well as probabilistic Boolean networks. This package is available at http://cran.r-project.org/web/packages/BoolNet/index.html.

2.4. Analyzing the Dynamics of the System

After selecting the model for state transitions (or time implementation), one can proceed with determining the long-term behavior of the system. To this end, first, the initial conditions for the system should be set. For a Boolean network with N nodes, there are 2^N possible initial states. Setting initial states of a model can be done according to prior biological knowledge. When sufficient information is not available, one can sample over a large set of initial conditions. Since the signal transduction processes are initiated once the input signal is active, it is reasonable to set the state of the signal at ON and that of the output node at OFF initially. However, in order to figure out whether the signal is necessary for observing the output, one may consider the cases where the signal is OFF as well. The state of intermediate nodes can be set similarly, either based on biological evidence or randomly.

Due to the finite number of states for a Boolean network, after a large enough number of time steps, the system must necessarily return to a previously visited state. As a result, the system eventually will converge to a finite set of attractors. Attractors of synchronous and deterministic asynchronous models include fixed points (steady states) and limit cycles (i.e., a set of states that are repeated regularly). The length (period) of a limit cycle is the number of states in the cycle. In random asynchronous models, including random order and general asynchronous, the fixed points are the same as deterministic models; however, limit cycles are replaced with the so-called loose attractors, where the system's trajectory travels irregularly over a set of states (22). For each attractor, the basin of attraction is defined as the set of initial conditions that lead to that particular attractor.

Let us find all possible attractors for the simple signaling network in Example 1 using BooleanNet (24) and the approaches introduced in Subheading 2.3. We start with illustrating the synchronous update method. Since the network has four nodes, totally $2^4 = 16$ initial conditions are possible. Let us set the initial condition of the network at $A_0 = C_0 = 1$ (ON) and $B_0 = D_0 = 0$ (OFF), which we denote by the vector (1010). To obtain the state of the system at the first time step using the synchronous method, one needs to plug in the initial states of the nodes in the corresponding Boolean functions given in Fig. 1b. This results in $A_1 = 1$ (since nothing affects the signal), $B_1 = 0$ (since NOT $C_0 = 0$), $C_1 = 0$ (since both $B_0 = 0$ and NOT $A_0 = 0$), and $D_1 = 0$ (since $B_0 = 0$). Therefore, the state of the system at $t = 1$ will be (1000). Following this procedure leads to a repeating sequence of four states, i.e., (1100), (1110), (1011), and (1000). This is a limit cycle of length four which has the initial state (1010) in its basin of attraction. Starting from another initial condition, where $A_0 = C_0 = D_0 = 0$ and $B_0 = 1$, leads to the system's state (0010) which remains unchanged with further updates. Thus, the initial state (0100) is in the basin of attraction of the fixed-point

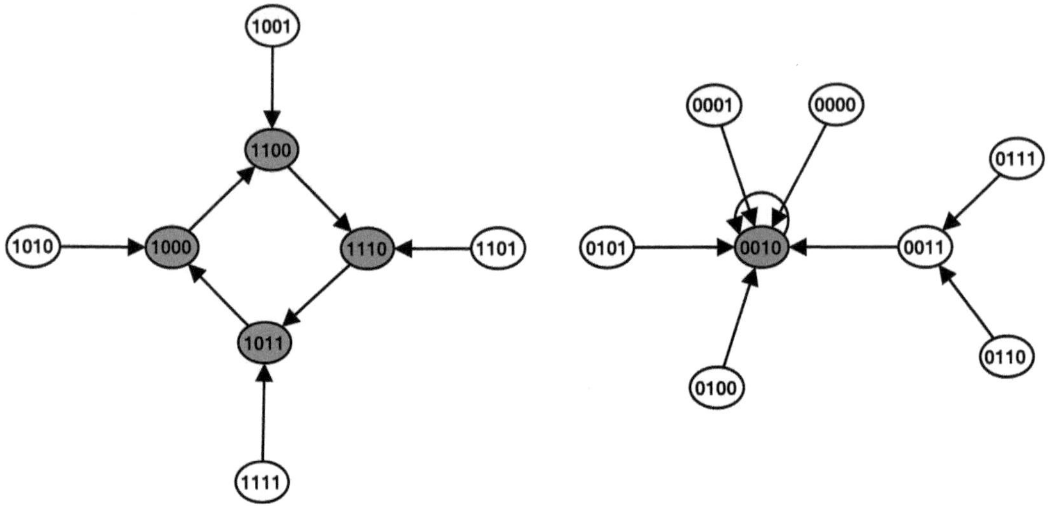

Fig. 2. The state transition graph of the network given in Example 1 obtained by the synchronous update method. The *binary digits* from left to right represent the state of the nodes A, B, C, and D, respectively. The four states represented with a *gray color* on the left form a limit cycle with four *white states* in its basin of attraction. The *gray state* with a self-loop on the right is the fixed point of the system which has seven other states in its basin of attraction.

attractor (0010). Synthesizing all state trajectories leads to a state transition graph depicted in Fig. 2. The binary digits represent the state of the nodes A, B, C, and D in order from left to right. The four gray states on the left form the limit cycle of the system, which has the four white states in its basin of attraction. The fixed point, which is represented with a gray color and a self-loop on it, has seven states plus itself in its basin of attraction.

Let us now consider the random order asynchronous updating scheme. To compute the system's state at $t = 1$ starting from the initial condition $A_0 = C_0 = 1$ (ON) and $B_0 = D_0 = 0$ (OFF) and the update order C–B–A–D, one needs to plug in the latest updated state of the nodes in the corresponding Boolean functions given in Fig. 1b. Based on the given update order, first, C is updated and becomes $C_1 =$ (NOT A_0) OR $B_0 = 0$, next B is updated and becomes $B_1 = A_0$ AND (NOT C_1) $= 1$, then A is updated and becomes $A_1 = 1$ (since nothing affects it), and finally D is updated and becomes $D_1 = B_1$ AND $C_1 = 0$. Therefore, the system's state at $t = 1$ is (1100). This state is part of a loose attractor, which is depicted with a gray color on the left in Fig. 3. Starting from initial condition $A_0 = B_0 = C_0 = 0$ (OFF) and $D_0 = 1$ (ON) and any random update order chosen from all $4! = 24$ possible permutations of four nodes, the system reaches a fixed point which is identical to the fixed point of the synchronous model.

Comparing the state transition graphs for the synchronous and random order asynchronous methods, we find that the fixed point is the same in both cases, as expected. This is explained by the fact that fixed points are update independent and can be obtained by taking

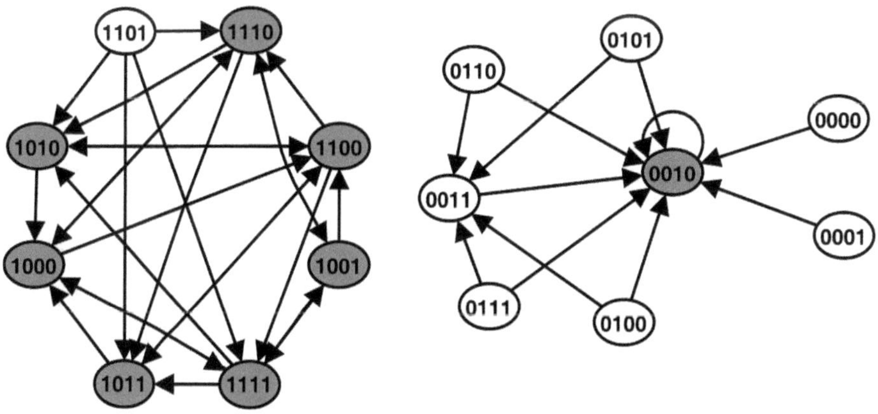

Fig. 3. The state transition graph of the network given in Example 1 obtained by the random order asynchronous update method. The *binary digits* from left to right represent the state of the nodes A, B, C, and D, respectively. The seven states represented with a *gray color* on the left form a loose attractor with a *white state* in its basin of attraction. The *gray state* with a self-loop on the right is the fixed point of the system which has seven other states in its basin of attraction.

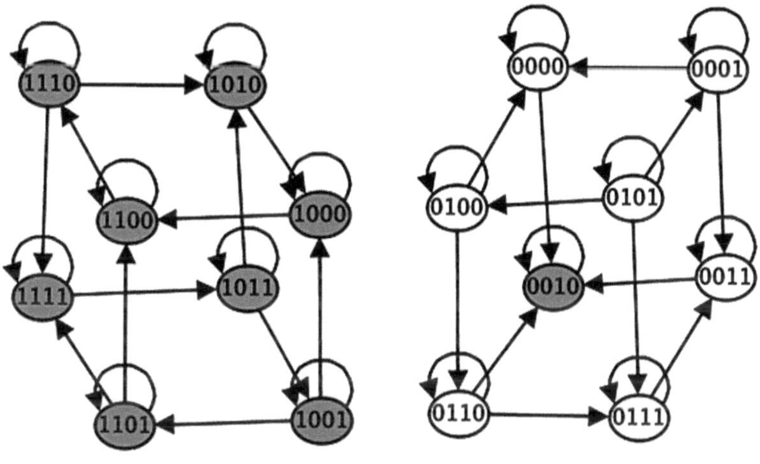

Fig. 4. The state transition graph of the network given in Example 1 obtained by the general asynchronous update method. The *binary digits* from left to right represent the state of the nodes A, B, C, and D, respectively. The eight states represented with a *gray color* on the left form a loose attractor. The *gray state* on the right is the fixed point of the system which has seven other states in its basin of attraction.

away the time dependencies of the Boolean functions and solving the resulting set of equations. We also observe that the state transition graph for the random order asynchronous method is denser than that of the synchronous method because each state can have multiple outgoing edges in random order asynchronous method, while in the synchronous method only one edge is going out of each state.

By using the general asynchronous method, the eight states in which the signal is ON form a loose attractor of the system depicted by a gray color on the left in Fig. 4. In this method, at each initial

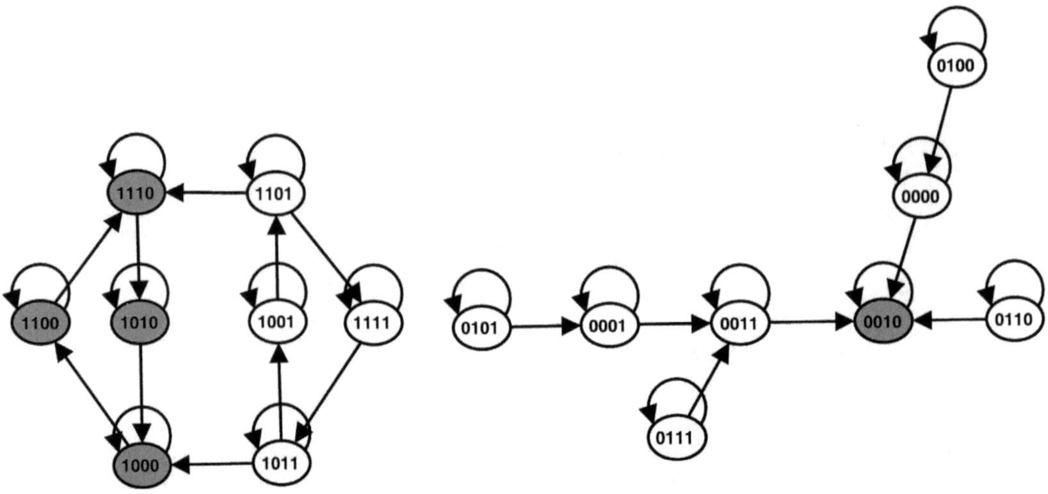

Fig. 5. The state transition graph of the network given in Example 1 obtained by the deterministic asynchronous update method. The *binary digits* from left to right represent the state of the nodes A, B, C, and D, respectively. The four states represented with a *gray color* on the left form a limit cycle that can be reached from the *four white* states. The *gray state* on the right is the fixed point of the system which has seven other states in its basin of attraction.

condition, there are four possibilities of choosing a node randomly. For example, starting from the initial condition $A_0 = B_0 = C_0 = 0$ (OFF) and $D_0 = 1$ (ON) and selecting node D for update, the system reaches the state (0000). Then, from this state and updating node C, the system will converge to the only fixed point of the system, i.e., the state (0010). As can be seen in Fig. 4, there are more self-loops in the state transition graph of the general asynchronous model since it is quite possible that updating the state of a particular node does not change the state of the system.

Finally, we illustrate the deterministic asynchronous method with the preselected time units as $t_A = 1$, $t_B = 2$, $t_C = 3$, and $t_D = 6$. This choice of time units implies that node A is updated at all time instants (but its state will not change because it has no input node), node B is updated at multiples of two, node C is updated at multiples of three, and finally node D is updated at multiples of six. For example, starting from the initial condition $A_0 = B_0 = D_0 = 1$ (ON) and $C_0 = 0$ (OFF), at $t = 1$ only A is updated but its state will be unchanged and as a result a self-loop appears at the (1101) state. At $t = 2$, nodes A and B are updated but their states remain unchanged (because nothing affects A, and $B_2 = A_1$ AND (NOT C_1) = 1). At $t = 3$, nodes A and C are updated. While node A remains unchanged, node C is turned ON (because C_3 = (NOT A_2) OR $B_2 = 1$). At $t = 6$, all the nodes are updated, and after traversing through some transient states the system eventually reaches a limit cycle of length four. The state transition graph corresponding to the deterministic asynchronous model is depicted in Fig. 5. The limit cycle is different from that of the synchronous method, but the fixed point is the same as that of the other updating schemes. It should be noted that a change in the time units may result in observing different attractors. Thus, this method is suitable for modeling signaling networks for which the time units are known beforehand.

In the particular example that we considered here, an oscillatory behavior was obtained both with synchronous and asynchronous updating schemes. It is known that synchronous models may generate some artifacts, such as spurious limit cycles (26, 27). As a result, it is possible that in some cases oscillations observed in the synchronous model disappear in asynchronous models.

2.5. Testing the Validity and Robustness of the Dynamic Model

One also needs to assess the validity and robustness of the constructed model and the dynamic behavior exhibited by it. To address the validity, the ability of the model in replicating the available experimental observations should be tested. If there is experimental evidence for a certain biological outcome which cannot be obtained using the dynamic model, then the model needs a revision. For instance, when there is experimental evidence for an ON state of the output node under a certain initial condition while the model shows the opposite or when the dynamic model indicates that there is no fixed point for the system while in reality the system eventually approaches a steady state, then the dynamic model needs to be refined. Revision in the dynamic model may include changing some of the Boolean functions (e.g., changing the AND operator to OR and vice versa) or incorporating possible additional components to the model.

To address the robustness, the ability of the model in maintaining the wild-type response under small topological perturbations can be tested. To this end, one can add or delete a random edge, rewire edges in the network randomly (e.g., changing any pair of parallel edges to cross-talk edges), or make an inhibitory interaction into activation or vice versa. Currently, comprehensive models of signal transduction systems are rare and it is quite possible that some new components or interactions are added to the model as new experimental evidence is found. Thus, probing the robustness of a model provides the opportunity to take into account the situation when such a new node or edge will be added to the reconstructed network and to assess its corresponding effects on the dynamic behavior.

2.6. Analyzing the Effect of Node Perturbations

In order to evaluate the robustness of a real biological system, one can perturb certain biologically significant components in the network, for example by knocking them out or over-expressing them. These perturbations can be modeled by fixing the state of a node, i.e., setting it at OFF to simulate knockout or at ON to simulate over- or constitutive expression. Checking the effect of node disruptions on the dynamic behavior, such as attractors, of the system leads to generating experimentally testable predictions and to refining the dynamic model even further.

Let us elaborate more on node perturbations using the simple network given in Example 1 and considering synchronous and the first two asynchronous methods. We omit the deterministic asynchronous updating scheme since the results would be highly dependent on the time units. As we have seen in Subheading 2.4, when the signal node A is OFF, the system will always converge to the fixed point (0010) in which the output node D stabilizes in the OFF state. This implies that activation of the signal is required for observing the output of the signaling process. On the other hand, when the signal is constitutively ON, the state transition graph possesses a complex attractor (either a limit cycle or a loose attractor) that contains sustained oscillations of the output node.

The effects of perturbations of intermediary nodes in Example 1 are the same for synchronous, random order asynchronous, and general asynchronous update methods. Knocking out the intermediary node B leads to two fixed points (0010) and (1000), where in both cases the output node settles into the OFF state. Conversely, over-expression of B results in two other fixed points, namely, (0111) and (1111), where the output node is stabilized in the ON state. As a result, when the signal is absent, the over-expression of node B would be sufficient for activation of the output in the long-term behavior. With eliminating node C, the state transition graph converges to either (0000) or (1100) fixed point, whereas with its over-expression the system reaches (0010) or (1010) fixed point. Therefore, neither knockout nor over-expression of node C suffices for the ON state of the output node D.

3. Application Examples

Boolean networks have been widely employed in modeling many signaling pathways, including T-cell receptor signaling (28), neurotransmitter signaling pathway (29), human cholesterol regulatory pathway (30), as well as glucose repression signaling pathways in *Saccharomyces cerevisiae* (31). In the following, we describe two examples of Boolean dynamic models of signal transduction networks, one corresponding to drought response in plants (15) and the other related to a disease in humans (17).

3.1. Abscisic Acid (ABA) Signal Transduction Network

Plants take up carbon dioxide for photosynthesis and lose water by transpiration through stomata (pores in the epidermis of leaves). The stomatal size is determined by a surrounding pair of guard cells. During drought conditions, the plant hormone ABA is synthesized to induce stomatal closure (15). This hormone serves as a signal to a process that leads to the closure of stomata.

A consistent network model of this signaling process has been assembled by Li et al. (15) through an extensive literature search. The 54 nodes of this network contain proteins, ion channels, and secondary messengers, as well as a few conceptual nodes such as stomatal closure and plasma membrane depolarization. ABA acts as the input to network, whereas closure is considered as the output node. The edges of the network represent protein–protein interactions, chemical reactions, and indirect regulatory relationships between two components. The Boolean functions governing the state of the regulated nodes (39 out of 54) were also suggested in the paper.

Li et al. (15) also developed a predictive dynamic model for the ABA signaling network using the random order asynchronous update. Since they focused on the behavior of stomatal closure, the output of the model was chosen as the percentage of simulations, involving different initial conditions for nodes other than ABA and different update orders, that attained the ON state for the node closure at any given time step (called the percentage of closure). Li et al. (15) identified the essential components of the ABA model by knocking out certain nodes and node combinations and determining whether the percentage of closure in response to ABA changes. The response of the resulting perturbed system was then categorized into five types based on the percentage of closure time course. According to their classification, the responses that are identical or very close to the wild-type response are marked as having normal sensitivity. A response is said to have reduced sensitivity if the percentage of closed stomata stabilizes at a positive value less than 100%. Additionally, insensitivity refers to those responses for which the percentage of closed stomata decreases to zero after the first few steps. Finally, a response for which the time course of reaching 100% of closed stomata has below-wild-type values is referred as hyposensitivity, whereas hypersensitivity corresponds to ABA-induced closure that is faster than wild-type (15). Their knockout study resulted in several biological predictions and implications, some of which have been validated by wet-bench experiments. We have recently carried out a thorough comparative study on the dynamic behavior of this network using the synchronous and three different asynchronous approaches delineared in Subheading 2.4 (32). Our analysis revealed that the wild-type system possesses an update-independent fixed point, and an additional complex attractor is present if strict constraints regarding the timing of certain processes and the initial state of the system are satisfied. Interestingly, in the case of disruption of a particular node, all models lead to a complex attractor (32).

3.2. T-LGL Leukemia Survival Signaling Network

Cytotoxic T lymphocytes (CTLs) are a type of T-cells that are capable of eliminating virus-infected cells as well as some tumor cells. T-cell large granular lymphocyte (T-LGL) leukemia is a blood cancer that exhibits an abnormal expansion of antigen-primed mature CTLs potentially driven by chronic virus infection. Unlike normal CTLs which are eliminated through activation-induced cell death after they successfully fight the virus, leukemic T-LGL cells remain long-term competent (17).

A network model of T-LGL survival signaling was constructed by Zhang et al. (17) through an extensive curation of the available experimental literature and also using NET-SYNTHESIS (19). This network consists of 60 nodes and 142 regulatory edges. The nodes of the network contain proteins, mRNAs, and some small molecules, such as lipids. The node "Stimuli," which represents antigen stimulation, serves as the input to the network, and the main output node is "Apoptosis," which denotes programmed cell death. It was shown by Zhang et al. (17) that all known signaling abnormalities in leukemic T-LGL can be reproduced by only keeping two proteins, platelet-derived growth factor (PDGF) and interleukin-15 (IL-15), constitutively expressed (ON). Based on a random asynchronous Boolean dynamic model of the assembled network, they identified potential therapeutic targets, such as NFκB, SPHK1, and BID, which stabilize in an ON or OFF state in T-LGL leukemia and the reversal of this state increases the frequency of the ON state of Apoptosis (17). Several of these predictions have been validated experimentally as well.

4. Conclusions

In this chapter, we first provided an introduction to network reconstruction of signaling processes from the available experimental data. Network representation can provide a foundation for developing computational approaches that are able to predict the dynamic behavior of a biological system in response to environmental stimuli and internal perturbations. We next carried out a detailed description of Boolean dynamic modeling, including synchronous and various types of asynchronous approaches, and of the ways in which they can be applied to decipher the dynamics of signaling networks. The successful applications of these approaches have demonstrated the power of Boolean dynamic modeling of signaling networks for such systems. Boolean models are most helpful when no detailed information on kinetic description of interactions for the biological processes involved in the signal transduction is available. Alternatively, when partial quantitative information about the biological interactions is available, piecewise linear differential equation models (33, 34), which provide a bridge between continuous and Boolean models, can be employed. The applicability of this approach has been shown before to model the *Drosophila* segment polarity gene network (23) as well as a pathogen–immune system interaction network (35).

Acknowledgment

This work was supported by NSF grant CCF-0643529.

References

1. Marks F, Klingmuller U, Muller-Decker K (2009) Cellular signal processing: an introduction to the molecular mechanisms of signal transduction: Garland sciences. Taylor and Francis Group, LLC, Philadelphia, PA
2. Chen KC, Csikasz-Nagy A, Gyorffy B, Val J, Novak B et al (2000) Kinetic analysis of a molecular model of the budding yeast cell cycle. Mol Biol Cell 11:369–391
3. Kauffman SA (1969) Metabolic stability and epigenesis in randomly constructed genetic nets. J Theor Biol 22:437–467
4. Thomas R (1973) Boolean formalization of genetic control circuits. J Theor Biol 42:563–585
5. Mendoza L, Alvarez-Buylla ER (2000) Genetic regulation of root hair development in *Arabidopsis thaliana*: a network model. J Theor Biol 204:311–326
6. Sanchez L, Thieffry D (2001) A logical analysis of the *Drosophila* gap-gene system. J Theor Biol 211:115–141
7. Sanchez L, van Helden J, Thieffry D (1997) Establishement of the dorso-ventral pattern during embryonic development of *Drosophila melanogasater*: a logical analysis. J Theor Biol 189:377–389
8. May RM (1976) Simple mathematical models with very complicated dynamics. Nature 261:459–467
9. Song M, Ouyang Z, Liu ZL (2009) Discrete dynamical system modelling for gene regulatory networks of 5-hydroxymethylfurfural tolerance for ethanologenic yeast. IET Syst Biol 3:203–218
10. Chaouiya C (2007) Petri net modelling of biological networks. Brief Bioinform 8:210–219
11. Peterson JL (1981) Petri Net Theory and the modeling of systems. Prentice Hall PTR, Upper Saddle River, NJ
12. Sackmann A, Heiner M, Koch I (2006) Application of Petri net based analysis techniques to signal transduction pathways. BMC Bioinformatics 7:482
13. Kauffman S (1993) Origins of order: self-organization and selection in evolution. Oxford University Press, Oxford
14. Papin JA, Hunter T, Palsson BO, Subramaniam S (2005) Reconstruction of cellular signalling networks and analysis of their properties. Nat Rev Mol Cell Biol 6:99–111
15. Li S, Assmann SM, Albert R (2006) Predicting essential components of signal transduction networks: a dynamic model of guard cell abscisic acid signaling. PLoS Biol 4:e312
16. Thakar J, Pilione M, Kirimanjeswara G, Harvill ET, Albert R (2007) Modeling systems-level regulation of host immune responses. PLoS Comput Biol 3:e109
17. Zhang R, Shah MV, Yang J, Nyland SB, Liu X et al (2008) Network model of survival signaling in large granular lymphocyte leukemia. Proc Natl Acad Sci USA 105:16308–16313
18. Albert R, DasGupta B, Dondi R, Kachalo S, Sontag E et al (2007) A novel method for signal transduction network inference from indirect experimental evidence. J Comput Biol 14:927–949
19. Kachalo S, Zhang R, Sontag E, Albert R, DasGupta B (2008) NET-SYNTHESIS: a software for synthesis, inference and simplification of signal transduction networks. Bioinformatics 24:293–295
20. Albert R, Barabasi AL (2002) Statistical mechanics of complex networks. Rev Mod Phys 74:47–97
21. Chaves M, Albert R, Sontag ED (2005) Robustness and fragility of Boolean models for genetic regulatory networks. J Theor Biol 235:431–449

22. Harvey I, Bossomaier T (1997) Time out of joint: attractors in asynchronous random Boolean networks. In: Husbands P, Harvey I, editors; Proceedings of the Fourth European Conference on Artificial Life, MIT Press; Cambridge, pp 67–75
23. Chaves M, Sontag ED, Albert R (2006) Methods of robustness analysis for Boolean models of gene control networks. Syst Biol (Stevenage) 153:154–167
24. Albert I, Thakar J, Li S, Zhang R, Albert R (2008) Boolean network simulations for life scientists. Source Code Biol Med 3:16
25. Mussel C, Hopfensitz M, Kestler HA (2010) BoolNet – an R package for generation, reconstruction, and analysis of Boolean networks. Bioinformatics 26:1378–1380
26. Faure A, Naldi A, Chaouiya C, Thieffry D (2006) Dynamical analysis of a generic Boolean model for the control of the mammalian cell cycle. Bioinformatics 22:e124–e131
27. Mochizuki A (2005) An analytical study of the number of steady states in gene regulatory networks. J Theor Biol 236:291–310
28. Saez-Rodriguez J, Simeoni L, Lindquist JA, Hemenway R, Bommhardt U et al (2007) A logical model provides insights into T cell receptor signaling. PLoS Comput Biol 3:e163
29. Gupta S, Bisht SS, Kukreti R, Jain S, Brahmachari SK (2007) Boolean network analysis of a neurotransmitter signaling pathway. J Theor Biol 244:463–469
30. Kervizic G, Corcos L (2008) Dynamical modeling of the cholesterol regulatory pathway with Boolean networks. BMC Syst Biol 2:99
31. Christensen TS, Oliveira AP, Nielsen J (2009) Reconstruction and logical modeling of glucose repression signaling pathways in *Saccharomyces cerevisiae*. BMC Syst Biol 3:7
32. Saadatpour A, Albert I, Albert R (2010) Attractor analysis of asynchronous Boolean models of signal transduction networks. J Theor Biol 266:641–656
33. Glass L, Kauffman SA (1973) The logical analysis of continuous, non-linear biochemical control networks. J Theor Biol 39:103–129
34. Glass L (1975) Classification of biological networks by their qualitative dynamics. J Theor Biol 54:85–107
35. Thakar J, Saadatpour-Moghaddam A, Harvill ET, Albert R (2009) Constraint-based network model of pathogen–immune system interactions. J R Soc Interface 6:599–612

Chapter 13

Analytic Methods for Modeling Stochastic Regulatory Networks

Aleksandra M. Walczak, Andrew Mugler, and Chris H. Wiggins

Abstract

Recent single-cell experiments have revived interest in the unavoidable or *intrinsic* noise in biochemical and genetic networks arising from the small number of molecules of the participating species. That is, rather than modeling regulatory networks in terms of the deterministic dynamics of concentrations, we model the dynamics of the probability of a given copy number of the reactants in single cells. Most of the modeling activity of the last decade has centered on stochastic simulation, i.e., Monte Carlo methods for generating stochastic time series. Here we review the mathematical description in terms of probability distributions, introducing the relevant derivations and illustrating several cases for which analytic progress can be made either instead of or before turning to numerical computation. Analytic progress can be useful both for suggesting more efficient numerical methods and for obviating the computational expense of, for example, exploring parametric dependence.

Key words: Stochastic gene regulation, Feedback, Gene regulation, Bursts, Birth-death process, Spectral method, Master equation, Fokker-Planck approximation, Linear noise approximation, Langevin equation, Generating function, Analytical methods

1. Introduction

Theoretical descriptions of biochemical kinetic processes rely on considering the concentration changes of the molecules involved in the reactions, and specifying the types of regulatory interactions between them. The theoretical tools used to describe many types of biochemical reactions share many similarities. In this chapter, we give an overview of the analytical approaches to study biochemical kinetics using the example of small gene regulatory networks.

The regulation of genes by transcription factor proteins is an intrinsically stochastic process, owing to the small numbers of copies of molecules involved. With the development of imaging techniques in molecular biology, we are able to observe directly the

Xuedong Liu and Meredith D. Betterton (eds.), *Computational Modeling of Signaling Networks*, Methods in Molecular Biology, vol. 880, DOI 10.1007/978-1-61779-833-7_13,

fluctuations in the concentrations of proteins and mRNAs and, by measuring the intensity profiles of fluorescence markers, measure full probability distributions (1–3). Experiments over the last decade have shown that in fact gene regulation is a noisy process, and noise can propagate in gene networks (4).

Many methods for solving the resulting stochastic equations rely on computer simulations. The efficiency of these methods has been greatly advanced in the last several years (5–16). However, numerical simulations are naturally limited to a specific choice of parameters, and changing the parameters requires a completely new calculation. Furthermore, simulation suffers from the curse of dimensionality: the computational runtime grows prohibitively as the number of species increases. These problems can be bypassed by developing analytical approaches, which often require certain approximations.

This chapter is intended as a tutorial on theoretical descriptions of biochemical kinetics. For clarity, we start by introducing the simplified kinetic description of the production of proteins. We first consider the deterministic description of the system and then introduce noise. After familiarizing the reader with different levels of description, we discuss models of regulation. We also present a wide spectrum of analytic tools used to find the steady state probability distributions of protein copy numbers in the cell. We point out that while for concreteness we focus on the case of gene regulation, the methods presented in this chapter are very general.

The results presented in this review are not new; some can be found in textbooks (17–19), while others have been derived more recently in the context of gene regulation (3, 20–38). The goal of this review is to give the reader who is not familiar with analytical methods for modeling stochastic gene regulation an overview of the mathematical tools used in the field. Naturally, we are not able to cover all the developments in the field, but we hope to give the reader a useful starting point. For concreteness we also limit our discussion to the case of small gene networks and do not discuss approximations used to describe larger networks, which is currently an active area of research (10, 11, 15, 16, 28, 29).

2. Model for Protein Production and Degradation: Simple Birth–Death Process

In this section, we describe a simple birth death process for one species with constant production rate, i.e., absent of regulation. We first remind the reader of the deterministic description given by chemical kinetics. Next we introduce the full probabilistic description (given by the master equation obeyed by the probability distribution $p_n(t)$), and, restricting our discussion to the simplest case, show how the deterministic equations arise as the dynamics of the mean $\langle n \rangle \equiv \sum_n n p_n$. After calculating the variance, we introduce the generating function formalism, useful in solving the master equation. This formalism may also be described using raising and lowering operators familiar to physicists from quantum mechanics. Finally, we relate this description to the Fokker–Plank equation, the analogous master equation for continuous state variables (e.g., real-valued coordinates rather than the integer-valued copy number). These results will be used in later sections when we introduce autoregulation and regulation among different species.

2.1. Deterministic Description: the Kinetic Rate Equation

In the simplest case, the number of copies n of a protein species X can change either due to the production of a protein, which occurs at a constant rate $\tilde{g}$, or due to degradation of a protein, which occurs at a constant rate r:

$$\emptyset \underset{r}{\overset{\tilde{g}}{\rightleftharpoons}} X. \tag{1}$$

Here we condense the complicated molecular machinery of transcription, translation, and protein modification into a single constant production rate. Similarly, we do not specify the molecular processes which have led to degradation; for simplicity we imagine either dilution due to cell division or active degradation with a constant rate.

When n is large, the dynamics for the mean $\langle n \rangle$ are well approximated by the continuous, deterministic description provided by the kinetic rate equation for the concentration $c \equiv \langle n \rangle / V$ in a volume V:

$$|G(t)\rangle = \sum_j G_j(t)|j\rangle, \tag{2}$$

The solution of Eq. 2 is

$$c(t) = \frac{\tilde{g}}{Vr} + \mathrm{e}^{-rt}\left[c(0) - \frac{\tilde{g}}{Vr}\right], \tag{3}$$

where $c(0)$ is the concentration of proteins at initial time $t = 0$. In steady state, the mean number of proteins is simply the ratio of the production and degradation rates, $\langle n \rangle = cV = \tilde{g}/r$, which is easily seen by taking either $\mathrm{d}c/\mathrm{d}t = 0$ in Eq. 2 or $t \to \infty$ in Eq. 3.

2.2. Introducing Noise: The Master Equation

The general probabilistic description of chemical reactions, respecting the finite copy number of the reactants, is the master equation, which specifies the rate of change of $p_{\mathbf{n}}$, the probability that there are $\mathbf{n} = (n_1, n_2, \ldots, n_L)$ copies of the L reactants. The macroscopic, deterministic description in terms of chemical kinetics is recovered by considering the dynamics for the mean concentration of these reactants, i.e., the vector $\langle \mathbf{n} \rangle / V = (\langle n_1 \rangle / V, \langle n_2 \rangle / V, \ldots, \langle n_L \rangle / V) = (c_1, c_2, \ldots, c_L)$. Such a description in terms of a summary statistic necessarily ignores a great amount of information, including all information about fluctuations about these mean values.

Given the transition matrices $w_{nn'}$, with dimensions of rates, the general form of the master equation (cf. Appendix A), set by conservation of total probability, is

$$\dot{p}_{\mathbf{n}} = \sum_{\mathbf{n}'} [w_{\mathbf{n}\mathbf{n}'}(t)p_{\mathbf{n}'} - w_{\mathbf{n}'\mathbf{n}}(t)p_{\mathbf{n}}], \tag{4}$$

For the case of only one species, we need specify only the dynamics of $\mathbf{n} = n$. Restricting further to the case of a simple birth–death process, there are only two nonzero contributions from the transition matrices $w_{nn'}$, given by

$$w_{n+1,n}(t) = \tilde{g}, \tag{5}$$

$$w_{n-1,n}(t) = rn, \tag{6}$$

all other values of $w_{nn'}$ being 0. Under this restriction Eq. 4 reduces to the familiar

$$\dot{p}_n = -\tilde{g}p_n - rnp_n + \tilde{g}p_{n-1} + r(n+1)p_{n+1}. \tag{7}$$

Qualitatively, the four terms on the right-hand side (RHS) represent how the probability of having n proteins can either (1) decrease in time, if there are n initially and one is either produced (first term) or degraded (second term), or (2) increase in time, if there are either $n - 1$ initially and one is produced (third term) or $n + 1$ initially and one is degraded (fourth term).

The dynamics of the mean number of proteins $\langle n \rangle \equiv \sum_{n=0}^{\infty} np_n$ are readily obtained (see Appendix B), as

$$\frac{\mathrm{d}\langle n \rangle}{\mathrm{d}t} = \tilde{g} - r\langle n \rangle. \tag{8}$$

Equation 8 shows that the dynamics of the mean of the protein distribution reproduces the kinetic rate equation, Eq. 2.

From here on, we will rescale time t by the degradation rate r, such that $rt \to t$, and define $g = \tilde{g}/r$, making the master equation

$$\dot{p}_n = -gp_n - np_n + gp_{n-1} + (n+1)p_{n+1}. \tag{9}$$

2.2.1. Steady State Solution

The master equation can be rewritten in terms of shift operators E^+ and E^- which increase and decrease the number of proteins by one, respectively (17), i.e.

$$E^+ f_n = f_{n+1}, \tag{10}$$

$$E^- f_n = f_{n-1}, \tag{11}$$

for any function f_n. We begin by writing Eq. 9 in terms of only E^+ to make clear that, as is the case for any one-dimensional master equation, its steady state can be found iteratively. In terms of E^+,

$$\dot{p}_n = (E^+ - 1)(np_n - gp_{n-1}). \tag{12}$$

Setting $\dot{p}_n = 0$ for steady state requires that the second term in parentheses vanish, giving for $n > 0$ the recursive relation

$$p_n = p_{n-1}\frac{g}{n}. \tag{13}$$

Starting with $n = 1$ and computing the first few terms reveals the pattern

$$p_n = \frac{g^n}{n!}p_0, \tag{14}$$

where $p_0 = e^{-g}$ is set by normalization:

$$1 = \sum_{n=0}^{\infty} p_n = p_0 \sum_{n=0}^{\infty} \frac{g^n}{n!} = p_0 e^{g}. \tag{15}$$

Thus the steady state probability of having n proteins is the Poisson distribution,

$$p_n = \frac{g^n}{n!}e^{-g}, \tag{16}$$

with parameter g, the ratio of production to degradation rates.

We remind the reader that the variance $\sigma^2 = \langle n^2\rangle - \langle n\rangle^2$ of the Poisson distribution is equal to its mean $\langle n\rangle$ (Appendix C). Therefore the standard deviation σ over the mean falls off like

$$\frac{\sigma}{\langle n\rangle} = \frac{\sqrt{\langle n\rangle}}{\langle n\rangle} = \frac{1}{\sqrt{\langle n\rangle}}, \tag{17}$$

demonstrating that the relative effect of fluctuations diminishes for large protein number. In Appendix D we show that in the limit of large protein number the steady state asymptotes to a Gaussian distribution with mean and variance equal to g.

For comparison with the other representations of the master equation described below, we now write Eq. 9 in terms of both shift operators E^+ and E^-:

$$\dot{p}_n = (E^+ - 1)(n - gE^-)p_n. \tag{18}$$

Equation 12 can be rewritten slightly by inserting between the parenthetic terms the unit operator $1 = E^-E^+$ and distributing the E^- to the left and the E^+ to the right, giving

$$\dot{p}_n = -(E^- - 1)[(n+1)E^+ - g]p_n. \tag{19}$$

where the negative sign has been factored out of the first parenthetic term.

2.2.2. Generating Function Representation

Not all master equations are solvable by straightforward iteration. A more generalizable way to solve a master equation is by the introduction of a generating function (17). Here we demonstrate the generating function approach on the birth–death master equation,

Eq. 9. The utility of the generating function for solving a master equation is that it turns an infinite set of ordinary differential equations (e.g., Eq. 9) into a single partial differential equation.

The generating function G is defined as

$$G(x, t) \equiv \sum_{n=0}^{\infty} p_n(t) x^n, \tag{20}$$

a power-series expansion in a continuous variable x whose coefficients are the probabilities p_n. Since x is a variable we introduce, we note that defining $x = e^{i\varphi}$ makes clear that the generating function is equivalent to the Fourier transform of p_n in protein number (this point is further developed in Appendix E). The probability distribution may be recovered via the inverse transform

$$p_n(t) = \frac{1}{n!} \partial_x^n G(x, t)|_{x=0}. \tag{21}$$

Additionally, we note that the ℓth moment may be generated by

$$\langle n^\ell \rangle = \left[(x\partial_x)^\ell G \right]_{x=1}, \tag{22}$$

(which is the reason for the generating function's name).

To obtain an easily solvable partial differential equation for the generating function, we multiply Eq. 9 by x^n and sum over n to obtain (see Appendix B for the detailed derivation):

$$\dot{G} = -(x - 1)(\partial_x - g)G. \tag{23}$$

We see immediately from comparison of Eqs. 19 and 23 that the representations of operators E^- and $(n + 1)E^+$ in x space are x and ∂_x, respectively.

In steady state ($\dot{G} = 0$) Eq. 23 must satisfy $(\partial_x - g)G = 0$, which is solved by

$$G(x) = G(0)e^{gx}, \tag{24}$$

where $G(0) = e^{-g}$ is set by normalization:

$$G(0)e^{g} = G(1) = \sum_{n=0}^{\infty} p_n(1)^n = \sum_{n=0}^{\infty} p_n = 1. \tag{25}$$

The steady state distribution is recovered via inverse transform:

$$p_n = \frac{1}{n!} \partial_x^n [e^{g(x-1)}]_{x=0} = e^{-g} \frac{g^n}{n!}, \tag{26}$$

a Poisson distribution with parameter g, as before (Eq. 16).

The time-dependent solution to Eq. 23 may be obtained using the method of characteristics, in which one looks for characteristic curves along which the partial differential equation becomes an ordinary differential equation. The curves are parameterized by s, such that $G(x,t) = G(x(s),t(s)) = G(s)$, and we look for the solution to the ordinary differential equation for $G(s)$, having eliminated x and t from the problem. First Eq. 23 can be rewritten more explicitly as

$$(x-1)gG = \frac{\partial G}{\partial t} + \frac{\partial G}{\partial x}(x-1). \tag{27}$$

Using the chain rule on $G(s)$ gives

$$\frac{dG}{ds} = \frac{\partial G}{\partial t}\frac{dt}{ds} + \frac{\partial G}{\partial x}\frac{dx}{ds}. \tag{28}$$

Consistency of Eqs. 27 and 28 requires

$$\frac{dt}{ds} = 1, \tag{29}$$

$$\frac{dx}{ds} = x-1, \tag{30}$$

$$\frac{dG}{ds} = (x-1)gG. \tag{31}$$

Equation 29 implies $s = t$ (we set $t_0 = 0$ without loss of generality). Therefore, defining $y \equiv x - 1$, Eq. 30 implies

$$y = y_0 e^t. \tag{32}$$

Finally, straightforward integration of Eq. 31 yields

$$\begin{aligned} G &= G_0 \exp\left[g y_0 \int_0^t dt' e^{t'} \right] = G_0 \exp[g y_0 (e^t - 1)] \\ &= F(y_0)\exp[g y_0 e^t]. \end{aligned} \tag{33}$$

where we define $F(y_0) \equiv G_0 e^{-g y_0}$. We may expand $F(y_0)$ as a function of its argument, i.e., $F(y_0) = \sum_{j=0}^{\infty} A_j y_0^j$ for some coefficients A_j, such that

$$G = \sum_j A_j y_0^j \exp[g y_0 e^t]. \tag{34}$$

Inserting $y_0 = ye^{-t} = (x-1)e^{-t}$ (Eq. 32), we obtain

$$G(x,t) = \sum_j A_j e^{-jt}(x-1)^j e^{g(x-1)}. \tag{35}$$

For $t \to \infty$ only the $j = 0$ term survives and we recover the steady state generating function (Eq. 24),

$$G(x) = c_0 e^{g(x-1)}, \tag{36}$$

where $c_0 = 1$ by normalization. Equation 35 constitutes the full time-dependent solution to the birth–death process; the time-dependent distribution can be retrieved using the inverse transform, Eq. 21, and the coefficients A_j are computable from the initial distribution $p_n(0)$, which will be made explicit in the next section.

Since Eq. 23 is linear in G, one may arrive at its solution in a second, elegant way by expanding in the eigenfunctions of the x-dependent operator. That is, writing Eq. 23 as $\dot{G} = -\mathcal{L}G$, where

$$\mathcal{L} = (x-1)(\partial_x - g), \tag{37}$$

we expand G in eigenfunctions $\varphi_j(x)$ with time-dependent expansion coefficients $G_j(t)$,

$$G(x,t) = \sum_j G_j(t)\,\varphi_j(x), \tag{38}$$

where the $\varphi_j(x)$ satisfy

$$\mathcal{L}\varphi_j = \lambda_j \varphi_j. \tag{39}$$

for eigenvalues λ_j. Substituting Eq. 38 into Eq. 23 gives $\sum_j \dot{G}_j \varphi_j = -\sum_j \lambda_j G_j \varphi_j$, which, by orthogonality of the φ_j, yields the set of ordinary differential equations $\dot{G}_j = -\lambda_j G_j$, solved by

$$G_j(t) = A_j e^{-\lambda_j t}. \tag{40}$$

for some constants A_j. Comparing the result,

$$G(x,t) = \sum_j A_j e^{-\lambda_j t} \varphi_j(x), \tag{41}$$

with Eq. 35 reveals the forms of the eigenvalues

$$\lambda_j = j \in \{0, 1, 2, \ldots\}, \tag{42}$$

and the eigenfunctions

$$\varphi_j(x) = (x-1)^j e^{g(x-1)}, \tag{43}$$

facts that we will confirm in the next section using operator methods.

2.2.3. Operator Representation

We now introduce a representation of the master equation in terms of raising and lowering operators. This representation makes the solution more elegant, yielding a simple algebra which allows calculation of projections between the spaces of protein numbers and eigenfunctions without explicit computation of overlap integrals; it also lays the formal groundwork for solving models of multidimensional regulatory networks. The formalism was first used for diffusion by Doi (39) and Zel'dovich (40) and later developed by Peliti (41).

As before the generating function is defined as an expansion in a complete basis indexed by protein number n in which the expansion coefficients are the probabilities p_n. Within the operator formalism, this basis is represented as the set of states $|n\rangle$, i.e.

$$|G(t)\rangle = \sum_{n=0}^{\infty} p_n(t)|n\rangle. \tag{44}$$

Here we have adopted state notation commonly used in quantum mechanics (42); the previous representation (Eq. 20) is recovered by projecting onto this equation the conjugate states $\langle x|$ with the provisions

$$\langle x|G(t)\rangle \equiv G(x,t), \tag{45}$$

$$\langle x|n\rangle \equiv x^n. \tag{46}$$

In Appendix E we show how the orthonormality of the $|n\rangle$ states,

$$\langle n|n'\rangle = \delta_{nn'}, \tag{47}$$

dictates the form of the conjugate state:

$$\langle n|x\rangle = \frac{1}{x^{n+1}}. \tag{48}$$

Projecting Eq. 44 onto $\langle n|$ and using Eq. 47 gives the inverse transform:

$$\langle n|G(t)\rangle = p_n(t). \tag{49}$$

The equation of motion in the operator representation is obtained by summing the master equation (Eq. 9) over n against $|n\rangle$ (see Appendix B), giving

$$|\dot{G}\rangle = -\hat{\mathcal{L}}|G\rangle, \tag{50}$$

where

$$\hat{\mathcal{L}} \equiv (\hat{a}^+ - 1)(\hat{a}^- - g). \tag{51}$$

Just as in the operator treatment of the quantum harmonic oscillator (42), the operators $\hat{a}^+$ and $\hat{a}^-$ here raise and lower the protein number by 1 respectively, i.e.,

$$\hat{a}^+|n\rangle = |n+1\rangle, \tag{52}$$

$$\hat{a}^-|n\rangle = n|n-1\rangle, \tag{53}$$

(note, however, that the prefactors here, 1 and n, are different than those conventionally used for the harmonic oscillator, $\sqrt{n+1}$ and $\sqrt{n}$, respectively). Comparison of Eqs. 19, 23, and 51 makes clear the following correspondences among the master equation, generating function, and operator representations respectively:

$$E^- \leftrightarrow x \leftrightarrow \hat{a}^+, \tag{54}$$

$$(n+1)E^+ \leftrightarrow \partial_x \leftrightarrow \hat{a}^-. \tag{55}$$

While it might seem strange that the downshift operator E^- corresponds to the raising operator $\hat{a}^+$ (and the upshift operator to the lowering operator), in Appendix F we show that, just as with the quantum harmonic oscillator, $\hat{a}^+$ and $\hat{a}^-$ lower and raise protein number, respectively, when operating to the left, i.e.,

$$\langle n|\hat{a}^+ = \langle n-1|, \tag{56}$$

$$\langle n|\hat{a}^- = (n+1)\langle n+1|, \tag{57}$$

which makes the correspondence more directly apparent. Finally, again as with the quantum harmonic oscillator, the raising and lowering operators enjoy the commutation relation (see Appendix F)

$$[\hat{a}^-, \hat{a}^+] = 1, \tag{58}$$

and $\hat{a}^+\hat{a}^-$ acts as a number operator, i.e., $\hat{a}^+\hat{a}^-|n\rangle = n|n\rangle$.

Equation 51 shows that the full operator $\hat{\mathcal{L}}$ factorizes, suggesting the definition of the shifted operators

$$\hat{b}^+ \equiv \hat{a}^+ - 1, \tag{59}$$

$$\hat{b}^- \equiv \hat{a}^- - g. \tag{60}$$

Since $\hat{b}^+$ and $\hat{b}^-$ differ from $\hat{a}^+$ and $\hat{a}^-$, respectively, by scalars, they obey the same commutation relation, i.e.,

$$[\hat{b}^-, \hat{b}^+] = 1. \tag{61}$$

Since the equation of motion (Eq. 50) is linear, it will benefit from expansion of $|G\rangle$ in the eigenfunctions $|\lambda_j\rangle$ of the operator $\hat{\mathcal{L}}$, where

$$\hat{\mathcal{L}}|\lambda_j\rangle = \lambda_j|\lambda_j\rangle, \tag{62}$$

for eigenvalues λ_j. In Appendix G we show that the commutation relation (Eq. 61) and the steady state solution $\langle x|G\rangle = e^{g(x-1)}$, for which $\hat{\mathcal{L}}|G\rangle = 0$, completely define the eigenfunctions and eigenvalues of $\hat{\mathcal{L}}$; we summarize the results of Appendix G here. The eigenvalues are nonnegative integers j,

$$\lambda_j = j \in \{0, 1, 2, \ldots\}, \tag{63}$$

as in Eq. 42. The eigenvalue equation now reads

$$\hat{\mathcal{L}}|j\rangle = \hat{b}^+\hat{b}^-|j\rangle = j|j\rangle, \tag{64}$$

which is consistent with interpretation of $\hat{b}^+\hat{b}^-$ as a number operator for the eigenstates $|j\rangle$. The eigenfunctions are (in x space)

$$\langle x|j\rangle = (x-1)^j e^{g(x-1)}, \tag{65}$$

as in Eq. 43 with $\varphi_j(x) \equiv \langle x|j\rangle$. The conjugate eigenfunctions are

$$\langle x|j\rangle = \frac{e^{-g(x-1)}}{(x-1)^{j+1}}. \tag{66}$$

The operators $\hat{b}^+$ and $\hat{b}^-$ raise and lower the eigenstates $|j\rangle$, respectively, as $\hat{a}^+$ and $\hat{a}^-$ do the $|n\rangle$ states, i.e.,

$$\hat{b}^{+}|j\rangle = |j+1\rangle, \tag{67}$$

$$\hat{b}^{-}|j\rangle = j|j-1\rangle, \tag{68}$$

$$\langle j|\hat{b}^{+} = \langle j-1|, \tag{69}$$

$$\langle j|\hat{b}^{-} = (j+1)\langle j+1|. \tag{70}$$

Now employing the expansion of $|G\rangle$ in the eigenstates $|j\rangle$,

$$|G(t)\rangle = \sum_j G_j(t)|j\rangle, \tag{71}$$

the equation of motion (Eq. 50) gives a trivial equation for the expansion coefficients

$$\dot{G}_j = -jG_j, \tag{72}$$

which is solved by

$$G_j(t) = A_j e^{-jt}, \tag{73}$$

for some constants A_j, as in Eq. 40. The inverse of Eq. 71 (obtained by projecting $\langle j|$ and using $\langle j|j'\rangle = \delta_{jj'}$) is

$$\langle j|G(t)\rangle = G_j(t) \tag{74}$$

The probability distribution is retrieved by inverse transform (Eq. 49),

$$p_n(t) = \langle n|G(t)\rangle = \langle n|\sum_j G_j(t)|j\rangle = \sum_j A_j e^{-jt}\langle n|j\rangle, \tag{75}$$

where the coefficients A_j are computed from the initial distribution $p_n(0)$:

$$A_j = G_j(0) = \langle j|G(0)\rangle = \sum_n p_n(0)\langle j|n\rangle. \tag{76}$$

Equations 75 and 76 give the full time-dependent solution to the birth–death process as expansions in the eigenmodes $\langle n|j\rangle$ and conjugate eigenmodes $\langle j|n\rangle$. Because we have decomposed the problem using the eigenbasis, or spectrum, of the underlying operator, we will refer to this as the spectral solution.

The quantities $\langle n|j\rangle$ and $\langle j|n\rangle$ are overlaps between the protein number basis $|n\rangle$ and the eigenbasis $|j\rangle$. Both $\langle n|j\rangle$ and $\langle j|n\rangle$ are readily computed by contour integration or more efficiently by recursive updating; these techniques are presented in Appendix H. Notable special cases are

$$\langle n|0\rangle = e^{-g}\frac{g^n}{n!}, \tag{77}$$

$$\langle 0|n\rangle = 1, \tag{78}$$

which confirm that Eq. 75 describes a Poisson distribution in the steady state ($j = 0$).

2.3. Fokker–Planck Approximation

The previous sections discuss several methods for calculating the steady state and time-dependent solutions of the master equation for the birth–death process. For many larger systems full solution of the master equation is not possible. In some of these cases one can make progress by deriving an approximate equation which is valid when protein numbers are large. In the limit of large protein numbers the master equation can be expanded to second order to yield the Fokker–Planck equation. In this section, we derive the Fokker–Planck equation for a general one-dimensional master equation with arbitrary production and degradation rates; the method is easily generalizable to models with regulation.

For arbitrary production rate g_n and degradation rate r_n, the master equation reads

$$\partial_t p_n = -(g_n + r_n)p_n + g_{n-1}p_{n-1} + r_{n+1}p_{n+1}; \tag{79}$$

setting $g_n = g$ and $r_n = n$ recovers the simple birth–death process with time rescaled by the degradation rate, as in Eq. 9.

2.3.1. Large Protein Number

The Fokker–Plank equation is derived under the assumption that the typical protein number is large ($n \gg 1$), such that n can be approximated as a continuous variable, and a change of 1 protein can be treated as a small change. We will use parentheses when treating n as continuous and a subscript when treating n as discrete. Under this approximation, the function $f(n \pm 1)$, where $f(n) \in \{g(n)p(n), r(n)p(n)\}$, can be expanded to second order as

$$f(n \pm 1) = f(n) \pm \partial_n f(n) + \frac{1}{2}\partial_n^2 f(n). \tag{80}$$

Inserting the results of the expansion into Eq. 79, we obtain the Fokker–Planck equation

$$\partial_t p(n) = -\partial_n[v(n)p(n)] + \frac{1}{2}\partial_n^2[D(n)p(n)], \tag{81}$$

where $v(n) \equiv g(n) - r(n)$, an effective drift velocity, recovers the RHS of the deterministic equation, and $D(n) \equiv g(n) + r(n)$, an effective diffusion constant, sets the scale of the fluctuations in protein number. The drift term plays the role of an effective force. The diffusion coefficient plays the role of an effective temperature: the larger it is, the larger the excursions a single trajectory (of particle number versus time) takes from the mean.

The steady state solution of Eq. 81 is easily obtained by noticing that the Fokker–Planck equation is a continuity equation of the form

$$\partial_t p = -\partial_n j, \tag{82}$$

where $j(n) \equiv v(n)p(n) - (1/2)\partial_n[D(n)p(n)]$ is the current of the probability. In steady state ($\partial_t p = 0$) the current is constant, and since $p(n) \to 0$ and $\partial_n p(n) \to 0$ as $n \to \infty$ (typically more quickly than $v(n)$ or $D(n)$ diverges), the current vanishes for all n. The steady state distribution is then found by direct integration, i.e.

$$p(n) = \frac{1}{\mathcal{Z}D(n)} \exp\left[2\int_0^n \mathrm{d}n' \frac{v(n')}{D(n')}\right], \tag{83}$$

where $\mathcal{Z}$ is a normalization constant ensuring $\int_0^\infty \mathrm{d}n\, p(n) = 1$.

In the simple birth–death process, for which $v(n) = g - n$ and $D(n) = g + n$, Eq. 83 evaluates to

$$p(n) = \frac{1}{\mathcal{Z}g}\left(1 + \frac{n}{g}\right)^{4g-1} \mathrm{e}^{-2n}. \tag{84}$$

In Appendix D, we show that the exact steady state (the Poisson distribution, Eq. 16) and Eq. 84 have the same asymptotic limit for large protein number: a Gaussian distribution with mean and variance equal to g, the ratio of production to degradation rates.

2.3.2. Small Noise

In addition to the approximation that the typical protein number is large, one may make the further approximation that the noise is small. This is often referred to as the "linear noise approximation" (17, 37, 38) or "small noise approximation" (32, 33). Specifically, one assumes that the fluctuations η in n are small around the mean $\bar{n}$, i.e.

$$n = \bar{n} + \eta, \tag{85}$$

where $\bar{n} \equiv \langle n \rangle = \sum_n n p_n$ (for brevity we use bar notation in this and several subsequent sections). Since $\mathrm{d}n/\mathrm{d}\eta = 1$ we have $p(n) = p(\eta)$, and the master equation (Eq. 81) becomes an equation in η:

$$\partial_t p(\eta) = -\partial_\eta[v(\bar{n} + \eta)p(\eta)] + \frac{1}{2}\partial_\eta^2[D(\bar{n} + \eta)p(\eta)], \tag{86}$$

We use the approximation that η is small to expand the drift and diffusion terms to first nonzero order in η:

$$v(\bar{n} + \eta) = v(\bar{n}) + \eta v'(\bar{n}) + \cdots \approx \eta v'(\bar{n}), \tag{87}$$

$$D(\bar{n} + \eta) = D(\bar{n}) + \eta D'(\bar{n}) + \cdots \approx D(\bar{n}), \tag{88}$$

where prime denotes differentiation with respect to n, and the last step in Eq. 87 recalls the fact that $v(\bar{n}) = g(\bar{n}) - r(\bar{n}) = 0$ in the steady state, as given by the kinetic rate equation (e.g., Eq. 8). Equation 86 is now

$$\partial_t p(\eta) = -v'(\bar{n})\partial_\eta[\eta p(\eta)] + \frac{1}{2}D(\bar{n})\partial_\eta^2[p(\eta)]. \tag{89}$$

As with Eq. 83, the steady state of Eq. 89 is found by direct integration:

$$p(\eta) = \frac{1}{\mathcal{Z}D(\bar{n})} \exp\left[\frac{2v'(\bar{n})}{D(\bar{n})}\int_0^\eta \mathrm{d}\eta' \eta'\right] = \frac{1}{\sqrt{2\pi\sigma^2}}\mathrm{e}^{-\eta^2/2\sigma^2}, \tag{90}$$

where

$$\sigma^2 \equiv \frac{D(\bar{n})}{-2v'(\bar{n})} \tag{91}$$

(note that $v'(\bar{n})$ is negative for stable fixed points). Equation 90 is a Gaussian distribution, meaning that under the linear noise approximation, the steady state probability distribution is a Gaussian centered at the exact mean $\bar{n}$ and with width determined by mean birth and death rates according to Eq. 91. We note that Eq. 90 can be equivalently derived by expanding the integrand in Eq. 83 about its maximum (or "saddle point") $\bar{n}$.

The linear noise approximation is stricter than the large protein number approximation made in the previous section (Subheading 2.3.1). While the previous approximation makes no assumption about the form of the probability distribution, the linear noise approximation assumes that the distribution is unimodal and sharply peaked around its mean. In practice, $\bar{n}$ and σ^2 are obtained by finding the steady state(s) (i.e., the stable fixed point(s)) of the corresponding deterministic rate equation. However, it is easily possible (for processes more complicated than the simple birth–death process) for the deterministic equation to have more than one stable fixed point (see Fig. 2). Although one may make a Gaussian expansion around each fixed point in turn, the linear noise approximation does not describe how these Gaussians might be weighted in a multimodal distribution. In these cases it is most accurate to use (if solvable) either the large protein number approximation (Eq. 83) or the original master equation (Eq. 79).

In the simple birth–death process, for which $v(n) = g - n$, $D(n) = g + n$, and $\bar{n} = g$, Eq. 91 gives $\sigma^2 = 2g/[-2(-1)] = g$, and therefore Eq. 90 reproduces the asymptotic behavior derived in Appendix D.

2.4. Langevin Approximation

We now consider a second stochastic approximation: the Langevin equation. The advantage of the Langevin approach is that a large amount can be learned about the process without finding the full distribution, but instead by considering the correlation functions of the concentration, which are readily computed. Starting from the Fokker–Planck equation we can calculate the equation for the mean of the distribution $\langle n \rangle = \sum_{n=0}^{\infty} np_n$. We arrive at the kinetic rate equation given in Eq. 2.

We can think about the change in concentration in time as a trajectory in n space. Each realization of a birth–death process, will be described by a certain $n(t)$, and averaging over many experiments, we obtain $\langle n(t) \rangle$, as given by Eq. 2. If we consider the change of concentrations on time scales longer than the characteristic timescales of the particular reactions, we can assume that there are many birth and death processes in each interval and that the fluctuations of each realization of $n(t)$ around the mean are Gaussian distributed:

$$P[\eta] \sim e^{-\int dn \int dt \frac{\eta^2(n,t)}{4D(n)}}. \tag{92}$$

We can include fluctuations around the mean values by considering an additional noise term $\eta(t)$, such that the equation for the change of the concentration of proteins n becomes:

$$\frac{dn}{dt} = v(n) + \eta(t) = g - n + \eta(t). \tag{93}$$

We require of the noise that:

$$\langle \eta(t) \rangle = 0, \tag{94}$$

$$\langle \eta(t')\eta(t) \rangle = \delta(t - t')D(n) = \delta(t - t')(g + n), \tag{95}$$

where $D(n)$ is in general the diffusion coefficient in the Fokker–Planck equation. In Appendix I we show the equivalence of the Fokker–Planck and Langevin descriptions.

In the general case when there are many types of proteins in the system, we can define a time-dependent correlation function

$$C_{ij}(t) = \langle \delta N_i(0) \delta N_j(t) \rangle, \tag{96}$$

where the average implies a time average and $\delta N_i(t) = N_i(t) - \langle N_i(t) \rangle$ are the variances of each type of protein. In one dimension δn^2 is the variance σ^2 (in subsequent sections we use δn^2 and σ^2 interchangeably). We note that in the steady state the time average can be replaced by an ensemble average. For the case of the single species birth–death process we have already computed the means and variances in Subheading 2.2.1. In case of the birth–death process, we calculate the time-dependent correlation function to obey the equation:

$$\frac{dC(t)}{dt} = -C(t). \tag{97}$$

In the steady state the solution must reduce to the previously calculated variance (Appendix C): $\delta n^2 = \langle (n - \langle n \rangle)(n - \langle n \rangle) \rangle = \langle n^2 \rangle - \langle n \rangle^2 - g$. Therefore the solution of Eq. 97 for a simple birth–death process is

$$C(t) = g e^{-t}. \tag{98}$$

We can also consider the correlation functions in Fourier space

$$C_{ij}(\omega) = \int_0^{\infty} dt e^{-i\omega t} C_{ij}(t). \tag{99}$$

Often it is easier to calculate the Fourier transform of the correlation function directly from the Fourier transform of the Langevin equation, and then invert the transform back. In case of the simple birth–death process we can vary the Langevin equation around its mean values $n(t) = \langle n \rangle + \delta n(t)$ (that is simply linearize the equation around its mean) and consider the resulting equation in Fourier space to obtain

$$-i\omega\delta\tilde{n}(\omega) = -\delta\tilde{n}(\omega) + \tilde{\eta}(\omega), \tag{100}$$

where $\delta\tilde{n}(\omega) = \int_0^\infty \mathrm{d}t e^{-i\omega t}\delta n(t)$ and

$$\tilde{\eta}(\omega) = \int_0^\infty \mathrm{d}t e^{-i\omega t}\eta(t), \tag{101}$$

$$\langle \tilde{\eta}(\omega)\tilde{\eta}(\omega') \rangle = 2\pi\delta(\omega - \omega')(g + \langle n \rangle). \tag{102}$$

Solving Eq. 100 for $\delta\tilde{n}(\omega)$ we find

$$\delta\tilde{n}(\omega) = \frac{\tilde{\eta}(\omega)}{1 - i\omega}, \tag{103}$$

from which we may calculate the correlation function:

$$\langle \delta\tilde{n}^*(\omega)\delta\tilde{n}(\omega') \rangle = 2\pi\delta(\omega - \omega')\frac{g + \langle n \rangle}{\omega^2 + 1} = 2\pi\frac{2g}{\omega^2 + 1}, \tag{104}$$

where $\delta\tilde{n}^*(\omega)$ denotes the complex conjugate of $\delta\tilde{n}(\omega)$. Inverting the Fourier transform reproduces the real time correlation function.

$$C(t) = \int_0^\infty \mathrm{d}\omega e^{i\omega t}\frac{2g}{\omega^2 + 1} = 2\pi i e^{-t}\frac{2g}{2i} = g e^{-t}, \tag{105}$$

where we reproduce the result of Eq. 99. The real part of the auto correlation function in Fourier space is also called the power spectrum

$$\mathcal{N}(\omega) = \langle \delta\tilde{n}^*(\omega)\delta\tilde{n}(\omega') \rangle = \int_0^\infty \mathrm{d}t e^{-i\omega t} C(t), \tag{106}$$

because it tells us which frequency modes contribute most to the form of the noise. We note that the power spectrum $\mathcal{N}(\omega)$ may be written as integral of the correlation function in real time.

2.5. Comparison of the Descriptions

For the birth–death process, the steady state of the kinetic rate equation agrees with the mean of the steady state probability distribution (under both the exact and approximate stochastic descriptions): $\langle n \rangle = g$. In the limit of large protein number, the Poisson distribution is well approximated by a Gaussian, and the Fokker–Planck (and therefore the Langevin) approximation is a good description. To investigate the validity of the Fokker–Planck approximation at small and intermediate protein number, in Fig. 1 we compare for a range of protein numbers the probability distribution obtained directly from the master equation (i.e., the Poisson distribution, Eq. 16) and from the Fokker–Planck approximation (Eq. 83). We quantify the disagreement between two distributions using the Kullback–Leibler divergence,

$$D_{\mathrm{KL}} = \sum_n p_n \log\frac{p_n}{\tilde{p}_n}, \tag{107}$$

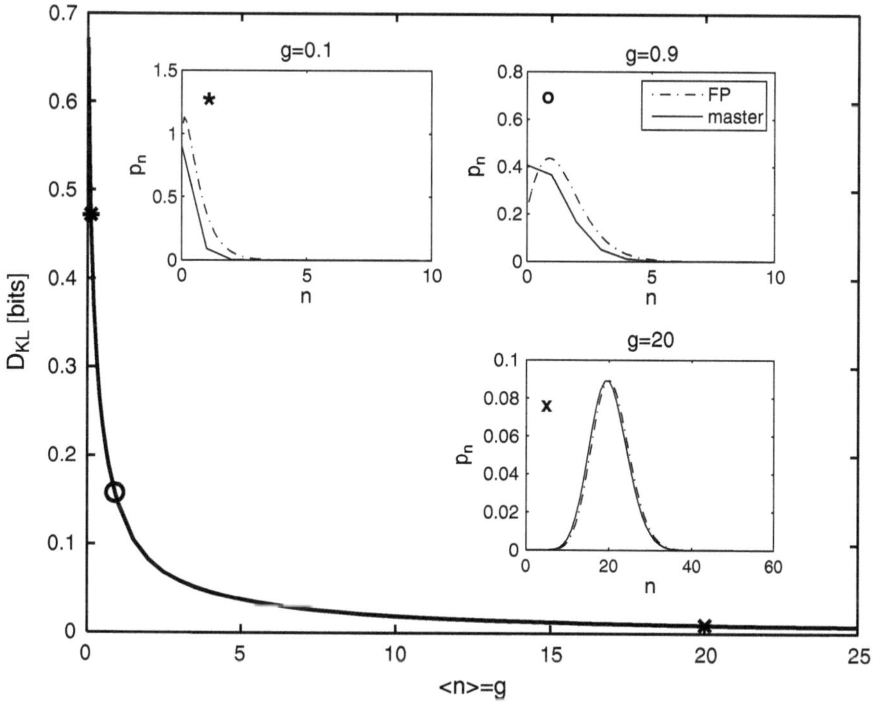

Fig. 1. Comparison of the distributions obtained from the master equation (Eq. 16) and from the Fokker–Planck approximation (Eq. 83), for the simple birth–death process. The Kullback–Leibler divergence D_{KL} (Eq. 107) between the two distributions is plotted as a function of mean protein number $\langle n \rangle = g$. *Insets* show distributions obtained from the master equation (*solid*) and from the Fokker–Planck equation (*dashed*) for $g = 0.1$ (*star*), $g = 0.9$ (*circle*), and $g = 20$ (*cross*), where the symbols correspond to points on the D_{KL} curve.

where p_n corresponds to the master equation and $\tilde{p}_n$ corresponds to the Fokker–Planck approximation. The Kullback–Leibler divergence is not symmetric and is therefore appropriate for a comparison between a "true" distribution (p_n here) and its approximation ($\tilde{p}_n$ here). As the mean protein number $\langle n \rangle = g$ increases, the accuracy of the approximation increases, and the divergence D_{KL} decreases. We plot explicitly three sample pairs of exact and approximate distributions, at small, intermediate, and large protein numbers. As expected the Fokker–Planck distribution deviates from the Poisson distribution at small protein number, and agrees well at large protein number.

3. Autoregulation

We now begin to turn our attention from the simple one-dimensional birth–death process to more realistic models of gene regulation (see refs. (43, 44) for a discussion of regulation functions of gene expression). We start with the description of autoregulation of one gene, in which its protein production rate is an arbitrary function of its own protein number.

3.1. Deterministic Model

As before the mean dynamics are captured by the kinetic rate equation. The kinetic rate equation for a birth–death process in which the production rate is an arbitrary function $g(\bar{n})$ of mean protein number $\bar{n}$ is

$$\frac{d\bar{n}}{dt} = g(\bar{n}) - \bar{n}. \tag{108}$$

The problem becomes potentially much harder, since the autoregulation function $g(\bar{n})$ can be nonlinear.

The form of the autoregulation function depends specifically on the molecular model of regulation which is being considered. For example, the Hill function,

$$g(n) = \frac{g_- K^h + g_+ n^h}{K^h + n^h}, \tag{109}$$

can be derived by considering a gene with two production rates g_+ and g_- (corresponding to the states in which a transcription factor protein is bound and unbound to the DNA, respectively) for which the rate of switching to the bound state depends on the protein number (see Subheading 4 and Appendix J). The derivation assumes that the rates of binding and unbinding are faster than the protein degradation rate (which sets the timescale for changes in protein number), such that equilibrium is reached, with equilibrium binding constant K. We remind the reader that, as we are modeling autoregulation, there is only one species of protein in this problem: the product of the gene acts as that gene's transcription factor.

The parameter h describes the cooperativity of protein binding. When $g_+ > g_-$, the case $h > 0$ corresponds to activation and the case $h < 0$ corresponds to repression; the special case $h = 0$ reproduces the simple birth–death process. For $|h| \geq 2$, Eq. 108 has two stable fixed points for certain parameter regimes; for $|h| \geq 2$ fixed points must be found numerically.

Although the Hill equation is often used in models of gene regulation, we note that other functional forms are derivable from other biochemical processes. Many of the following results are valid for arbitrary $g(n)$.

3.2. The Master Equation

The full stochastic description corresponding to the deterministic Eq. 108 is given by the birth–death master equation in which the production rate is described by the arbitrary autoregulation function g_n (recall that we replace parentheses with subscript when treating n as discrete):

$$\dot{p}_n = -g_n p_n - n p_n + g_{n-1} p_{n-1} + (n+1) p_{n+1}, \tag{110}$$

We may easily generalize the solution in Subheading 2.2.1 to find the steady state probability distribution,

$$p_n = \frac{p_0}{n!} \prod_{n'=0}^{n-1} g_{n'}, \tag{111}$$

with p_0 set by normalization. Except for special cases of the regulation function we cannot find a closed form solution for the distribution, but the product is easily evaluated.

The results in Subheading 2 for the simple birth–death process can be generalized in the case of autoregulation. We will pay particular attention to the generalization of the eigenfunction expansion in operator notation for arbitrary g_n, when discussing two genes in Subheading 5.

3.3. Bistability and Noise

Autoregulation can affect the statistics of the steady state distribution. Even before specifying the form of $g(n)$, we may obtain a general statistical result in the limit of large protein number by using the linear noise approximation (Subheading 2.3.2). The Fokker–Planck Eq. 91 describes the variance σ^2 of fluctuations around the steady state mean $\bar{n}$. Here we compute the ratio $\sigma^2/\bar{n}$ for autoregulation in order to compare with the Poisson distribution, for which $\sigma^2/\bar{n} = 1$.

For autoregulation the diffusion coefficient becomes $D(\bar{n}) = g(\bar{n}) + \bar{n} = 2\bar{n}$, where we have used the fact that $0 = g(\bar{n}) - \bar{n}$ at steady state. The derivative of the drift term evaluated at the mean is $v'(\bar{n}) = g'(\bar{n}) - 1$, where the prime denotes differentiation with respect to n. Thus Eq. 91 becomes

$$\frac{\sigma^2}{\bar{n}} = \frac{1}{1 - g'(\bar{n})}. \tag{112}$$

Equation 112 diverges when $g'(\bar{n}) = 1$. This corresponds to a bifurcation point for the self-activating gene, where there are two possible solutions to the steady state equation. In this case, the ratio of the variance to the mean is not very informative about the distribution because the distribution is bimodal. The self-repressing gene, in contrast, cannot be bistable, and the ratio of the variance to the mean is always a good description of noise. In Fig. 2 we demonstrate that the exact distribution becomes bimodal when the deterministic equation crosses a bifurcation point.

The ratio in Eq. 112 shows that self activation ($g'(\bar{n})>0$) leads to super-Poissonian noise ($\sigma^2/\bar{n}>1$), and self-repression ($g'(\bar{n})<0$) leads to sub-Poissonian noise ($\sigma^2/\bar{n}<1$); when there is no regulation we have $g'(\bar{n}) = 0$, and we recover the Poisson result, $\sigma^2/\bar{n} = 1$. Figure 3 demonstrates this behavior by computing the exact distribution (Eq. 111) for Hill function autoregulation (Eq. 109) with various values of the cooperativity parameter h.

Equation 112 can be equivalently obtained by computing the power spectrum from the linearized Langevin equation.

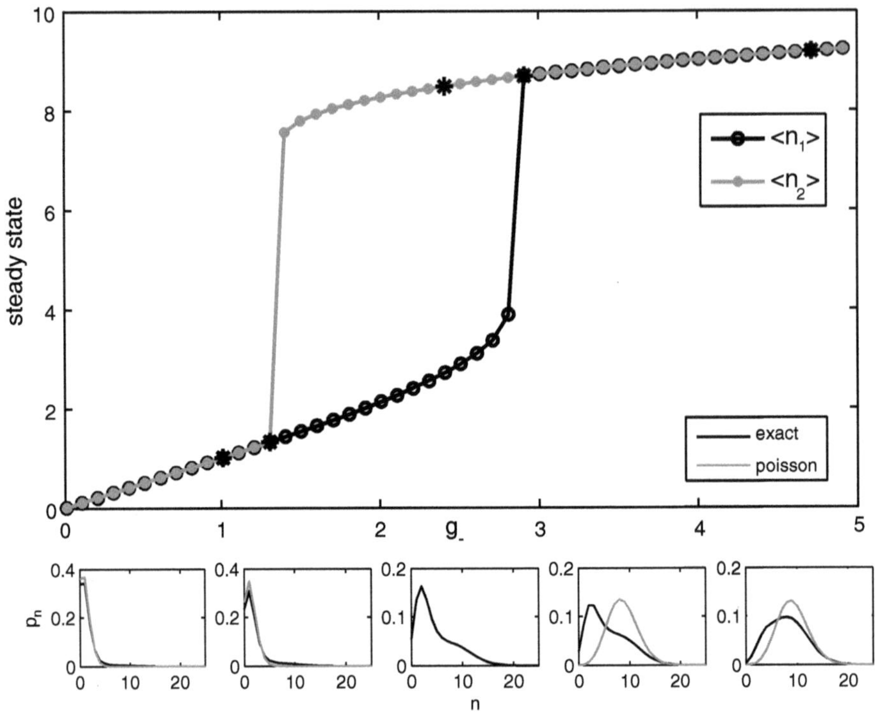

Fig. 2. The steady state solution(s) for a self-activating gene as a function of the basal rate g_-. The figure shows regimes which are monostable (high and low g_-) and bistable (intermediate g_-). The probability distributions below (*solid black lines*), obtained from Eq. 109, correspond consecutively to the points above marked by *stars*. In the monostable regime the distributions are compared to Poisson distributions (*solid gray lines*). The autoregulation function is a Hill function (Eq. 109) with $h = 4$, $g_+ = 10$, and $K = 6$.

Linearizing the Langevin equation and going to Fourier space we obtain

$$-i\omega\delta\tilde{n}(\omega) = g'(\bar{n})\tilde{n}(\omega) - \tilde{n}(\omega) + \tilde{\eta}(\omega). \tag{113}$$

The power spectrum becomes

$$\mathcal{N} = \frac{2\bar{n}}{\omega^2 + [1 - g'(\bar{n})]^2}. \tag{114}$$

The steady state autocorrelation function is therefore

$$\delta n^2 = \frac{\bar{n}}{|1 - g'(\bar{n})|}, \tag{115}$$

and the ratio of the variance to the mean is

$$\frac{\delta n^2}{\bar{n}} = \frac{1}{|1 - g'(\bar{n})|}. \tag{116}$$

We point the reader to the paper of Warren, Tanase-Nicola and ten Wolde (35) for a pedagogical discussion of power spectra in gene expression.

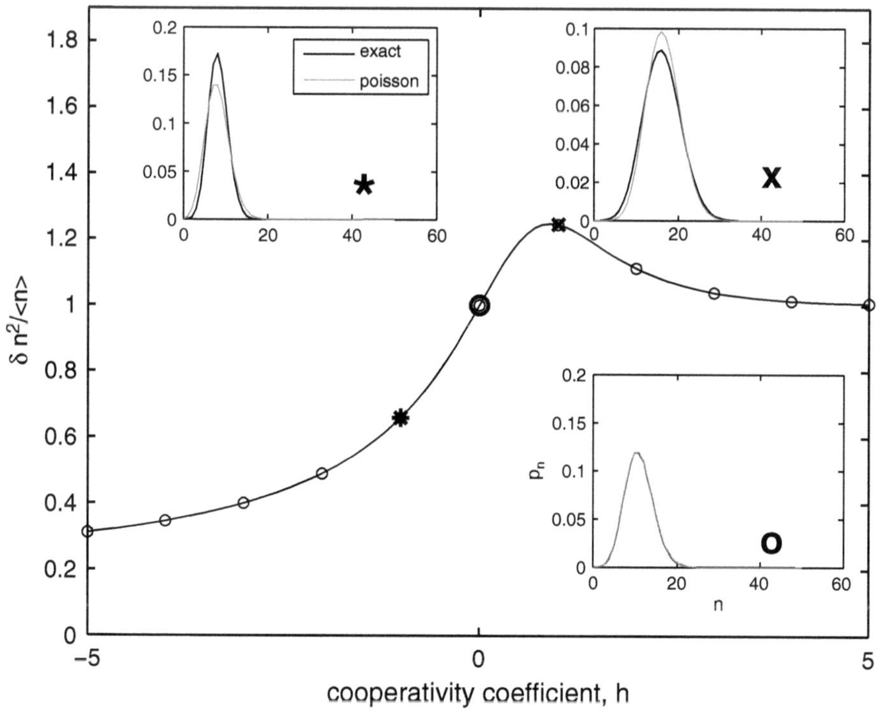

Fig. 3. The ratio of the variance δn^2 to the mean $\langle n\rangle$ as a function of the cooperativity coefficient *h*, for a gene undergoing Hill function (Eq. 109) autoregulation. For $h < 0$ the gene represses its own production; for $h > 0$ the gene activates its own production. The case $h = 0$ recovers the simple birth–death process result $\delta n^2/\langle n\rangle = 1$. Points are plotted for integer values of *h*; the *line* shows Eq. 109 for continuous *h*. *Insets* show probability distributions (*black solid lines*) from the master equation in Eq. 108, compared with Poisson distributions with the same mean (*gray solid lines*), at $h = -1$ (*star*), $h = 0$ (*circle*), and $h = 1$ (*cross*), where symbols correspond to points on the $\delta n^2/\langle n\rangle$ curve. Parameters are $g_- = 2$, $g_+ = 20$, and $K = 4$, which restrict solutions to the monostable regime, yielding unimodal distributions.

4. Bursts

The previous section assumes that the processes of transcriptional and translational regulation can be described by one deterministic regulation function. Recent experiments have observed a feature of gene regulation that is not captured by such a model: protein production often occurs in bursts (3, 45, 46). Bursts can result from the gene transitioning among two or more states with different rates of protein production (modeled in Eq. 117, below), for example when a transcription factor is bound or unbound, or, in higher eukaryotes, when different production states are introduced by chromatin remodeling (45). Bursts can also arise during translation, in which a single mRNA transcript produces many copies of the protein in a short amount of time. Here we present simple models of the first type, which we call transcriptional bursts, and the second type, which we call translational bursts. We then discuss a model which combines bursting with autoregulation. We refer the reader to the literature for discussion of other burst models including those which capture both transcriptional and translational bursts (25, 47).

4.1. Transcriptional Bursts

In this section we consider a gene that can exist in multiple states, each with its own protein production rate. We first specialize to the case of two states, which is commonly used as a model for bursting (3, 34). We solve this case using a generating function. The conventional solution is limited to two states; for an arbitrary number of states we present the spectral solution using operator methods. Finally we present a solution to a simple model with both bursting and autoregulation.

4.1.1. The Master Equation

We begin by considering the general process in which a gene has Z different states. In each state z the protein production is described by a simple birth-death process (Subheading 2) with production rate g_z. Transitions between the states occur at rates given by the transition matrix $\Omega_{zz'}$, such that

$$\sum_{z=1}^{Z} \Omega_{zz'} = 0, \qquad (117)$$

which follows from conservation of probability. The rates are constant here; in Subheading 4.1.4 we extend the model such that the rates depend on protein number.

The system is described by a Z-state probability vector with elements given by p_n^z. The particular states evolve according to the coupled master equation

$$\dot{p}_n^z = -g_z p_n^z - n p_n^z + g_z p_{n-1}^z + (n+1) p_{n+1}^z + \sum_{z'} \Omega_{zz'} p_n^{z'}. \qquad (118)$$

where all but the last term describes the birth–death process for state z, and the last term describes transitions among states. The probabilities of being in state z regardless of the number of proteins are given by $\pi_z = \sum_n p_n^z$; summing the master equation over n in the steady state gives

$$\sum_{z'} \Omega_{zz'} \pi_{z'} = 0, \qquad (119)$$

and normalization requires

$$\sum_{z} \pi_z = 1. \qquad (120)$$

4.1.2. The Two-State Gene

We first consider the case in which the gene has only two states ($Z = 2$), an active state ($z = +$) and an inactive state ($z = -$):

$$p_n^z = \begin{pmatrix} p_n^+ \\ p_n^- \end{pmatrix}. \tag{121}$$

The transition matrix takes the form

$$\Omega_{zz'} = \begin{pmatrix} -\omega_- & \omega_+ \\ \omega_- & -\omega_+ \end{pmatrix}. \tag{122}$$

where ω_+ and ω_- are the transition rates to the active and to the inactive states, respectively.

As in Subheading 2.2.2, we define the generating function $G_\pm(x) = \sum_n p_n^\pm x^n$ and sum the master equation (Eq. 118) over x^n, yielding at steady state

$$0 = -y(\partial_y - g_\pm)G_\pm \pm \omega_+ G_- \mp \omega_- G_+, \tag{123}$$

where $y \equiv x - 1$. Writing $G_\pm(y) = e^{g_\pm y} H_\pm(y)$ yields a simpler equation for $H_\pm(y)$; writing out the + and − cases explicitly,

$$0 = -ye^{y\Delta}\partial_y H_+ + \omega_+ H_- - \omega_- e^{y\Delta} H_+, \tag{124}$$

$$0 = -ye^{-y\Delta}\partial_y H_- + \omega_- H_+ - \omega_+ e^{-y\Delta} H_-, \tag{125}$$

where $\Delta \equiv g_+ - g_-$. Equations 124–125 are a coupled pair of first-order ordinary differential equations; combining them yields a second-order differential equation. Specifically, solving Eq. 124 (125) for $H_-(H_+)$ and substituting it into Eq. 125 (124), one obtains

$$0 = u\partial_u^2 H_\pm + (\beta - u)\partial_u H_\pm - \alpha H_\pm, \tag{126}$$

where $u \equiv \mp y\Delta$, $\alpha \equiv \omega_\mp$, and $\beta \equiv \omega_+ + \omega_- + 1$. Equation 126 is the canonical equation for the confluent hypergeometric function. Only one of its two solutions satisfies $p_n \to 0$ as $n \to \infty$; it is $H_\pm(x) = N_\pm \Phi[\alpha, \beta; u]$, where

$$\Phi[\alpha, \beta; u] = \sum_{\ell=0}^{\infty} \frac{\Gamma(\ell + \alpha)}{\Gamma(\alpha)} \frac{\Gamma(\beta)}{\Gamma(\ell + \beta)} \frac{u^\ell}{\ell!} \tag{127}$$

is the confluent hypergeometric function of the first kind, and the scalars $N_\pm$ are found by normalization: $N_\pm = G_\pm(1) = \sum_n p_n^\pm (1)^n = \pi^\pm = \omega_\pm/(\omega_+ + \omega_-)$ (the last step uses Eqs. 119–120). Thus,

$$G_\pm(x) = \frac{\omega_\pm}{\omega_+ + \omega_-} e^{g_\pm(x-1)} \Phi[\omega_\mp, \omega_+ + \omega_- + 1; \mp(g_+ - g_-) \times (x - 1)]. \tag{128}$$

As before, the probability distribution can be recovered from $G(x) = \sum_\pm G_\pm(x)$ by differentiation (Eq. 21).

As shown in Appendix K, in the limit $g_- = 0$, Eq. 128 reduces to the slightly simpler expression

$$G(x) = \Phi[\omega_+, \omega_+ + \omega_-; g_+(x-1)], \tag{129}$$

for which the probability distribution has the explicit form

$$p_n = \frac{g_+^n}{n!} \frac{\Gamma(n+\omega_+)}{\Gamma(\omega_+)} \frac{\Gamma(\omega_+ + \omega_-)}{\Gamma(n+\omega_+ + \omega_-)} \Phi[\omega_+ + n, \omega_+ + \omega_- + n; -g_+], \tag{130}$$

in agreement with the results of Iyer-Biswas et al. (34) and Raj et al. (3).

4.1.3. The Spectral Solution

It is clear that the method presented in the previous section results in a *Z*th order ordinary differential equation, and therefore its utility is generally limited to $Z = 2$ states. In contrast, here we show that the spectral solution, in which we expand in eigenfunctions of the birth–death operator, can be used for an arbitrary number of states.

We begin as in Subheading 2.2.3 by defining $|G_z\rangle = \sum_n P_n^z |n\rangle$ and writing the master equation in terms of raising and lowering operators,

$$|\dot{G}_z\rangle = -\hat{b}^+ \hat{b}_z^- |G_z\rangle + \sum_{z'} \Omega_{zz'} |G_{z'}\rangle, \tag{131}$$

where now $\hat{b}_z^- = \hat{a}^- - g_z$ is z dependent. We expand the generating function in the z-dependent eigenfunctions of the birth–death process as

$$|G_z\rangle = \sum_j G_j^z |j_z\rangle, \tag{132}$$

where the eigenvalue relation reads

$$\hat{b}^+ \hat{b}_z^- |j_z\rangle = j |j_z\rangle. \tag{133}$$

This gives the following equation for the expansion coefficients:

$$\dot{G}_j^z = -j G_j^z + \sum_{z'} \Omega_{zz'} \sum_{j'} G_{j'}^{z'} \langle j_z | j'_{z'} \rangle. \tag{134}$$

Here the transition matrix $\Omega_{zz'}$ couples the otherwise independent birth–death processes, and

$$\langle j_z | j'_{z'} \rangle = \frac{(-\Delta_{zz'})^{j-j'}}{(j-j')!} \theta(j-j'), \tag{135}$$

where $\Delta_{zz'} = g_z - g_{z'}$, and the convention $\theta(0) = 1$ is used for the Heaviside function. Equation 135 is derived by evaluating the inner product using the x space representations of the eigenfunctions,

$$\langle x|j_z\rangle = (x-1)^j e^{g_z(x-1)}, \tag{136}$$

$$\langle j_z|x\rangle = \frac{e^{-g_z(x-1)}}{(x-1)^{j+1}}, \tag{137}$$

and Cauchy's theorem, as in Appendix E.

The Heaviside function makes Eq. 134 lower-triangular. This is explicitly clear, for example, in the steady state,

$$jG_j^z - \sum_{z'} \Omega_{zz'} G_j^{z'} = \sum_{z' \neq z} \Omega_{zz'} \sum_{j'<j} G_{j'}^{z'} \frac{(-\Delta_{zz'})^{j-j'}}{(j-j')!}, \tag{138}$$

where one observes that the jth component is a function only of components $j'<j$. The lower-triangular structure allows G_j^z to be computed iteratively, which makes the spectral solution numerically efficient. The lower-triangular structure is a consequence of rotating to the eigenbasis of the birth–death operator; this structure was not present in the original master equation. Other spectral decompositions are possible by exploiting other eigenbases; for example by expanding in a single eigenbasis parameterized by a constant (not z dependent) production rate, one obtains an equation that is sub-diagonal in j (29).

The probability distribution is recovered via inverse transform,

$$p_n^z = \sum_j G_j^z \langle n|j_z\rangle, \tag{139}$$

as in Subheading 2.3. The spectral solution is valid for any number of states Z, while direct solution of the differential equation is generally limited to $Z = 2$. As expected, for $Z = 2$ the spectral solution recovers the confluent hypergeometric function solution (Eq. 128), as shown in Appendix K.

Figure 4 demonstrates the spectral solution for $Z = 3$ states when the transition rate among states is (a) equal to, and (b) much less than, the degradation rate. In (a) the bursts give rise to a long tail in the distribution (compared to the Poisson distribution); in (b) the system spends long times in each production state before transitioning, resulting in a trimodal distribution.

4.1.4. Bursting and Autoregulation

Here we consider the simplest model that combines bursting with autoregulation (27). The solution follows that of Subheading 4.1.2. The gene has two states, an active state in which the regulatory protein is bound (p_n^+) and an inactive state in which the regulatory protein is unbound (p_n). The regulation is incorporated by making the rate at which the gene switches to the active state proportional to the protein number:

$$\Omega_{zz'} = \begin{pmatrix} -\omega_- & \omega_+ n \\ \omega_- & -\omega_+ n \end{pmatrix}. \tag{140}$$

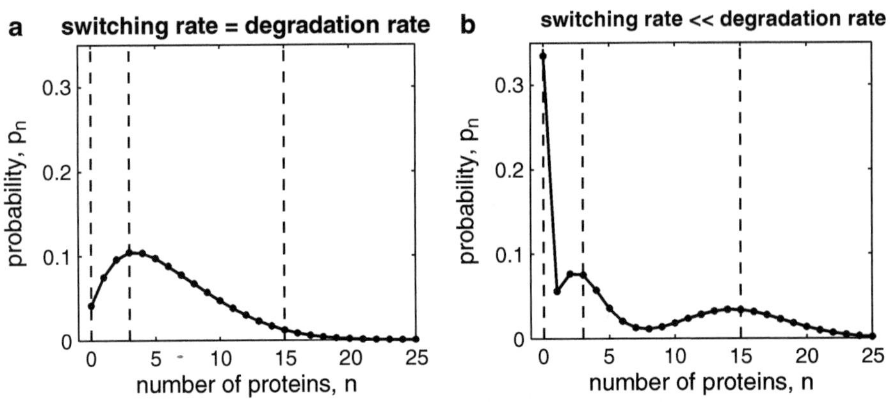

Fig. 4. Steady state probability distribution for transcriptional bursting with $Z = 3$ states, when the transition rate among states is (**a**) equal to the degradation rate ($\omega = 1$) and (**b**) much less than the degradation rate ($\omega = 0.01$). Production rates are $g_z = (0, 3, 15)$ and the rate of transition ω between any two states is equal (i.e., all off-diagonal elements of $\Omega_{zz'}$ are ω and all three diagonal elements are -2ω).

In this case the analog of Eq. 123 for the generating function is

$$0 = -(x-1)(\partial_x - g_\pm)G_\pm \pm \omega_+ x \partial_x G_- \mp \omega_- G_+, \qquad (141)$$

As in Subheading 4.1.2, the + and − equations can be combined to give a second-order differential equation, e.g. for $G_-(x)$,

$$0 = \partial_x^2 G_- + C_1(x)\partial_x G_- + C_0(x) G_-, \qquad (142)$$

with

$$C_1(x) \equiv \frac{g_- + g_+ + \omega_- + \omega_+ + 1 - x[g_+(1+\omega_+) + g_-]}{(1+\omega_+)x - 1}, \qquad (143)$$

$$C_0(x) \equiv \frac{g - g_+ x - g_-(g_+ + \omega_- + 1)}{(1+\omega_+)x - 1}, \qquad (144)$$

whose solution $G_-(x) = N_- e^{g_+(x-1)}\Phi[\alpha, \beta; u]$ is proportional to the confluent hypergeometric function of the first kind (Eq. 126) with

$$\alpha = 1 + \frac{\omega_-}{1+\omega_+}\left(1 + \frac{\omega_+ g_-}{g_- - (1+\omega_+)g_+}\right), \qquad (145)$$

$$\beta = 1 + \frac{\omega_-}{1+\omega_+} + \frac{\omega_+ g_-}{(1+\omega_+)^2}, \qquad (146)$$

$$u = -\frac{[g_+(1+\omega_+) - g_-][(1+\omega_+)x - 1]}{(1+\omega_+)^2}. \qquad (147)$$

Again N_- is set by normalization; $G_+(x)$ is computed from $G_-(x)$ using Eq. 141 (bottom signs).

In Appendix J we consider an extension of this model, in which the transition rate to the active state is proportional to n^h, where h is the cooperativity of the binding. We show that in the limit of fast transitions this model yields an effective autoregulation function of the Hill form (Eq. 109). Full solution of this model using the method presented here for $h = 0$ and $h = 1$ is of limited utility for $h > 1$, because the high-order differential equation is not solvable.

4.2. Translational Bursts

We now turn our attention to translational bursts, in which a single mRNA transcript produces many proteins in a short amount of time. In this case, the model includes just one protein production rate, but each production event results in an instantaneous burst of multiple proteins instead of one. This approach does not explicitly consider mRNAs; instead it models the translational step as an effective burst of proteins.

With production occurring in bursts of (integer) size B, the birth–death master equation becomes

$$\dot{p}_n = -gp_n - np_n + gp_{n-B} + (n+1)p_{n+1}. \tag{148}$$

This equation is easily solved by spectral decomposition. In terms of raising and lowering operators (Subheading 2.2.3), Eq. 148 reads

$$|\dot{G}\rangle = \left[-g - \hat{a}^+\hat{a}^- + g(\hat{a}^+)^B + \hat{a}^-\right]|G\rangle, \tag{149}$$

$$= (-\hat{b}^+\hat{b}^- + \Gamma)|G\rangle, \tag{150}$$

where $\hat{b}^+ = \hat{a}^+ - 1$ and $\hat{b}^- = \hat{a}^- - g$ as before, and $\Gamma \equiv (\hat{a}^+)^B - \hat{a}^+$. Just as in Subheading 2.2.3, Eq. 150 benefits from a spectral expansion in the eigenfunctions $|j\rangle$ of the simple birth–death process, i.e.,

$$|G\rangle = \sum_j G_j|j\rangle, \tag{151}$$

where

$$\hat{b}^+\hat{b}^-|j\rangle = j|j\rangle. \tag{152}$$

Equation 150 then becomes an equation for the expansion coefficients:

$$\dot{G}_j = -jG_j + \sum_{j'} \langle j|\Gamma|j'\rangle G_{j'}. \tag{153}$$

The simplification

$$\langle j|\Gamma|j'\rangle = \langle j|\left[(\hat{b}^+ + 1)^B - (\hat{b}^+ + 1)\right]|j'\rangle, \tag{154}$$

$$= \langle j|\left[\sum_{\ell=0}^{B} \binom{B}{\ell} (\hat{b}^+)^\ell - (\hat{b}^+ + 1)\right]|j'\rangle \tag{155}$$

$$= \langle j|\left[(B-1)\hat{b}^+ + \sum_{\ell=2}^{B} \binom{B}{\ell} (\hat{b}^+)^\ell\right]|j'\rangle \tag{156}$$

$$= (B-1)\delta_{j,j'-1} + \sum_{\ell=2}^{B} \binom{B}{\ell} \delta_{j,j'+\ell}. \tag{157}$$

gives

$$\dot{G}_j = -jG_j + (B-1)G_{j-1} + \sum_{\ell=2}^{B} \binom{B}{\ell} G_{j-\ell}, \tag{158}$$

The equation is lower-triangular, which makes a recursive solution simple. Note that setting $B = 1$ recovers Eq. 72 for the simple birth–death process. The probability distribution is retrieved by inverse transform,

$$p_n = \sum_j G_j \langle n|j \rangle. \tag{159}$$

with $\langle n|j \rangle$ computed as in Appendix H.

Translational bursts increase noise relative to the simple birth–-death process. For example, we can easily compute the variance and mean of the steady state distribution for translational bursts (see Appendix C); their ratio is

$$\frac{\sigma^2}{\bar{n}} = \frac{1+B}{2}, \tag{160}$$

which for $B>1$ is (potentially much) greater than the birth–death result $\sigma^2/\bar{n} = 1$.

We note that within the Langevin approach we can model translational bursts as:

$$\frac{\mathrm{d}n}{\mathrm{d}t} = Bg - n + \eta(t), \tag{161}$$

$$\langle \eta(t)\eta(t') \rangle = \delta(t-t')(Bg + n). \tag{162}$$

We obtain a power spectrum of the form:

$$\mathcal{N} = \frac{B+1}{\omega^2 + 1}, \tag{163}$$

which again is a signature of a more noisy process.

5. Two Genes

At this point we have discussed models of a single gene, including the simple birth–death process, autoregulation, and bursts. Here we extend the discussion to include multiple genes with regulation. We focus on the case of two genes, and we describe how the method may be extended to larger regulation networks. We refer the reader to the literature for more detailed discussion of the application of the spectral method to cascades (28) and to systems with both regulation and bursts (29).

Here we consider two genes, the first with protein number n, and the second with protein number m. The first gene regulates its own expression via arbitrary function g_n, and it regulates the expression of the second gene via arbitrary function q_n. This system can be thought of as the simplest possible regulatory network; it is also the minimum system necessary for completely describing a regulatory cascade of arbitrary length (28). Because the model is general, the two degrees of freedom could correspond to two species of proteins, or to a protein and an mRNA species.

The master equation for this system is

$$\dot{p}_{nm} = -g_n p_{nm} - n p_{nm} + g_{n-1} p_{n-1,m} + (n+1) p_{n+1,m} + \rho[-q_n p_{nm} - m p_{nm} + q_n p_{n,m-1} + (m+1) p_{n,m+1}], \quad (164)$$

where time is rescaled by the degradation rate of the first gene and ρ is the ratio of the degradation rate of the first gene to that of the second.

5.1. Deterministic Model

As always, the kinetic rate equations can be obtained by deriving the mean dynamics from the master equation. Summing the master equation over both indices against either n or m (and performing index shifts as in Appendix B) gives

$$\frac{d\bar{n}}{dt} = g(\bar{n}) - \bar{n}, \quad (165)$$

$$\frac{d\bar{m}}{dt} = q(\bar{n}) - \bar{m}, \quad (166)$$

respectively. Here we have assumed that $\sum_n g_n p_n \approx g(\bar{n})$ and $\sum_n q_n p_n \approx q(\bar{n})$ by taking $n = \bar{n} + \eta$ and keeping the first terms of the Taylor expansions $g(n) = g(\bar{n}) + \eta g'(\bar{n}) + \cdots$ and $q(n) = q(\bar{n}) + \eta q'(\bar{n}) + \cdots$, respectively.

Solving for the steady state of the deterministic system requires finding the roots of two coupled nonlinear equations, which can only be done numerically. Solving the deterministic system is essential for finding the power spectrum and calculating the correlation functions in the Langevin approach. However, as we show in the next section, the full stochastic description can be solved quite efficiently using the spectral method, i.e., by expanding in the eigenfunctions of the simple birth–death process.

5.2. The Spectral Method

Although it is usually not possible to find an exact solution at either the master and Fokker–Planck equation levels, here we show that efficient numerical computation of the full steady state distribution is possible by expanding in the eigenfunctions of the simple birth–death process (Subheading 2.2.3). This approach is general and does not depend on the algebraic forms of the regulation functions.

As in Subheading 2.2.3 we define the generating function $|G\rangle = \sum_{nm} p_{nm}|n, m\rangle$, which makes the master equation

$$|\dot{G}\rangle = -\hat{\mathcal{L}}|G\rangle, \tag{167}$$

where the full linear operator

$$\hat{\mathcal{L}} = \hat{b}_n^+ \hat{b}_n^-(n) + \rho \hat{b}_m^+ \hat{b}_m^-(n) \tag{168}$$

is written in terms of raising and lowering operators defined analogously to Eqs. 59–60:

$$\hat{b}_n^+ = \hat{a}_n^+ - 1, \tag{169}$$

$$\hat{b}_m^+ = \hat{a}_m^+ - 1, \tag{170}$$

$$\hat{b}_n^-(n) = \hat{a}_n^- - \hat{g}(n), \tag{171}$$

$$\hat{b}_m^-(n) = \hat{a}_m^- - \hat{q}(n). \tag{172}$$

The difference here is that both lowering operators are n dependent because of the regulation functions. This suggests that we separate the full operator as $\hat{\mathcal{L}} = \hat{\mathcal{L}}_0 + \hat{\mathcal{L}}_1(n)$, where we have defined a constant part,

$$\hat{\mathcal{L}}_0 = \hat{b}_n^+ \hat{b}_n^- + \rho \hat{b}_m^+ \hat{b}_m^-, \tag{173}$$

with $\hat{b}_n^- \equiv \hat{a}_n^- - \bar{g}$ and $\hat{b}_m^- \equiv \hat{a}_m^- - \bar{q}$ and an n-dependent part

$$\hat{\mathcal{L}}_1(n) = \hat{b}_n^+ \hat{\Gamma}(n) + \rho \hat{b}_m^+ \hat{\Delta}(n), \tag{174}$$

with $\hat{\Gamma}(n) \equiv \bar{g} - \hat{g}(n)$ and $\hat{\Delta}(n) \equiv \bar{q} - \hat{q}(n)$. The operator $\hat{\mathcal{L}}_0$ describes two independent birth–death processes with constant production rates $\bar{g}$ and $\bar{q}$, respectively, and the operator $\hat{\mathcal{L}}_1$ captures how the regulated processes differ from the constant rate processes. The constants $\bar{g}$ and $\bar{q}$ act as gauges: they can be chosen freely, and the final solution is independent of this choice.

We have already solved for the eigenfunctions $|j, k\rangle$ of the constant rate birth–death processes in Subheading 2.2.3. The spectral method exploits this fact by expanding the solution of the full problem in these eigenfunctions:

$$|G\rangle = \sum_{jk} G_{jk}|j, k\rangle. \tag{175}$$

Equation 167 then gives an algebraic relation among the expansion coefficients:

$$\dot{G}_{jk} = -(j + \rho k)G_{jk} - \sum_{j'} \Gamma_{j-1,j'} G_{j'k} - \rho \sum_{j'} \Delta_{jj'} G_{j',k-1}, \tag{176}$$

where

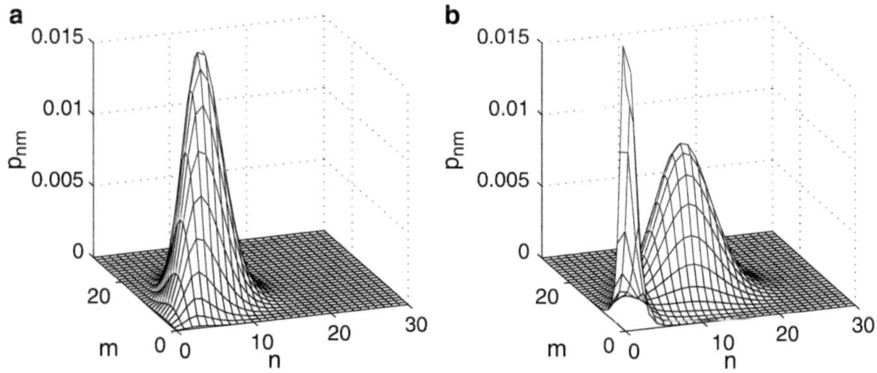

Fig. 5. Demonstration of the spectral method with two genes. Regulation is of Hill form, $q_n = (q_- K^h + q_+ n^h)/(K^h + n^h)$, with $q_- = 1$, $q_+ = 12$, $K = 4$, and $h = 4$. In (**a**) the first gene undergoes a simple birth–death process with $g_n = 7$; in (**b**) the first gene regulates itself with $g_n = q_n$. Degradation rates are equal: $\rho = 1$.

$$\Gamma_{jj'} = \langle j|\hat{\Gamma}(n)|j'\rangle = \sum_n \langle j|n\rangle(\bar{g} - g_n)\langle n|j'\rangle, \tag{177}$$

$$\Delta_{jj'} = \left\langle j|\hat{\Delta}(n)|j'\right\rangle = \sum_n \langle j|n\rangle(\bar{q} - q_n)\langle n|j'\rangle. \tag{178}$$

Equation 176 enjoys the property that it is sub-diagonal in the second gene's eigenvalue k. This feature comes from the fact that both regulation functions are dependent on the first gene's protein number n (and not m). The sub-diagonal structure allows the matrix G_{jk} to be computed iteratively by columns in k, which makes the method numerically efficient. The initial condition

$$G_{j0} = \langle j, k = 0|G\rangle = \sum_{nm} p_{nm}\langle j|n\rangle\langle 0|m\rangle = \sum_n p_n\langle j|n\rangle \tag{179}$$

requires the marginal probability distribution of the first gene p_n, which can be found recursively (Eq. 111) as described in Subheading 3.2 on one-dimensional auto regulation. The joint probability distribution is retrieved by inverse transform:

$$p_{nm} = \sum_{jk} G_{jk}\langle n|j\rangle\langle m|k\rangle. \tag{180}$$

To demonstrate the capability of the spectral method, Fig. 5 plots the steady state joint probability distribution for a particular choice of regulation function, for cases (a) without and (b) with autoregulation. To demonstrate the accuracy of the spectral method, Fig. 6 compares steady state marginal distributions obtained by the spectral method with those obtained both by direct iterative solution of Eq. 164 and by standard Gillespie simulation (48). The spectral method attains excellent agreement with both methods in orders of magnitude less runtime than either other method; full details on the efficiency and accuracy of the spectral method are available in (28).

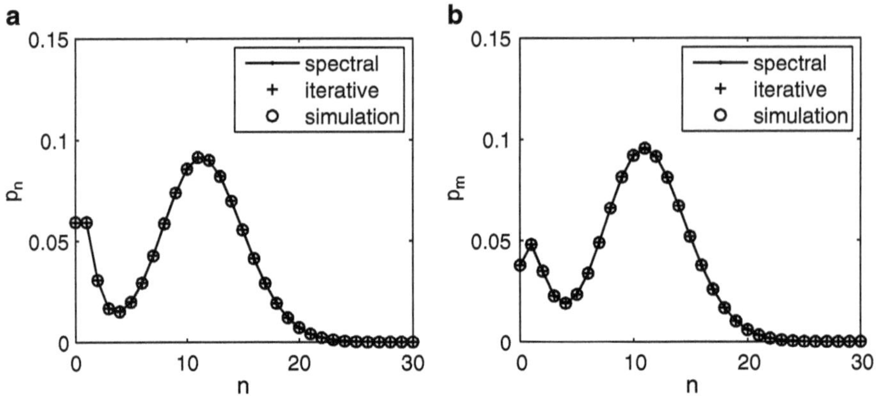

Fig. 6. Marginal distributions for the (**a**) first and (**b**) second gene in a two-gene network, summed from the joint distribution in Fig 5b. Agreement is demonstrated among distributions obtained by the spectral method (*solid*), by direct iterative solution of the master equation (Eq. 16; *plus signs*), and by standard Gillespie simulation (48) (*circles*).

The spectral decomposition presented here constitutes an expansion in only one of many possible eigenbases: that with constant production rates. Other spectral expansions are possible, as discussed in detail in (29); we do not repeat the discussion here.

6. Summary

In this pedagogical chapter, we have attempted to introduce the reader to the phenomenon of intrinsic noise and to the simplest possible mathematical tools for modeling. We have avoided detailed discussion of a vast literature in Monte-Carlo simulation of such phenomena in favor of simple calculations which can be performed analytically or which are useful in numerical solutions of the master equation itself (rather than random generation of sample trajectories).

The discussion also hopefully illustrates how different problem settings motivate different analytic methods, many of which have close parallels in the physics canon. These include approximation of differences with derivatives, yielding a description very similar to that used in Fokker–Planck descriptions; as well as the algebraic relations among the different eigenfunctions of the birth–death process itself, yielding an algebra of "ladder operators" similar to that used in quantum mechanics.

While the literature in simulation of intrinsic noise has grown immensely since the introduction of time-varying Monte-Carlo methods over 30 years ago (48), it is our hope that for model systems these calculations will help point out how the linearity of master equations may be exploited to yield either solutions, numerical methods, or analytical approximations which give more direct insight into the design principles of stochastic regulatory networks.

Appendix A. Deriving the Master Equation

Here we derive the master equation in one dimension from the laws of probability. The probability of having n proteins at a time $t + \tau$ (where τ is small) is equal to the probability of having had n' proteins at time t multiplied by the (possibly time dependent) probability of transitioning from n' to n in time τ, summed over all n':

$$p_n(t+\tau) = \sum_{n'} p_{n'}(t) p_{n|n'}(t,\tau). \tag{181}$$

The transition probability $p_{n|n'}(t,\tau)$ has two contributions: (1) the probability of transitioning from state $n' \neq n$ to state n, equal to a transition rate $w_{nn'}(t)$ times the transition time τ, and (2) the probability of starting and remaining in state $n' = n$, which we call $\pi_n(t)$:

$$p_{n|n'}(t,\tau) = (1 - \delta_{nn'}) w_{nn'}(t)\tau + \delta_{nn'}\pi_n(t). \tag{182}$$

Applying the normalization condition $\sum_n p_{n|n'}(t,\tau) = 1$ to Eq. 182, we obtain

$$\pi_n(t) = 1 - \tau \sum_{n' \neq n} w_{n'n}(t), \tag{183}$$

making Eq. 181

$$p_n(t+\tau) = \sum_{n'} p_{n'}(t) \times \left[(1 - \delta_{nn'}) w_{nn'}(t)\tau + \delta_{nn'} \left(1 - \tau \sum_{n'' \neq n} w_{n''n}(t) \right) \right], \tag{184}$$

$$= \tau \sum_{n'} p_{n'}(t) w_{nn'}(t) - \tau p_n(t) w_{nn} + p_n(t) - \tau p_n(t) \sum_{n'' \neq n} w_{n''n}(t), \tag{185}$$

$$= p_n(t) + \tau \left[\sum_{n'} p_{n'}(t) w_{nn'}(t) - p_n(t) \sum_{n'} w_{n'n}(t) \right]. \tag{186}$$

Taking the limit $\tau \to 0$, we obtain

$$\frac{\mathrm{d}p_n}{\mathrm{d}t} = \sum_{n'} p_{n'} w_{nn'}(t) - p_n \sum_{n'} w_{n'n}(t), \tag{187}$$

as in Eq. 4.

Appendix B. Summation of the Birth–Death Master Equation

Here we sum the one-dimensional birth–death master equation over n against (1) n, to derive the mean dynamics, (2) x^n, to derive the equation of motion in generating function space, and (3) $|n\rangle$, to find the equation of motion in operator notation.

First we sum the master equation (Eq. 9) over n against n to derive the dynamics of the mean $\langle n\rangle = \sum_n np_n$:

$$\partial_t \sum_n np_n = -g\sum_n np_n - \sum_n n^2 p_n + g\sum_n np_{n-1} + \sum_n n(n+1)p_{n+1}. \tag{188}$$

The left-hand side (LHS) is the time derivative of the mean. On the RHS, recalling that n runs from 0 to ∞, we may define for the third term $n' \equiv n-1$ and for the fourth term $n'' \equiv n+1$, giving

$$\partial_t\langle n\rangle = -g\sum_n np_n - \sum_n n^2 p_n + g\sum_{n'=-1}^{\infty}(n'+1)p_{n'} + \sum_{n''=1}^{\infty}(n''-1)n'' p_{n''}. \tag{189}$$

The lower limits on all sums can be set to 0 once more without changing the expression (in the third term we impose $p_{-1} \equiv 0$ since protein number cannot be negative); simplifying gives

$$\partial_t\langle n\rangle = -g\sum_n np_n - \sum_n n^2 p_n + g\sum_{n'} n' p_{n'} + g\sum_{n'} p_{n'} + \sum_{n''} n''^2 p_{n''} - \sum_{n''} n'' p_{n''}, \tag{190}$$

$$= g\sum_{n'} p_{n'} - \sum_{n''} n'' p_{n''} \tag{191}$$

$$= g - \langle n\rangle, \tag{192}$$

as in Eq. 8.

Next we sum the master equation (Eq. 9) over n against x^n to derive the equation of motion of the generating function $G(x,t) = \sum_n p_n(t)x^n$:

$$\partial_t \sum_n p_n x^n = -g\sum_n p_n x^n - \sum_n np_n x^n + g\sum_n p_{n-1}x^n + \sum_n (n+1)p_{n+1}x^n. \tag{193}$$

The LHS and the first term on the RHS reduce directly to functions of G. The third and fourth terms on the RHS benefit from the same index shifts applied to obtain Eq. 189, giving

$$\partial_t G = -gG - \sum_n p_n n x^n + g \sum_{n'} p_{n'} x^{n'+1} + \sum_{n''} n'' p_{n''} x^{n''-1}. \tag{194}$$

We may now eliminate bare appearances of n in the second and fourth terms on the RHS using $nx^n = x\partial_x x^n$ and $nx^{n-1} = \partial_x x^n$ respectively, giving

$$\partial_t G = -gG - \sum_n p_n x \partial_x x^n + g \sum_{n'} p_{n'} x^{n'+1} + \sum_{n''} p_{n''} \partial_x x^{n''}, \tag{195}$$

$$= -gG - x\partial_x G + gxG + \partial_x G, \tag{196}$$

$$= -(x-1)(\partial_x - g)G. \tag{197}$$

as in Eq. 23.

Finally, we sum the master equation (Eq. 9) over n against $|n\rangle$ to derive the equation of motion of the generating function in operator notation $|G\rangle = \sum_n p_n |n\rangle$:

$$\partial_t \sum_n p_n |n\rangle = -g \sum_n p_n |n\rangle - \sum_n n p_n |n\rangle + g \sum_n p_{n-1} |n\rangle + \sum_n (n+1) p_{n+1} |n\rangle. \tag{198}$$

Again we may reduce the LHS and the first term on the RHS directly, and we may apply index shifts to the third and fourth terms on the RHS:

$$\partial_t |G\rangle = -g|G\rangle - \sum_n n p_n |n\rangle + g \sum_{n'} p_{n'} |n'+1\rangle + \sum_{n''} n'' p_{n''} |n''-1\rangle. \tag{199}$$

Now defining operators $\hat{a}^+$ and $\hat{a}^-$ that raise and lower the protein number by 1, respectively, i.e.,

$$\hat{a}^+ |n\rangle \equiv |n+1\rangle, \tag{200}$$

$$\hat{a}^- |n\rangle \equiv n|n-1\rangle, \tag{201}$$

Equation 198 can be written

$$\partial_t |G\rangle = g|G\rangle - \sum_n p_n \hat{a}^+ \hat{a}^- |n\rangle + g \sum_{n'} p_{n'} \hat{a}^+ |n'\rangle + \sum_{n''} p_{n''} \hat{a}^- |n''\rangle, \tag{202}$$

$$= -g|G\rangle - \hat{a}^+ \hat{a}^- |G\rangle + g\hat{a}^+ |G\rangle + \hat{a}^- |G\rangle, \tag{203}$$

$$= -(\hat{a}^+ - 1)(\hat{a}^- - g)|G\rangle \tag{204}$$

as in Eq. 50.

Appendix C. Statistics of the Birth–Death Distribution

In this appendix we calculate the means of a birth–death distribution. In the simplest case, starting directly from the solution in Eq. 16 we obtain:

$$\langle n\rangle = \sum_n np_n = \sum_n n\frac{g^n}{n!}\mathrm{e}^{-g} = \mathrm{e}^{-g}g\partial_g\sum_n\frac{g^n}{n!} = g. \tag{205}$$

Similarly we can calculate

$$\langle n(n-1)\rangle = \sum_n n(n-1)\frac{g^n}{n!}\mathrm{e}^{-g} = \mathrm{e}^{-g}g^2\partial_g^2\sum_n\frac{g^n}{n!} = g^2, \tag{206}$$

which gives us the variance

$$\langle n^2\rangle - \langle n\rangle^2 = g^2 + g - g^2 = g. \tag{207}$$

These results can also be obtained in the operator formalism, and the fact that for steady state $|G\rangle = |j=0\rangle$. For example:

$$\begin{aligned}\langle n\rangle &= \langle x|\hat{a}^+\hat{a}^-|j=0\rangle\big|_{x=1} = \langle x|\hat{a}^+\hat{a}^-|G\rangle\big|_{x=1}\\ &= \langle x|\hat{a}^+\hat{a}^-\sum_n p_n|n\rangle|_{x=1} = \sum_n np_nx^n|_{x=1} = \sum_n np_n.\end{aligned} \tag{208}$$

To use the number operator on the $|j=0\rangle$ state we need to rewrite it in terms of $\hat{a}^+ = \hat{b}^+ + 1$ and $\hat{a}^- = \hat{b}^- + g$. Acting to the left we obtain:

$$\begin{aligned}\langle n\rangle &= \langle x|(\hat{b}^+\hat{b}^- + g + \hat{b}^- + g\hat{b}^+)|0\rangle_{x=1}\\ &= \langle x|g|0\rangle_{x=1} + \langle x|g|1\rangle_{x=1} = g.\end{aligned} \tag{209}$$

We can also calculate the mean of the bursty process. Multiplying Eq. 148 by n and summing over n, we use,

$$\sum_{n=0}^{\infty} np_{n-B} = \sum_{n=0}^{\infty}(n+B)p_n = \langle n\rangle + B, \tag{210}$$

to obtain:

$$g(\langle n\rangle + B) - g\langle n\rangle + \langle n^2\rangle - \langle n\rangle - \langle n^2\rangle = 0, \tag{211}$$

which is solved by:

$$\langle n\rangle = gB. \tag{212}$$

The second moment $\langle n^2\rangle = \sum n^2p_n$ is calculated the same way:

$$\langle n^2\rangle = \frac{1}{2}\left(2g\langle n\rangle B + \langle n\rangle + gB^2\right) = (gB)^2 + \frac{gB^2 + gB}{2}, \tag{213}$$

which yields the variance as

$$\delta n^2 = \langle n^2\rangle - \langle n\rangle^2 = \frac{gB}{2}(1+B). \tag{214}$$

Appendix D. Asymptotic Distributions for Large Protein Number

Here we show that in the limit of large protein number n, the steady state of the birth–death process (the Poisson distribution, Eq. 16) approaches a Gaussian distribution with mean and variance equal to g, the ratio of production to degradation rates. We then show that in the same limit, the steady state solution to the birth–death Fokker–Planck equation (Eq. 84) also approaches a Gaussian distribution with mean and variance equal to g.

The large-n limit of the Poisson distribution is most conveniently evaluated by first taking the log, which allows one to make use of Stirling's approximation, $\log n! \approx n \log n - n + \log \sqrt{2\pi n}$ for large n:

$$\log p(n) \approx -g + n \log g - n \log n + n - \log \sqrt{2\pi n}. \tag{215}$$

The derivative

$$\partial_n \log p(n) = \log \frac{g}{n} + \mathcal{O}(1/n) \tag{216}$$

vanishes at the maximum

$$n = g \tag{217}$$

about which we Taylor expand to second order:

$$\log p(n) \approx \log p(g) + \frac{1}{2}(n-g)^2 \partial_n^2 [\log p(n)]_g, \tag{218}$$

$$= -\log \sqrt{2\pi g} - \frac{(n-g)^2}{2g}. \tag{219}$$

The result is a Gaussian distribution with mean and variance equal to g:

$$p(n) = \frac{1}{\sqrt{2\pi g}} e^{-(n-g)^2/(2g)}. \tag{220}$$

The large-n limit of the Fokker–Planck distribution is similarly evaluated. the log and its derivatives are

$$\log p(n) = \log \frac{1}{Z_g} + (4g-1) \log\left(1 + \frac{n}{g}\right), \tag{221}$$

$$\partial_n \log p(n) = \frac{4g-1}{n+g} \quad 2, \tag{222}$$

$$\partial_n^2 \log p(n) = \frac{1-4g}{(n+g)^2}. \tag{223}$$

The first derivative vanishes at $n = g - 1/2 \approx g$ for large g, at which the second derivative evaluates to $\partial_n^2[\log p(n)]_g = -1/g + 1/(4g^2) \approx -1/g$. Taylor expanding to second order and exponentiating then give

$$p(n) = \frac{1}{\sqrt{2\pi g}} \mathrm{e}^{-(n-g)^2/(2g)}, \tag{224}$$

where $\mathcal{Z}$ is eliminated by normalization.

Appendix E. Orthonormality and the Inner Product

Here we show that interpreting the generating function as a Fourier transform motivates a particular choice of inner product and conjugate state in the protein number basis.

We have defined the generating function in terms of a continuous variable x as $G(x) = \sum_n p_n x^n$. We could equivalently write $x \equiv \mathrm{e}^{i\varphi}$ and define

$$G(\varphi) = \sum_n p_n \mathrm{e}^{in\varphi}, \tag{225}$$

which makes clear that the generating function is simply the Fourier transform of the probability distribution in protein number n. In the state notation commonly used in quantum mechanics (42), Eq. 225 is written $\langle\varphi|G\rangle = \sum_n p_n\langle\varphi|n\rangle$, where $\langle\varphi|G\rangle \equiv G(\varphi)$ and

$$\langle\varphi|n\rangle = \mathrm{e}^{in\varphi}. \tag{226}$$

Equation 226 is the representation of $|n\rangle$ in φ space. In order to compute projections of $|n\rangle$ onto other states or itself using this representation, we must define a conjugate state $\langle n|\varphi\rangle$ and a consistent inner product. With complex exponentials, it is common to make the conjugate state the complex conjugate,

$$\langle n|\varphi\rangle \equiv \langle\varphi|n\rangle^* = \mathrm{e}^{-in\varphi}, \tag{227}$$

and the inner product the integral

$$\langle f|h\rangle \equiv \int_0^{2\pi} \frac{\mathrm{d}\varphi}{2\pi} \langle f|\varphi\rangle\langle\varphi|h\rangle, \tag{228}$$

for any f and h. Given the definition of conjugate state, this choice of inner product ensures the orthonormality of the $|n\rangle$ states:

$$\langle n|n'\rangle = \int_0^{2\pi} \frac{\mathrm{d}\varphi}{2\pi} \langle n|\varphi\rangle\langle\varphi|n'\rangle = \int_0^{2\pi} \frac{\mathrm{d}\varphi}{2\pi} \mathrm{e}^{i(n'-n)\varphi} = \delta_{nn'}. \tag{229}$$

We may now reinterpret these definitions in terms of our original variable x. Since $x = \mathrm{e}^{i\varphi}$ we have $\mathrm{d}x = i\mathrm{e}^{i\varphi}\mathrm{d}\varphi = ix\mathrm{d}\varphi$, and in the orthonormality condition (Eq. 229) the integration from 0 to 2π along the real φ line segment becomes contour integration along the unit circle in the complex x plane:

$$\begin{aligned}\langle n|n'\rangle &= \int_0^{2\pi} \frac{\mathrm{d}\varphi}{2\pi} \mathrm{e}^{-in\varphi}\mathrm{e}^{in'\varphi} = \oint \frac{\mathrm{d}x}{2\pi i x} x^{-n} x^{n'} \\ &= \oint \frac{\mathrm{d}x}{2\pi i} \frac{1}{x^{n+1}} x^{n'} = \delta_{nn'}. \end{aligned} \tag{230}$$

Since we have defined

$$\langle x|n\rangle = x^n, \tag{231}$$

Equation 230 suggests the definition of conjugate state

$$\langle n|x\rangle = \frac{1}{x^{n+1}} \tag{232}$$

and inner product

$$\langle f|h\rangle = \oint \frac{\mathrm{d}x}{2\pi i} \langle f|x\rangle\langle x|h\rangle \tag{233}$$

in x space.

With these definitions, we will find Cauchy's theorem,

$$\oint \frac{\mathrm{d}x}{2\pi i} \frac{f(x)}{(x-a)^{n+1}} = \frac{1}{n!} \partial_x^n [f(x)]_{x=a} \theta(n), \tag{234}$$

(with the convention $\theta(0) = 1$ for the Heaviside function) very useful in evaluating projections. For example, we may immediately use it to confirm the orthonormality condition,

$$\langle n|n'\rangle = \oint \frac{\mathrm{d}x}{2\pi i} \frac{1}{x^{n+1}} x^{n'} = \frac{1}{n!} \partial_x^n [x^{n'}]_{x=0} \theta(n) = \delta_{nn'}, \tag{235}$$

and in Appendix H we use it to compute analytic expressions for the projections between protein number states $|n\rangle$ and birth–death eigenstates $|j\rangle$.

Appendix F. Properties of the Raising and Lowering Operators

Just as in the operator treatment of the quantum harmonic oscillator (42), in this paper we have defined operators $\hat{a}^+$ and $\hat{a}^-$ that act on $|n\rangle$ states by raising and lowering the protein number n by, 1 respectively, i.e.,

$$\hat{a}^+|n\rangle = |n+1\rangle, \tag{236}$$

$$\hat{a}^-|n\rangle = n|n-1\rangle. \tag{237}$$

(note, however, that the prefactors here, 1 and n, are different than those conventionally used for the harmonic oscillator, $\sqrt{n+1}$ and $\sqrt{n}$, respectively). In this appendix, we derive the actions of these operators when acting to the left, as well as their commutation relation.

The actions of $\hat{a}^+$ and $\hat{a}^-$ to the left can be found by projecting onto Eqs. 236 and 237 the conjugate state $\langle n'|$:

$$\langle n'|\hat{a}^+|n\rangle = \langle n'|n+1\rangle = \delta_{n',n+1} = \delta_{n'-1,n} = \langle n'-1|n\rangle, \quad (238)$$

$$\begin{aligned}\langle n'|\hat{a}^-|n\rangle &= \langle n'|n|n-1\rangle = n\delta_{n',n-1} = (n'+1)\delta_{n'+1,n} \\ &= (n'+1)\langle n'+1|n\rangle.\end{aligned} \quad (239)$$

The first and last expressions in each case imply

$$\langle n'|\hat{a}^+ = \langle n'-1|, \quad (240)$$

$$\langle n'|\hat{a}^- = (n'+1)\langle n'+1|, \quad (241)$$

as in Eqs. 56–57.

The commutation relation between $\hat{a}^-$ and $\hat{a}^+$ is defined as $[\hat{a}^-, \hat{a}^+] \equiv \hat{a}^-\hat{a}^+ - \hat{a}^+\hat{a}^-$. It is evaluated by considering its action on a state $|n\rangle$:

$$\begin{aligned}[\hat{a}^-, \hat{a}^+]|n\rangle &= \hat{a}^-\hat{a}^+|n\rangle - \hat{a}^+\hat{a}^-|n\rangle \\ &= \hat{a}^-|n+1\rangle - \hat{a}^+ n|n-1\rangle \\ &= (n+1)|n\rangle - n|n\rangle = |n\rangle.\end{aligned} \quad (242)$$

The first and last expressions imply

$$[\hat{a}^-, \hat{a}^+] = 1, \quad (243)$$

just as with the quantum harmonic oscillator.

Appendix G. Eigenvalues and Eigenfunctions in the Operator Representation

Here we derive the eigenfunctions $|\lambda_j\rangle$ (also called eigenstates in the operator representation) and eigenvalues λ_j of the operator $\hat{\mathcal{L}} = \hat{b}^+\hat{b}^-$, for which

$$\hat{\mathcal{L}}|\lambda_j\rangle = \lambda_j|\lambda_j\rangle, \quad (244)$$

as well as the actions of the individual operators $\hat{b}^+$ and $\hat{b}^-$ on the eigenstates. We will find that the commutation relation

$$\left[\hat{b}^-, \hat{b}^+\right] = 1 \quad (245)$$

and the existence of the steady state solution $\langle x|G\rangle = e^{q(x-1)}$, for which

$$\hat{\mathcal{L}}|G\rangle = 0, \quad (246)$$

are all that are necessary to completely define the eigenvalues and eigenstates of $\hat{\mathcal{L}}$. The treatment here finds many parallels with the derivation of the eigenvalue spectrum of the quantum harmonic oscillator (42).

First it is useful to compute the commutation relation of $\hat{\mathcal{L}}$ with each of its components $\hat{b}^+$ and $\hat{b}^-$:

$$\left[\hat{b}^+, \hat{\mathcal{L}}\right] = \left[\hat{b}^+, \hat{b}^+\hat{b}^-\right] = \hat{b}^+\left[\hat{b}^+, \hat{b}^-\right] + \left[\hat{b}^+, \hat{b}^+\right]\hat{b}^- = -\hat{b}^+, \quad (247)$$

$$\left[\hat{b}^-, \hat{\mathcal{L}}\right] = \left[\hat{b}^-, \hat{b}^+\hat{b}^-\right] = \hat{b}^+\left[\hat{b}^-, \hat{b}^-\right] + \left[\hat{b}^-, \hat{b}^+\right]\hat{b}^- = -\hat{b}^-. \quad (248)$$

(here we have used Eq. 245 and the properties $[f, gh] = g[f, h] + [f, g]h$ and $[f, f] = 0$ for any f, g, and h). We now consider a particular eigenvalue λ and evaluate the action of $\hat{\mathcal{L}}\hat{b}^+$ on $|\lambda\rangle$:

$$\begin{aligned}\hat{\mathcal{L}}\hat{b}^+|\lambda\rangle &= \hat{b}^+\hat{\mathcal{L}}|\lambda\rangle - \left[\hat{b}^+, \hat{\mathcal{L}}\right]|\lambda\rangle = \hat{b}^+\lambda|\lambda\rangle + \hat{b}^+|\lambda\rangle \\ &= (\lambda + 1)\hat{b}^+|\lambda\rangle. \end{aligned} \quad (249)$$

The equality of the first and last expressions,

$$\hat{\mathcal{L}}\left(\hat{b}^+|\lambda\rangle\right) = (\lambda + 1)\left(\hat{b}^+|\lambda\rangle\right) \quad (250)$$

reveals two results: (1) the existence of a state with eigenvalue λ implies the existence of a state with eigenvalue $\lambda + 1$, and (2) the state with eigenvalue $\lambda + 1$ is proportional to $\hat{b}^+|\lambda\rangle$. By induction, the first result means that the eigenvalues are spaced by 1; moreover, the existence of a state with eigenvalue zero (the steady state solution; Eq. 246) anchors the eigenvalues to the integers, i.e.,

$$\lambda_j = j \quad (251)$$

for integer j. The second result can now be written explicitly

$$\hat{b}^+|j\rangle = |j + 1\rangle, \quad (252)$$

where we are free by normalization to set the proportionality constant to 1. Equation 252 demonstrates that $\hat{b}^+$ is a raising operator for the eigenstates $|j\rangle$.

Evaluating the action of $\hat{\mathcal{L}}\hat{b}^-$ on eigenstate $|j\rangle$ similarly reveals that $\hat{b}^-|j\rangle$ is proportional to $|j - 1\rangle$ (cf. Eq. 249); consistency with the eigenvalue equation (Eq. 244) requires that the proportionality constant be j, giving

$$\hat{b}^-|j\rangle = j|j - 1\rangle. \quad (253)$$

Equation 253 demonstrates that $\hat{b}^-$ is a lowering operator for the eigenstates $|j\rangle$. The operators $\hat{b}^+$ and $\hat{b}^-$ raise and lower $|j\rangle$ as $\hat{a}^+$ and $\hat{a}^-$ do $|n\rangle$; therefore they act to the left as (cf. Eqs. 238–239)

$$\langle j|\hat{b}^+ = \langle j - 1|, \quad (254)$$

$$\langle j|\hat{b}^- = (j+1)\langle j+1|. \tag{255}$$

Equation 253 imposes a floor on the eigenvalue spectrum, since $\hat{b}^-|0\rangle = 0$, and no lower eigenstates can be generated. Therefore the eigenvalues are limited to nonnegative integers:

$$j \in \{0, 1, 2, 3, \ldots\}. \tag{256}$$

Equation 252 describes how any state can be obtained from the $j = 0$ state:

$$|j\rangle = (\hat{b}^+)^j|0\rangle. \tag{257}$$

Recalling that $\hat{b}^+ = \hat{a}^+ - 1$ and that $|0\rangle$ is the steady state solution, Eq. 257 can be used to derive the representation of the eigenfunctions in x space, $\langle x|j\rangle$. Projecting $\langle x|$ onto Eq. 257 and recalling that $\hat{a}^+$ corresponds to x (Eq. 54) and that $\langle x|0\rangle = e^{g(x-1)}$ (Eq. 24), we obtain

$$\langle x|j\rangle = (x-1)^j e^{g(x-1)}. \tag{258}$$

We may now define the conjugate state $\langle j|x\rangle$ such that the $|j\rangle$ are orthonormal under the inner product defined in Eq. 233:

$$\begin{aligned} \delta_{jj'} = \langle j|j'\rangle &= \oint \frac{dx}{2\pi i}\langle j|x\rangle\langle x|j'\rangle \\ &= \oint \frac{dx}{2\pi i}\langle j|x\rangle(x-1)^{j'}e^{g(x-1)} = \oint \frac{dy}{2\pi i}y^{j'}f_{j(y)}, \end{aligned} \tag{259}$$

where $y \equiv x - 1$ and $f_j(x-1) \equiv e^{g(x-1)}\langle j|x\rangle$. Equation 259 is equivalent to Eq. 230, from which we identify $f_j(y) = 1/y^{j+1}$ and thus

$$\langle j|x\rangle = \frac{e^{-g(x-1)}}{(x-1)^{j+1}}. \tag{260}$$

We have shown that the operator $\hat{\mathcal{L}} = \hat{b}^+\hat{b}^-$ has nonnegative integer eigenvalues j and eigenfunctions (in x space) given by Eqs. 258 and 260.

Appendix H. Overlaps Between Protein Number States and Eigenstates

Here we describe two methods for computing the overlaps $\langle n|j\rangle$ and $\langle j|n\rangle$ between the protein number states $|n\rangle$ and the eigenstates $|j\rangle$: by contour integration and by recursive updating.

The first method evaluates the overlaps using the inner product defined in Eq. 233 and Cauchy's theorem (Eq. 234). Recalling the representations in x space of the protein number states and eigenstates (Eqs. 46, 48, and 65–66), the first overlap becomes

$$\langle n|j\rangle = \oint \frac{dx}{2\pi i}\langle n|x\rangle\langle x|j\rangle = \oint \frac{dx}{2\pi i}\frac{e^{g(x-1)}(x-1)^j}{x^{n+1}}$$
$$= \frac{1}{n!}\partial_x^n[e^{g(x-1)}(x-1)^j]_{x=0}. \quad (261)$$

Repeated derivatives of a product follow a binomial expansion, i.e.,

$$\langle n|j\rangle = \frac{1}{n!}\sum_{\ell=0}^{n}\frac{n!}{\ell!(n-\ell)!}\partial_x^{n-\ell}[e^{g(x-1)}]_{x=0}\partial_x^{\ell}[(x-1)^j]_{x=0}, \quad (262)$$

$$= \sum_{\ell=0}^{n}\frac{n!}{\ell!(n-\ell)!}[g^{n-\ell}e^{-g}]\left[\frac{j!}{(j-\ell)}(-1)^{j-\ell}\theta(j-\ell)\right], \quad (263)$$

$$= (-1)^j e^{-g} g^n j!\xi_{nj}, \quad (264)$$

where we define

$$\xi_{nj} \equiv \sum_{\ell-0}^{\min(n,j)}\frac{1}{\ell!(n-\ell)!(j-\ell)!(-g)^{\ell}}. \quad (265)$$

Following similar steps, the conjugate overlap evaluates to

$$\langle j|n\rangle = n!(-g)^j\xi_{nj}, \quad (266)$$

with ξ_{nj} as in Eq. 265. For the special case $j = 0$, Eqs. 264 and 266 reduce to Eqs. 77–78.

It is more computationally efficient to compute the overlaps recursively using rules that can be derived from the raising and lowering operations (Eqs. 52–53, 56–57, and 67–70). For example, using the raising operators,

$$\langle n|j+1\rangle = \left\langle n|\hat{b}^+|j\right\rangle = \langle n|(\hat{a}^+ - 1)|j\rangle$$
$$= \langle n-1|j\rangle - \langle n|j\rangle, \quad (267)$$

which can be initialized using $\langle n|0\rangle = e^{-g}g^n/n!$ (Eq. 264) and updated recursively in j. Equation 267 makes clear that in n space the $(j+1)$th mode is simply the (negative of the) discrete derivative of the jth mode. The lowering operators give an alternative update rule,

$$(n+1)\langle n+1|j\rangle = \langle n|\hat{a}^-|j\rangle = \left\langle n|(\hat{b}^- + g)|j\right\rangle$$
$$= j\langle n|j-1\rangle + g\langle n|j\rangle, \quad (268)$$

which can be initialized using $\langle 0|j\rangle = (-1)^j e^{-g}$ (Eq. 264) and updated recursively in n. One may similarly derive recursion relations for $\langle j|n\rangle$, i.e.,

$$\langle j|n+1\rangle = \langle j-1|n\rangle + \langle j|n\rangle, \quad (269)$$
$$(j+1)\langle j+1|n\rangle = n\langle j|n-1\rangle - g\langle j|n\rangle, \quad (270)$$

initialized with $\langle j|0\rangle = (-g)^j/j!$ or $\langle 0|n\rangle = 1$, respectively (Eq. 266) and updated recursively in n or j respectively. Two-term recursion relations can be similarly derived from the full operator $\hat{b}^+\hat{b}^-$ (29).

Appendix I. The Equivalence of the Fokker–Planck and Langevin Descriptions

In this appendix, we start from the Langevin equation defined in Eqs. 93 and 94 and derive the Fokker–Planck equation in Eq. 81. As described in Subheading 4, the observed trajectory is one realization $r_\eta(t)$ of the random stochastic process with Gaussian noise as presented in Eq. 92:

$$p_n(t) = \int \mathcal{D}\eta\delta(n - r_\eta(t))P[\eta] \equiv \langle\delta(n - r_\eta(t))\rangle. \tag{271}$$

Consider the evolution of the probability distribution:

$$p_n(t + \Delta t) - p_n(t) = \langle\delta(n - r_\eta(t + \Delta t)) - \delta(n - r_\eta(t))\rangle, \tag{272}$$

and expand the increment in the trajectory as $r(t + \Delta t) = r(t) + \Delta r(t)$. We can now Taylor expand the difference in delta functions:

$$p_n(t + \Delta t) - p_n(t) = \left\langle(-\Delta r(t))\delta'(n - r_\eta(t)) + \frac{1}{2}(-\Delta r(t))^2\delta''(n - r_\eta(t))\right\rangle, \tag{273}$$

$$= -\partial_n\langle\Delta r(t)\delta(n - r_\eta(t))\rangle + \frac{1}{2}\partial_n^2\left\langle(\Delta r(t))^2\delta(n - r_\eta(t))\right\rangle, \tag{274}$$

where the primes denote derivatives in n. Using the Langevin equation to calculate the increments in the trajectories:

$$\Delta r(t) = v[r(t)]\Delta t + \eta(t)\Delta t, \tag{275}$$

we obtain:

$$\langle\Delta r(t)\delta(n - r_\eta(t))\rangle = \langle v[r(t)]\Delta t\delta(n - r_\eta(t))\rangle + \langle\eta(t)\Delta t\delta(n - r_\eta(t))\rangle, \tag{276}$$

$$= v[r(t)]\Delta t\langle\delta(n - r_\eta(t))\rangle = v(v)\Delta t p_n(t). \tag{277}$$

$$\begin{aligned}\left\langle(\Delta r(t))^2\delta(n - r_\eta(t))\right\rangle = (\Delta t)^2&\left\langle(v[r(t)])^2\delta(n - r_\eta(t))\right\rangle \\ &+ 2\langle v[r(t)]\eta(t)\delta(n - r_\eta(t))\rangle \\ &+ \langle\eta^2(t)\delta(n - r_\eta(t))\rangle,\end{aligned} \tag{278}$$

$$= (\Delta t)^2 (v(n))^2 p_n(t) + \langle \eta^2(t) \rangle p_n(t), \tag{279}$$

where we have used (assuming a discretized process):

$$\langle \eta(t) \delta(n - r_\eta(t)) \rangle = \langle \eta(t) \rangle \langle \delta(n - r_\eta(t)) \rangle = 0. \tag{280}$$

We identify

$$D(n) = \frac{\Delta t}{2} \langle \eta^2(t) \rangle. \tag{281}$$

Putting together all elements and keeping only leading terms in Δt:

$$\frac{p_n(t + \Delta t) - p_n(t)}{\Delta t} = -\partial_n [u(n) \Delta t p_n(t)] + \partial_n^2 [D(n) p_n(t)]. \tag{282}$$

Taking the limit $\Delta t \to 0$ we recover the Fokker–Planck equation:

$$\partial_t p_n(t) = -\partial_n [u(n) \Delta t p_n(t)] + \partial_n^2 [D(n) p_n(t)]. \tag{283}$$

Appendix J. Derivation of the Hill Function

The Hill function can be derived as an effective production rate for an autoregulating gene with two production states (Subheading 4.1.4). The gene is found either in an inactive state (−), in which the production rate is a constant g_-, or in an active state (+), in which the production rate $w_+ n^h$ depends on the number of proteins n, which incorporates the autoregulation; h describes the cooperativity, with $h>0$ corresponding to activation and $h<0$ corresponding to repression. The probability $p_n^\pm$ of the gene being in a given state + or − and there being n proteins evolves in time according to the master equation in Eq. 118 with $\Omega_{zz'}$ as in Eq. 140; in the steady state:

$$0 = -\mathcal{L}^\pm p_n^\pm \pm \omega_+ n^h p_n^- \mp \omega - p_n^+, \tag{284}$$

where $\mathcal{L}^\pm$ describes simple birth–death terms with constant production rates $g_\pm$ in each of the two states. We define moments of the master equation as

$$\pi^\pm \equiv \sum_n p_n^\pm, \tag{285}$$

$$\pi^+ \mu_\ell^\pm = \sum_n p_n^\pm n^\ell \quad \text{for } \ell \geq 1. \tag{286}$$

with

$$\pi^+ + \pi^- = 1 \tag{287}$$

by normalization.

Summing Eq. 284 (top signs) over n (and recalling that the birth–death terms sum to zero) gives

$$0 = \omega_+ \sum_n n^h p_n^- - \omega - \sum_n p_n^+, \tag{288}$$

$$= \omega_+ \pi^- \mu_h^- - \omega_- \pi^+ \tag{289}$$

which, with Eq. 287, becomes an expression for π^+, the probability of being in the active state:

$$\pi^+ = \frac{\mu_h^-}{\mu_h^- + \omega_+/\omega_-}. \tag{290}$$

We solve for μ_h^- using two approximations. The first is that higher moments can be decoupled, i.e.,

$$\mu_{h+1}^- \approx \mu_h^- \mu_1^- \quad \text{for } h \geq 1, \tag{291}$$

which implies that

$$\mu_h^- \approx (\mu_1^-)^h. \tag{292}$$

This approximation allows one to simplify the mean equation in the + state (obtained by summing the Eq. 284, top signs, against n over n):

$$0 = \pi^+ g_+ - \sum_n n p_n^+ + \omega_+ \sum_n n^{h+1} p_n^- - \omega - \sum_n n p_n^+, \tag{293}$$

$$= \pi^+ g_+ - \pi^+ \mu_1^+ + \omega + \pi^- \mu_{h+1}^- - \omega_- \pi^+ \mu_1^+, \tag{294}$$

$$\approx \pi^+ g_+ - \pi^+ \mu_1^+ + \omega_+ \pi^- \mu_h^- \mu_1^- - \omega_- \pi^+ \mu_1^+, \tag{295}$$

$$= \pi^+ g_+ - \pi^+ \mu_1^+ + \omega_- \pi^+ (\mu_1^- - \mu_1^+), \tag{296}$$

where the last step uses Eq. 289. The second approximation is that transitions between states are fast, i.e., $\omega_+ \sim \omega_- \gg 1$. This approximation allows us to neglect the first two terms of Eq. 296 compared to the third term, which implies that $\mu_1^- \approx \mu_1^+$. Summing Eq. 286 over $\pm$ for $\ell = 1$ then gives

$$\pi^+ \mu_1^+ + \pi_- \mu_1^- \approx (\pi^+ + \pi^-) \mu_1^- = \mu_1^- = \sum_n p_n n = \bar{n}, \tag{297}$$

which shows that μ_1^- approximates the mean of the distribution. Therefore with Eq. 292 the probability that the gene is active (Eq. 290) can be written

$$\pi^+ = \frac{\bar{n}^h}{\bar{n}^h + K}, \tag{298}$$

with equilibrium constant $K \equiv \omega_+/\omega_-$. The effective production rate is the sum of the production rates in each state times the corresponding probabilities of being in each state:

$$g(\bar{n}) = g_- \pi^- + g_+ \pi^+, \tag{299}$$

$$= g_-\left(1 - \frac{\bar{n}^h}{\bar{n}^h + K}\right) + g_+ \frac{\bar{n}^h}{\bar{n}^h + K}, \quad (300)$$

$$= \frac{g_- K + g_+ \bar{n}^h}{\bar{n}^h + K}, \quad (301)$$

which is the Hill function (Eq. 109).

Appendix K. Limiting Case of the Two-State Gene

Here we show that the spectral solution to transcriptional bursting (Eq. 138) reduces to the hypergeometric form (Eq. 128) in the limit of $Z = 2$ production states. We also derive the slightly simpler expression for the special case of zero production in the inactive state.

In the case of two states (Eqs. 121–122; $z = \pm$), Eq. 138 becomes

$$jG_j^{\pm} + \omega_{\mp} G_j^{\pm} - \omega_{\pm} G_j^{\mp} = \omega_{\pm} \sum_{j'<j} G_{j'}^{\mp} \frac{(\mp\Delta)^{j-j'}}{(j-j')!}, \quad (302)$$

where $\Delta = \Delta_{+-} = -\Delta_{-+}$. Initializing with $G_0^{\pm} = \omega_{\pm}/(\omega_+ + \omega_-)$ and computing the first few terms reveals the pattern

$$G_j^{\pm} = \frac{\omega_{\pm}}{\omega_+ + \omega_-} \frac{(\mp\Delta)^j}{j!} \frac{\prod_{j'=0}^{j-1} (j' + \omega_{\mp})}{\prod_{j''=0}^{j-1} (j'' + \omega_+ + \omega_- + 1)}, \quad (303)$$

$$= \frac{\omega_{\pm}}{\omega_+ + \omega_-} \frac{(\mp\Delta)^j}{j!} \frac{\Gamma(j + \omega_{\mp})}{\Gamma(\omega_{\mp})} \frac{\Gamma(\omega_+ + \omega_- + 1)}{\Gamma(j + \omega_+ + \omega_- + 1)}, \quad (304)$$

where in the second line the products are written in terms of the Gamma function. Writing the total generating function $|G\rangle = \sum_{\pm} |G_{\pm}\rangle$ in position space recovers the hypergeometric form (Eq. 128):

$$G(x) = \sum_{\pm} \langle x | G_{\pm} \rangle, \quad (305)$$

$$= \sum_{\pm} \sum_j \langle x | j_{\pm} \rangle \langle j_{\pm} | G_{\pm} \rangle, \quad (306)$$

$$= \sum_{\pm} \sum_j (x-1)^j e^{g_{\pm}(x-1)} G_j^{\pm}, \quad (307)$$

$$= \sum_{\pm} \frac{\omega_{\pm}}{\omega_+ + \omega_-} e^{g_{\pm}(x-1)} \Phi[\omega_{\mp}, \omega_+ + \omega_- + 1; \mp\Delta(x-1)], \quad (308)$$

where

$$\Phi[\alpha, \beta; u] = \sum_{j=0}^{\infty} \frac{\Gamma(j+\alpha)}{\Gamma(\alpha)} \frac{\Gamma(\beta)}{\Gamma(j+\beta)} \frac{u^j}{j!} \quad (309)$$

is the confluent hypergeometric function of the first kind.

In the limit $g_- = 0$, Eq. 308 reads

$$G(x) = \frac{\omega_+}{\omega_+ + \omega_-} e^u \Phi[\omega_-, \omega_+ + \omega_- + 1; -u] + \frac{\omega_-}{\omega_+ + \omega_-} \Phi[\omega_+, \omega_+ + \omega_- + 1; u], \quad (310)$$

where $u \equiv g + (x - 1)$. Using the fact that (49)

$$e^u \Phi[\alpha, \beta; -u] = \Phi[\beta - \alpha, \beta; u], \quad (311)$$

Equation 310 can be written

$$G(x) = \frac{\omega_+}{\omega_+ + \omega_-} \Phi[\omega_+ + 1, \omega_+ + \omega_- + 1; u] + \frac{\omega_-}{\omega_+ + \omega_-} \Phi[\omega_+, \omega_+ + \omega_- + 1; u], \quad (312)$$

or, noting Eq. 309 and the fact that $\Gamma(s + 1) = s\Gamma(s)$ for any s,

$$G = (x) \sum_j \left(\frac{\omega_+}{\omega_+ + \omega_-} \frac{\Gamma(j + \omega_+ + 1)}{\Gamma(\omega_+ + 1)} + \frac{\omega_-}{\omega_+ + \omega_-} \frac{\Gamma(j + \omega_+)}{\Gamma(\omega_+)} \right) \times \frac{\Gamma(\omega_+ + \omega_- + 1)}{\Gamma(j + \omega_+ + \omega_- + 1)} \frac{u^j}{j!}, \quad (313)$$

$$= \sum_j \left(\frac{\omega_+}{\omega_+ + \omega_-} \frac{(j + \omega_+)\Gamma(j + \omega_+)}{\omega_+ \Gamma(\omega_+)} + \frac{\omega_-}{\omega_+ + \omega_-} \frac{\Gamma(j + \omega_+)}{\Gamma(\omega_+)} \right) \frac{(\omega_+ + \omega_-)\Gamma(\omega_+ + \omega_-)}{(j + \omega_+ + \omega_-)\Gamma(j + \omega_+ + \omega_-)} \frac{u^j}{j!}, \quad (314)$$

$$= \sum_j \frac{\Gamma(j + \omega_+)}{\Gamma(\omega_+)} \frac{\Gamma(\omega_+ + \omega_-)}{\Gamma(j + \omega_+ + \omega_-)} \frac{u^j}{j!}, \quad (315)$$

$$= \Phi[\omega_+, \omega_+ + \omega_-; u], \quad (316)$$

as in Eq. 129. The marginal p_n is obtained by $p_n = \partial_u^n [G(x)]_0 / n!$; using Eq. 316 and the derivative of the confluent hypergeometric function,

$$\partial_u^n \Phi[\alpha, \beta; u] = \frac{\Gamma(n + \alpha)}{\Gamma(\alpha)} \frac{\Gamma(\beta)}{\Gamma(n + \beta)} \Phi[\alpha + n, \beta + n; u], \quad (317)$$

one obtains

$$p_n = \frac{g_+^n}{n!} \frac{\Gamma(n + \omega_+)}{\Gamma(\omega_+)} \frac{\Gamma(\omega_+ + \omega_-)}{\Gamma(n + \omega_+ + \omega_-)} \Phi[\omega_+ + n, \omega_+ + \omega_- + n; -g_+], \quad (318)$$

as in Eq. 130.

References

1. Elowitz MB, Levine AJ, Siggia ED, Swain PS (2002) Stochastic gene expression in a single cell. Science 297(5584):1183–6
2. Ozbudak EM, Thattai M, Kurtser I, Grossman AD, van Oudenaarden A (2002) Regulation of noise in the expression of a single gene. Nat Genet 31(1):69–73
3. Raj A, Peskin CS, Tranchina DA, Vargas DY, Tyagi S (2006) Stochastic mrna synthesis in mammalian cells. Plos Biol 4(10):e309
4. Pedraza JM, van Oudenaarden A (2005) Noise propagation in gene networks. Science 307 (5717):1965–9
5. van Zon JS, ten Wolde PR (2005) Simulating biochemical networks at the particle level and in time and space: Green's function reaction dynamic. Phys Rev Lett 94:128103
6. van Zon JS, ten Wolde PR (2005) Green's function reaction dynamics: a particle-based approach for simulating biochemical networks in time and space. J Chem Phys 123:234910
7. Allen RJ, Frenkel D, ten Wolde PR (2006) Forward flux sampling-type schemes for simulating rare events: efficiency analysis. J Chem Phys 124:194111
8. Valeriani C, Allen R, Morelli M, Frenkel D, ten Wolde PR (2007) Computing stationary distributions in equilibrium and non-equilibrium systems with Forward Flux Sampling. J Chem Phys 127:114109
9. Morelli MJ, Allen RJ, Tanase-Nicola S, ten Wolde PR (2008) Eliminating fast reactions in stochastic simulations of biochemical networks: a bistable genetic switch. J Chem Phys 128:045105
10. Munsky B, Khammash M (2006) The finite state projection algorithm for the solution of the chemical master equation. J Chem Phys 124:044104
11. Munsky B, Khammash M (2007) A multiple time interval finite state projection algorithm for the solution to the chemical master equation. J Comput Phys 226:818–835
12. El-Samad H, Khammash M, Petzold L, Gillespie D (2005) Stochastic modeling of gene regulatory networks. Int J Robust Nonlinear Control 15:691–711
13. Lampoudi S, Gillespie DT, Petzold LR (2009) The multinomial simulation algorithm for discrete stochastic simulation of reaction-diffusion systems. J Chem Phys 130:094104–16
14. Gillespie DT, Roh M, Petzold LR (2009) Refining the weighted stochastic simulation algorithm. J Chem Phys 130:174103
15. Chevalier M, El-Samad H (2009) A rigorous framework for multi-scale simulation of stochastic cellular networks. J Chem Phys 131:054102
16. Didier F, Henzinger TA, Mateescu M, Wolf V (2009) Fast adaptive uniformization of the chemical master equation. Proceedings of the High Performance Computational Systems Biology Workshop, IEEE Computer Society, pp 118–127
17. van Kampen NG (1992) Stochastic processes in physics and chemistry. North-Holland, Amsterdam
18. Gardiner CW (2004) Handbook of stochastic methods: for physics, chemistry and the natural sciences, 3rd edn. Springer, New York
19. Zwanzig R (2001) Nonequilibrium statistical mechanics, 1st edn. Oxford University Press, Oxford, USA
20. Kepler TB, Elston TC (2001) Stochasticity in transcriptional regulation: origins, consequences, and mathematical representations. Biophys J 81:3116
21. Swain PS, Elowitz MB, Siggia ED (2002) Intrinsic and extrinsic contributions to stochasticity in gene expression. Proc Natl Acad Sci USA 99(20):12795–800
22. Hasty J, Issacs F, Dolnik M, McMillen D, Collins JJ (2001) Designer gene networks: towards fundamental cellular control. Chaos 11:207
23. Hasty J, Pradines J, Dolnik M, Collins JJ (2000) Noise-based switches and amplifiers for gene expression. Proc Natl Acad Sci USA 97:2075
24. Bialek W (2001) Stability and noise in biochemical switches. In: Leen TK, Dietterich TG, Tresp V (eds) Advances in neural information processing 13, 13th edn. MIT Press, Cambridge, pp 103–109
25. Mehta P, Mukhopadhyay R, Wingreen NS (2008) Exponential sensitivity of noise-driven switching in genetic networks. Phys Biol 5:026005
26. Thattai M, van Oudenaarden A (2001) Intrinsic noise in gene regulatory networks. Proc Natl Acad Sci USA 98(15):8614–9
27. Hornos JEM, Schultz D, Innocentini GCP, Wang J, Walczak AM, Onuchic JN, Wolynes PG (2005) Self-regulating gene: an exact solution. Phys Rev E Stat Nonlin Soft Matter Phys 72:051907
28. Walczak AM, Mugler A, Wiggins CH (2009) A stochastic spectral analysis of transcriptional

regulatory cascades. Proc Natl Acad Sci USA 106(16):6529
29. Mugler A, Walczak AM, Wiggins CH (2009) Spectral solutions to stochastic models of gene expression with bursts and regulation. Phys Rev E Stat Nonlin Soft Matter Phys 80 (4):041921
30. Mugler A, Ziv E, Nemenman I, Wiggins CH (2008) Quantifying evolvability in small biological networks. IET Syst Biol 3 (5):379–387
31. Bialek W, Setayeshgar S (2005) Physical limits to biochemical signaling. Proc Natl Acad Sci USA 102(29):10040–5
32. Tkacik G, Walczak AM, Bialek W (2009) Optimizing information flow in small genetic networks. Phys Rev E Stat Nonlin Soft Matter Phys 80:031920
33. Walczak AM, Tkacik G, Bialek W (2010) Optimizing information flow in small genetic networks II: feed forward networks. Phys Rev E Stat Nonlin Soft Matter Phys 81:041905
34. Iyer-Biswas S, Hayot F, Jayaprakash C (2009) Stochasticity of gene products from transcriptional pulsing. Phys Rev E Stat Nonlin Soft Matter Phys 79:31911
35. Warren PB, Tanase-Nicola S, ten Wolde PR (2006) Exact results for noise power spectra in linear biochemical reaction networks. J Chem Phys 125:1449041–11
36. Tanase-Nicola S, Warren PB, ten Wolde PR (2006) Signal detection, modularity, and the correlation between extrinsic and intrinsic noise in biochemical networks. Phys Rev Lett 97(6):068102
37. Paulsson J (2004) Summing up the noise in gene networks. Nature 427(6973):415–8
38. Elf J, Ehrenberg M (2003) Fast evaluation of fluctuations in biochemical networks with the linear noise approximation. Genome Res 13:2475–2484
39. Doi M (1976) Second quantization representation for classical many-particle system. J Physics A Math Gen 9:1465
40. ZelDovich YB, Ovchinnikov AA (1978) The mass action law and the kinetics of chemical reactions with allowance for thermodynamic fluctuations of the density. Sov J Exp Theor Phys 47:829
41. Peliti L (1986) Renormalisation of fluctuation effects in the a+a to a reaction. J Phys A Math Gen 19:L365
42. Sakurai JJ (1985) Modern quantum mechanics. Pearson Education India, Delhi
43. Bintu L, Buchler NE, Garcia HG et al (2005) Transcriptional regulation by the numbers: models. Curr Opin Genet Dev 15:116
44. Bintu L, Buchler NE, Garcia HG et al (2005) Transcriptional regulation by the numbers: applications. Curr Opin Genet Dev 15:125
45. Raj A, van Oudenaarden A (2008) Nature, nurture, or chance: stochastic gene expression and its consequences. Cell 135(2):216–226
46. Golding I, Paulsson J, Zawilski SM, Cox EC (2005) Real-time kinetics of gene activity in individual bacteria. Cell 123(6):1025–36
47. Walczak AM, Sasai M, Wolynes PG (2005) Self-consistent proteomic field theory of stochastic gene switches. Biophys J 88:828
48. Gillespie DT (1977) Exact stochastic simulation of coupled chemical reactions. J Phys Chem 81(25):2340–2361
49. Koepf W (1998) Hypergeometric summation: an algorithmic approach to summation and special function identities. Vieweg, Braunschweig, Germany

Index

Xuedong Liu and Meredith D. Betterton (eds.), *Computational Modeling of Signaling Networks*, Methods in Molecular Biology, vol. 880, DOI 10.1007/978-1-61779-833-7,

MIX
Papier aus verantwortungsvollen Quellen
Paper from responsible sources
FSC® C105338

If you have any concerns about our products,
you can contact us on
ProductSafety@springernature.com

In case Publisher is established outside the EU,
the EU authorized representative is:
Springer Nature Customer Service Center GmbH
Europaplatz 3, 69115 Heidelberg, Germany

Printed by Libri Plureos GmbH
in Hamburg, Germany